GLYCOBIOLOGY AND MEDICINE

ADVANCES IN EXPERIMENTAL MEDICINE AND BIOLOGY

Recent Volumes in this Series

Volume 526
TAURINE 5: Beginning the 21st Century
Edited by John B. Lombardini, Stephen W. Schaffer, and Junichi Azuma

Volume 527
DEVELOPMENTS IN TRYPTOPHAN AND SEROTONIN METABOLISM
Edited by Graziella Allegri, Carlo V. L. Costa, Eugenio Ragazzi, Hans Steinhart, and Luigi Varesio

Volume 528
ADAMANTIADES-BEHÇET'S DISEASE
Edited by Christos C. Zouboulis

Volume 529
THE GENUS *YERSINIA*: Entering the Functional Genomic Era
Edited by Mikael Skurnik, José Antonio Bengoechea, and Kaisa Granfors

Volume 530
OXYGEN TRANSPORT TO TISSUE XXIV
Edited by Jeffrey F. Dunn and Harold M. Swartz

Volume 531
TROPICAL DISEASES: From Molecule to Bedside
Edited by Sangkot Marzuki, Jan Verhoef, and Harm Snippe

Volume 532
NEW TRENDS IN CANCER FOR THE 21ST CENTURY
Edited by Antonio Llombart-Bosch and Vicente Felipo

Volume 533
RETINAL DEGENERATIONS: Mechanisms and Experimental Theory
Edited by Matthew M. LaVail, Joe G. Hollyfield, and Robert E. Anderson

Volume 534
TISSUE ENGINEERING, STEM CELLS, AND GENE THERAPIES
Edited by Y. Murat Elçin

Volume 535
GLYCOBIOLOGY AND MEDICINE
Edited by John S. Axford

GLYCOBIOLOGY AND MEDICINE

Edited by

John S. Axford

Director, The Sir Joseph Hotung Centre for Musculoskeletal Diseases
St Georges Hospital and Medical School
London, United Kingdom

Springer Science+Business Media, LLC

Library of Congress Cataloging-in-Publication Data

Jenner Glycobiology and Medicine Symposium (6th : 2002 : Seillac, France)
Glycobiology and medicine / edited by John S. Axford.
p. ; cm. -- (Advances in experimental medicine and biology, ISSN 0065-2598 ; v. 535)
"Proceedings of the Sixth Jenner Glycobiology and Medicine Symposium, held 14-17 September, 2002, in Seillac, France"--T.p. verso.
Includes bibliographical references and index.
ISBN 978-0-306-47782-9 ISBN 978-1-4615-0065-0 (eBook)
DOI 10.1007/978-1-4615-0065-0
1. Glycoconjugates--Pathophysiology--Congresses. 2. Glycoconjugates--Physiological effect--Congresses. I. Axford, John S. II. Title. III. Series.
[DNLM: 1. Glycosylation--Congresses. 2. Glycoproteins--physiology--Congresses. QU 55 J54g 2003]
QP702.G577J46 2002
612'.01578--dc21

2003047705

Proceedings of the Sixth Jenner Glycobiology and Medicine Symposium, held September 14–17, 2002, in Seillac, France

ISSN 0065-2598

ISBN 978-0-306-47782-9

Originally published by Kluwer Academic/Plenum Publishers, New York in 2003

http://www.wkap.nl/

10 9 8 7 6 5 4 3 2 1

A C.I.P. record for this book is available from the Library of Congress

PREFACE

Over the past decade significant advances in technology have opened up the field of glycobiology. In particular, improved methods for carbohydrate analysis have led to important biochemical observations demonstrating that sugars play crucial roles in human physiology. It is clear that many diseases are associated with characteristic changes in glycosylation and furthermore, the possibility of modulating glycan processing to treat disease is beginning to be realised. This volume summarises some of the important recent developments in "glycobiology and medicine."

We highlighted some of the numerous areas in which there are glycosylation dependant pathological mechanisms causing common diseases. The next decade will undoubtedly see novel diagnostic and therapeutic techniques originating from these observations. This will significantly enhance our ability to combat infection and diseases such as bacterial and viral infections, some cancers, glycolipid storage disorders, systemic autoimmune disease and disorders that involve cytokine related inflammatory mechanisms.

These topics were discussed at the 6th Jenner Glycobiology and Medicine meeting. This meeting received a European Commission Education grant (No. HPCF-2002-00250).

CONTENTS

Glycosylation Dependent Bacterial Infections

Glycosylation Dependent Viral Infections

Inflammation

GLYCOSYLATION DEPENDENT BACTERIAL INFECTIONS

1

A SWEET COATING—HOW BACTERIA DEAL WITH SUGARS

Anthony P. Corfield[1], Rebecca Wiggins[1,3], Cathryn Edwards[2], Neil Myerscough[1], Bryan F. Warren[2], Peter Soothill[3], Michael R. Millar[4], and Patrick Horner[5]

[1]Mucin Research Group
Univ. Div. Medicine
Bristol Royal Infirmary
Bristol BS2 8HW, UK
[2]Department of Cellular Pathology
John Radcliffe Hospital
Oxford OX3 9DU, UK
[3]Department of Obstetrics and Gynaecology
St Michael's Hospital
Bristol BS2 8EG, UK
[4]Department of Microbiology
Barts and the London NHS Trust, Smithfield
London EC1A 7BE, UK
[5]Milne Genito-Urinary Clinic, Bristol Royal Infirmary
Bristol BS2 8HW, UK

1. INTRODUCTION

The relationships between bacteria and their hosts are fundamental to our understanding of vital normal and pathogenic processes. They have attracted considerable interest recently with improved understanding of a variety of infectious diseases and normal developmental and adult processes at mucosal surfaces throughout the body (Tannock, 1999). An important part of the recognition process includes the glycobiology of the host mucosal cells and the bacteria themselves. Glycan structures and their manipulation are assuming increasing significance. This review serves to illustrate some of the current observations in this field.

Glycobiology and Medicine, edited by John S. Axford
Kluwer Academic / Plenum Publishers, New York, 2003

2. BACTERIAL POPULATIONS AT MUCOSAL SURFACES

Identification of the components of the complex ecosystems that are found at various body surfaces in man remains limited due to the failure to culture a high proportion of these organisms in the laboratory. The targets of interest are, the non-pathogens particularly in the gut, where the major populations of bacteria are found in man, and the pathogens that have usually attracted more attention. The non-pathogens are proving to be a focus of attention as their contribution to normal development and maintenance of stability and defence at mucosal surfaces is vital but still poorly defined and understood. The majority of the information available has been related to the gut (Deplancke and Gaskins, 2001; Hooper *et al.*, 1998, 2001; Hooper and Gordon, 2001a) and the principles discussed for this system are used in this general review.

The interactions that take place between the host and bacteria can be seen as a flexible and overlapping system encompassing symbiosis, commensalism and pathogenicity. The positive symbiotic and commensal interactions afforded by the non-pathogens have been described generally as mutualism (Hooper *et al.*, 2001). These dynamic systems are evolved as a result of exposure to the damaging pathogens and the need of the host to provide protection. These may take the form of reducing virulence and coexistence or processes that circumvent the innate and adaptive host defence (Hooper *et al.*, 2001). The need to adapt to these interactions provides the impetus for the coevolution of effective bacterial-host interactions. This is favored by the genetic diversity of bacteria, their capacity for rapid growth and high population density (Hooper *et al.*, 2001). A basic property of the successful mucosal defensive system is the ability to discriminate between pathogenic and beneficial bacteria (Hooper *et al.*, 2001; Philpott *et al.*, 2001).

There are features of the gut flora that gives an insight into the nature of the factors responsible for the formation of these systems.

Identification of bacterial strains at sites throughout the gastrointestinal tract have shown that certain species are commonly associated with particular mucosal surfaces throughout the tract (Boriello, 2002). Within the normal gut Lactobacilli is found in the stomach, and Bifidobacteria and *Escherichia coli*, in the large bowel. Furthermore distinct strains of bacteria can be found in the large intestine during the early stages of life (Tannock, 1994). Following colonization of the neonatal gut at birth the relative abundance of bacterial strains found through lactation weaning and the achievement of adulthood show characteristic patterns during normal growth.

A similar phenomenon can be identified in mucosal diseases and certain bacteria are well established with pathological conditions. Three examples of this are discussed in detail in other chapters in this volume. *Helicobacter pylori* has its major pathology in the stomach, being associated with gastritis and gastric cancer (Appelmelk *et al.*, 1996, Hooper and Gordon, 2001). *Pseudomonas aeruginosa* is found in the respiratory tract and has been studied in this organ with respect to cystic fibrosis (Ramphal and Arora, 2001; Scharfman *et al.*, 1999). Uropathogenic *Escherichia coli* (UPECs) are found associated with cystitis and pyelonephritis in the urinary tract (Mulvey *et al.*, 2000). These common patterns of behavior suggest the recognition of tissue specific binding sites for the bacteria. These sites may change in a specific and predictable manner during development, once again creating a series of targets for normal and pathogenic interactions.

3. GLYCOSYLATION DEPENDENCE OF BACTERIAL INFECTIONS AND HOST MUCOSAL CELL SYSTEMS

Molecular analysis of the factors that determine the nature of bacterial relationships with the host identify the repertoire of glycans present on the bacterial surface and those expressed on the host mucosal surface. Bacteria produce proteins that bind to glycans (adhesins/lectins), enzymes that degrade glycans, proteins for metabolic conversion of carbohydrates, and soluble factors capable of directing host mucosal cell surface glycosylation. It is clear that glycobiological processing is a major issue in these relationships.

The hypothesis that cell glycans play a role in the normal and pathological interactions with bacteria is supported at the host level by the variation of cell surface glycosylation at the sites where exposure to bacteria take place. The identification of the sites is related to the type of cell (cell lineage), the location within the cells (subcellular compartment) and the stage of growth of the host and the relevant mucosal surface (developmental stage). This evidence has been accumulating over the past 25–30 years leading to current knowledge of the wide variety of glycoconjugates expressed on cellular membranes and the demonstration of their biological role in cellular function (Cook, 1994; Dwek, 1995; Gabius and Gabius, 1997; Hakomori, 2002; Montreuil *et al.*, 1995; Rudd and Dwek, 1997; Rudd *et al.*, 2001; Taylor and Drickamer, 2003; Varki *et al.*, 1999).

Many different glycan structures have been identified associated with cell surface molecules, but demonstration of their association with well-defined biological functions has only recently started to accumulate. Examples of well-defined structure–function relationships include those shown in Table 1.

The formation of glycan structures does not arise directly from DNA sequence, as is the case with proteins. Oligosaccharide sequences are generated through the action of a family of glycosyltransferases, coded, transcribed and translated from the genome, that act on a variety of acceptors, which may be protein, lipid or saccharide based, and include the growing glycoconjugates themselves. Thus the fidelity of glycan sequence depends on the expression and specificity of a defined group of glycosyltransferases, the availability of an

Table 1. Glycan structures with defined functions (Taylor *et al.*, 2003).

Glycan structure	Binding protein	Carbohydrate Recognition Domain (CRD)	Function
Terminal β–Galactose N-glycans	Asialoglycoprotein receptor	C-type CRD	Clearance of circulating asialoglycoproteins and mediation of endocytosis
Terminal α–mannose N-glycans	Mannose binding protein	C-type CRD	Clearance of circulating high-mannose glycoproteins and mediation of endocytosis
GalNAc-4-SO_3	Mannose binding protein	R-type CRD	Clearance of circulating glycoprotein hormones
Sialyl-Le^x on O-glycans	P, L and E-Selectins	C-type CRD	Leukocyte trafficking
Mannose-6-P on N-glycans	Mannose-6-phosphate receptors	P-type CRD	Lysosomal enzyme targeting

activated sugar and the acceptor and is a posttranslational event. Furthermore, because of the extensive range of cellular glycoconjugates and their constant renewal this process demands a major energy commitment from the cell.

4. GENERATION OF GLYCAN DIVERSITY IN RESPONSE TO INTERACTION WITH BACTERIAL POPULATIONS

Overall the process of glycan formation and maintenance is highly programmed and regulated. This implies that during its evolution certain structures have been selected for certain biological functions. Part of the evolutionary pressure acting to generate a diverse library of glycan structures comes from the need to provide an effective and adaptive defensive system. The need to evade pathogenic microbes and create and sustain a symbiotic relationship with indigenous bacteria is a strong selective pressure to create the capacity for glycan diversity (Dennis *et al.*, 1999; Drickamer and Taylor, 1998; Gagneux and Varki, 1999; Varki, 1993).

The non-template nature of glycan biosynthesis provides a capacity to generate structural diversity in glycans in response to evolutionary pressure. It allows the host to accommodate for dynamic fluxes in the composition of the flora (particularly in the gut) and the rapid evolution of individual species.

Thus the repertoire of glycans found in and on the surfaces of cells illustrates the existing evolved glyco-programming and is integrated with the processes of normal development and continued function in adulthood. The functional interactions observed between the host and bacteria can be demonstrated in situations where the response to bacterial colonization can be controlled. The gastrointestinal tract of newborn animals can be kept sterile under germfree conditions. Comparison of these germfree animals with conventionally colonized individuals shows the interaction of the bacteria with the glycosylation mechanisms in the host gastrointestinal mucosa. Small intestinal glycoconjugates that terminate with Fucα1-2Galβ- show typical neonatal patterns with major expression during weaning, when large increases in gut flora are found. In contrast these patterns are not seen in germfree mice, but can be recovered by the introduction of intestinal microflora and is accompanied by the induction of an α1-2 fucosyltransferase activity (Bry *et al.*, 1996; Hooper and Gordon, 2001).

In addition, there is a typically increased mucus layer present in the gut of germfree mice which is thought to be caused by the absence of a mucinolytic action of the bacterial microflora rather than increased mucin production (Deplancke and Gaskins, 2001).

The degree of specificity and selectivity for this process can also be illustrated with the recognition of individual host based glycosylation by the colonising bacterial flora *in vivo*. Examination of the ability of the gut flora to degrade the mucus barrier in secretor positive individuals with different blood groups demonstrates an adaptation to regulated mucus turnover, without a level of destruction that would compromise the defensive role of the barrier (Hoskins and Boulding, 1976).

A further feature of the intestinal bacterial flora is its use of fermentable carbohydrates as an energy source. This is directly related to the abundance of glycoconjugates available in the defensive barrier as substrates for energy production. The tools to carry out this task include the glycosidases, endowing the ability to degrade glycoconjugates and to release monosaccharides (Hoskins *et al.*, 1985, 1992). It also creates an ecological advantage as the bacteria have direct access to a host derived energy source. The evolution and flexibility of the interaction is shown in that only some strains have all enzymes to degrade glycoconjugates (MOD strains). In doing so this population of bacteria generate monosaccharides and can recruit other organisms that will use them (Hoskins *et al.*, 1992).

Thus, there are several features that create conditions for the establishment of an ecosystem. Host glycans are present in abundance early in neonatal life, there is an individual adaptation (e.g., flora related to host blood group status) and there are members of the whole flora that can induce host glycosylation in response to nutritional requirements.

In a series of elegant experiments Gordon and his group have demonstrated a molecular basis for the glycans as legislators of host–microbial interactions. Using the strict anaerobe *Bacteroides thetaiotaomicron* they showed that the single bacterial strain could substitute for the whole intestinal flora in inducing the fucosylation patterns in germfree mice. A series of genes were identified that coded for proteins that catabolise L-fucose, released by the action of α-fucosidase action and taken into the bacterial cell by a specific L-fucose permease. These genes were shown to be under the control of a repressor gene acting as a sensor of fucose availability in the bacterial cell. The same repressor also coordinates the fucose degradation operon with a separate locus that produces a molecular signal inducing the synthesis of α1-2-fucosylated oligosaccharides in the host intestinal mucosa, one of the energy substrates for the bacterium. This mechanism creates a favorable environment for the establishment of this bacterium in an ecosystem (Bry *et al.*, 1996; Hooper *et al.*, 1999, 2000, 2001, 2002; Hooper and Gordon, 2001b).

Although the identity of the "signal" molecule has not yet been identified, direct demonstration of this concept has been reported experimentally in mice (Freitas *et al.*, 2002; Freitas and Cayuela, 2000).

These examples provide powerful evidence for the role of glycans in host–microbe interactions. In the case of pathogenic bacteria the intimate contact with host apical mucosal cell membranes is a crucial step in the infective process. The recognition of glycans by the host defensive mechanisms is a further area of current interest. In addition to the systems described above, further interest has focused on the toll receptors, particularly in the case of bacterial induced mucus hypersecretion in the lung (Basbaum *et al.*, 1999; Lemjabbar and Basbaum, 2002), the many examples of bacterial adhesins that recognize host cell surface glycans and the integration of bacteria into biofilms which themselves are glycoconjugates and are synthesized by the bacteria (Costerton, 1995).

5. EXAMPLES OF HOST BACTERIAL INTERACTIONS IN ACTION

Two examples from our own work serve to illustrate two aspects of host–bacterial interaction.

5.1. Sialidase in Bacterial Vaginosis (BV)—An Example of a Degrading Pathway

A wide variety of infectious diseases have been identified with individual bacterial strains targeting specific host organs and tissues. Among these there are well-defined examples where a bacterial sialidase is produced and plays a major role in the disease pathology. Some examples are shown in Table 2.

The action of the sialidases produced on host targets has been identified in some of these cases. Sialidases are often components of a broader group of enzymes designed to degrade the mucus barrier and provide access to the mucosal cell surface or to create binding sites for the bacteria within the mucus barrier itself. Under normal, non-pathological conditions these enzymes are responsible for the regulated turnover of the mucus barrier. The removal of "used" mucus is achieved while maintaining an effective protective barrier.

Under pathological conditions the sialidases act to remove the negative charge on adherent mucus glycoproteins and compromise their physiologically important viscoelastic properties. This allows access to the mucosal surface where the glycocalyx may also be degraded to generate sites for binding and colonization. Partial degradation of the mucins also may create binding sites which may give the bacteria an advantage in the infectious process.

In the female reproductive tract the normal vaginal and cervical mucus barrier is colonized largely by *Lactobacillus* strains. These maintain a characteristically acidic environment at approximately pH 4.0. A mucin degrading activity has not been identified in these bacteria but we have detected a low but reproducible sialidase activity which may play a role in cervical mucus turnover (Howe *et al.*, 1999; Wiggins *et al.*, 2001; Wiggins *et al.*, 2002). Changes in the cervical mucosal environment have been identified in bacterial vaginosis (BV) and these include an increase in pH and the appearance of typical bacterial strains (Table 3) with a loss of *Lactobacillus* sp.

Table 2. Bacterial sialidases in disease.

Disease	Bacteria	Site	Enzyme in blood
Gas gangrene	Clostridia	Wounds	Yes
Septicemia	Streptococcus ++	Various	Yes
Pneumonia	Streptococcus	Respiratory tract	Yes
Peritonitis	Clostridia ++	Peritoneum	Yes
Meningitis	Step. Gp B III	Brain	Yes
Otitis media	Streptococcus	Middle ear	No
Cholera	Vibrio cholerae	Gastrointestinal tract	No
Bacterial vaginosis	Prevotella ++	Reproductive tract	No

++, additional bacterial strains have been identified in these groups.

Table 3. Bacteria found in bacterial vaginosis.

Bacterium	Gram stain	Shape	Oxygen status	Sialidase
Gardnerella vaginalis	Negative	Bacillus	Microaerophilic	Yes
Mobiluncus sp.	Various	Curved	Anaerobe	?
Bacteroides	Negative	Bacillus	Anaerobe	Yes
Prevotella	Negative	Bacillus	Anaerobe	Yes
Porphyromonas	Negative	Bacillus	Anaerobe	?
Peptostreptococcus sp.	Positive	Coccus	Anaerobe	Yes
Mycobacterium hominis		No cell wall	Facultative or	?
Ureoplasma urealyticum		No cell wall	strict anaerobes	?

Table 4. Examples of physiological and synthetic sialidase substrates.

Substrate	Linkage	Occurrence
Physiological—Naturally occurring substrates		
Sialyl lactose	α2-3 or α2-6	Oligosaccharide
Colominic acid	α2-8	Polysaccharide
Fetuin	α2-3, α2-6	Glycoprotein (N & O)
α_1-Acid glycoprotein (AGP)	α2-6	Glycoprotein (N)
Salivary mucin (BSM)	α2-6 > α2-3	Glycoprotein (O)
Gangliosides	α2-3 > α2-6	Glycolipid
Synthetic substrates		
4-Nitrophenyl-α-Neu5Ac	*α	Synthetic
4-Methylumbelliferyl-α-Neu5Ac	*α	Synthetic
5-Bromo-4-chloro-3-indolyl-α-Neu5Ac	*α	Synthetic

*α-Glycosidic bond at C2 of sialic acid (Neu5Ac); N, N-linked oligosaccharide chains; O, O-linked oligosaccharide chains.

The routine clinical detection of bacterial vaginosis is carried out using the Gram stain and grades 1 (normal), 2 (intermediate) and 3 (bacterial vaginosis) are identified on the basis of the microflora identified using this stain. Together with this change in mucosal bacteria an increase in sialidase activity can be detected in vaginal swabs from these patients (Briselden *et al.*, 1992; Howe *et al.*, 1999; McGregor *et al.*, 1994). As shown in Table 3 the major strains associated with bacterial vaginosis are sialidase positive.

Sialidase are glycosidases that act on the wide variety of sialoglycoconjugates present in nature. Examination of their substrate specificity has led to the identification of discrete activities with specific subcellular location in mammals. However, bacterial sialidases show a much broader range of activity (Corfield, 1992). Specific sialoglycoconjugate substrates have been used to identify individual enzymes and assess the potential physiological substrates for the sialidases *in vivo*. In response to the increasing interest in sialidase activity several synthetic sialic acid glycosides were synthesized to allow rapid and sensitive screening. Examples of these substrates are shown in Table 4.

Closer examination of the substrate specificity of bacterial sialidases with synthetic substrates showed that some enzymes had only limited or no activity with these compounds. Due to their synthetic nature the aglycone gives no indication of the physiological nature of substrates cleaved by the enzyme. As a result each new sialidase activity requires assessment with both physiological and synthetic substrates to allow full characterization.

In spite of the identification of BV associated sialidase activity (Briselden *et al.*, 1992; McGregor *et al.*, 1994), no screening of substrate specificity was carried out until recently (Howe *et al.*, 1999). Testing with physiological sialidase substrates confirmed that these substrates are equally valid for detecting the disease (Table 5) and confirm that a basal level of sialidase activity is present in normal individuals.

On the basis of testing with different substrates the indoxyl-glycoside, 5-bromo-4-chloro-3-indolyl-α-Neu5Ac was tested as a rapid assay for the detection of BV.

Routine use of the Gram stain to test for BV requires more that one day to obtain a result that can be passed on to the patient and to allow assessment of the need for therapy. The short time required for a positive result in the sialidase assay provides a near patient test that could be used immediately in the clinic (Wiggins *et al.*, 2000). A considerable improvement in patient treatment can thus be achieved using this screening test.

Table 5. Sialidase activity in bacterial vaginosis.

	Activity (nmole/h/mg)				
Grade	AGP	p	BSM	p	n
1	0.69 ± 0.2		3.64 ± 0.5		46
2	2.54 ± 0.7	0.03	5.05 ± 0.9	0.02	22
3	4.67 ± 0.9	0.004	8.46 ± 0.7	0.0001	24

AGP—human α_1-acid glycoprotein; BSM—bovine submandibular gland mucin.

Table 6. Comparison of sialidase activity (spot test) with the Gram stain for BV at 20 weeks' gestational age in 99 patients.

	Sialidase		
Gram stain	Positive	Negative	Total
+	14	3	17
–	0	82	82

Fishers exact test: Exact mid-P one sided $P < 0.0001$, two sided $P < 0.0001$. Sensitivity: 100%, Specificity: 96.4%, PPV: 88.2%, NPV: 100%.

Examination of the 5-bromo-4-chloro-3-indolyl-α-Neu5Ac in a larger series of BV patients confirmed the value of this substrate in screening and showed very high specificity and selectivity (Table 6).

In addition to its value in the identification of BV positive patients the 5-bromo-4-chloro-3-indolyl-α-Neu5Ac has been used to identify sialidase positive clones of bacteria grown from the swabs taken from these patients. In this way the further characterization of BV related bacterial strains and their sialidases can be advanced.

Thus, BV is a further example of an infectious disease with major glycobiological involvement. The results reported above underline the importance of assessing substrate specificity in order to provide optimal screening and to assess the physiological significance of the disease mechanism. Current investigations are aimed at the characterization of the sialidase activity and identification of its physiological targets to enable improved therapy.

The glycobiological imbalance introduced by the pathogenic strains in BV is the sialidase activity. This knowledge now provides a series of possible mucosal targets for examination as part of the disease mechanism.

5.2. Fecal Bacterial Control of an O-Acetylated Sialic Acid Glycotope

The work of Gordon and his group have highlighted the close interaction and potential for signaling between bacteria and the host with regard to the glycobiology of mucosal defence (Hooper *et al.*, 1998, 2000, 2001; Hooper and Gordon, 2001a, 2001b). At present there is limited information on the manner in which this functions and how it is regulated.

We have been studying the interaction of the enteric bacterial flora with the gastrointestinal mucosa protective barrier in order to understand how this mutual, symbiotic-commensal system maintains effective and continuous defence while at the same time supporting a dynamic bacterial population that is able to resist the attack of pathogens.

We have adopted the study of a mucin linked poly-O-acetylated sialic acids glycotopes that are a marker for the human colorectal mucosa. Sialic acids have a diverse distribution in nature (Schauer *et al.*, 1997) and are especially abundant in mucins at many mucosal surfaces (Brockhausen, 1999; Corfield *et al.*, 2000, 2001; Roussel and Lamblin, 1996). O-Acetylated sialic acids are also common (Varki, 1992, 1997), but poly-O-acetylated sialic acids, having 2 and 3 O-acetyl esters at positions C7–C9 are found almost exclusively in the colorectal mucosa where they are major antigens (Corfield *et al.*, 1995, 1999, 2001; Filipe and Ramachandra, 1995; Jass and Roberton, 1994).

The biochemical measurement of these glycotopes is difficult due to their extreme lability under mild alkaline conditions and the significant losses associated with purification of the glycoconjugates and identification of the individual O-acetylated sialic acid forms.

Chemical analyses have demonstrated the presence of at least four forms of sialic acid in normal human colorectal mucins (Corfield *et al.*, 1995, 1999), Table 7.

The O-acetylated sialic acids have been demonstrated histochemically. The mild-periodic acid Schiff (mPAS) technique has proved particularly valuable and reproducible, using saponification to remove the esters (Campbell *et al.*, 1994; Filipe and Ramachandra, 1995; Jass and Roberton, 1994; Reid *et al.*, 1984; Roe *et al.*, 1989; Veh *et al.*, 1982). In addition, a number of saponification sensitive antibodies have been raised that show similar histological patterns to the mPAS detection of O-acetylated sialic acids and also localize to colonic goblet cell vesicles (Hughes *et al.*, 1986; Milton *et al.*, 1993; Richman and Bodmer, 1987; Smithson *et al.*, 1997). Although the precise nature of these antibodies has not yet been determined they are of considerable value in identifying changes occurring in inflammatory bowel disease (Milton *et al.*, 1993; Smithson *et al.*, 1997) and colorectal cancer (Corfield *et al.*, 1999; Hughes *et al.*, 1986; Milton *et al.*, 1993; Richman and Bodmer, 1987) with immunohistological methods.

We have used two of these antibodies, PR3A5 (Richman and Bodmer, 1987) and 6G4 (Smithson *et al.*, 1997) together with the mPAS method in the studies described here (Table 8). Comparison of the immuno-histological data from the mPAS compared with the

Table 7. Sialic acids in normal colorectal mucin (Corfield *et al.*, 1999).

Sialic acid	Abbreviation	% of total
N-Acetyl-neuraminic acid	Neu5Ac	47 ± 11
9-O-Acetyl-N-acetyl-neuraminic acid	Neu5,9Ac_2	15 ± 5
7,(8),9-Di-O-acetyl-N-acetyl-neuraminic acid	Neu5,7(8),9 Ac_3	18 ± 4
7,8,9-Tri-O-acetyl-N-acetyl-neuraminic acid	Neu5,7,8,9Ac_4	20 ± 6
Total O-acetylated		53 ± 15

Table 8. Detection of O-acetylated sialic acids in colorectal mucins.

Method	Type of assay	Direct assay	After saponification
mPAS	Chemical	Negative	Positive
PR3A5	Antibody	Positive	Negative
6G4	Antibody	Positive	Negative

Table 9. Patient groups and identification of non-O-acetylators.

Patient group	Total	Positive	Non O-Ac
Normal	19	16	3
Normal operated	15	15	0
IBD (UC + CD)	39	34	5
IBD operated	34	29	5

IBD, inflammatory bowel disease; UC, ulcerative colitis; CD, Crohn's disease.
All sections stained with mPAS with and without saponification.

Table 10. The impact of fecal flow on O-acetyl sialic acid expression in colonic mucins Tissue from patients in each group were stained with the mPAS method or using the antibodies PR3A5 and 6G4. The staining was scored as positive or negative and patterns for each group compared and tested statistically.

Group	Total	mPAS	PR3A5	*P*	6G4 positive	6G4 negative	*P*
Normal	16	16	16		13	3	
Normal div	15	15	15	NS	5	10	0.01
IBD	34	33	33		29	5	
IBD operated	29	29	29	NS	8	21	0.001

PR3A5 and 6G4 antibody results (Edwards *et al.*, 2000; Richman and Bodmer, 1987; Smithson *et al.*, 1997) show that the mPAS and PR3A5 antibody detect the total O-acetylated sialic acid population while the 6G4 antibody detects a subpopulation of O-acetylated sialic acids in the colonic goblet cell mucus. We therefore examined a series of patients (Table 9) who had undergone surgical diversion of the large bowel for inflammatory bowel disease or other reasons. These patients were compared with controls having inflammatory bowel disease but no surgery. The expression of O-acetylated sialic acids was compared within these groups to assess the influence of removal of the fecal flow and thus continuous exposure to the complete enteric bacterial flora. As approximately 10% of the normal population are non-O-acetylators, these individuals were identified using the mPAS assay, where a positive result was obtained without prior saponification (Table 9).

Examination of these groups showed that the removal of the fecal flow through surgical isolation resulted in a selective loss of the 6G4 epitope, while the pattern seen for mPAS and PR3A5 staining was retained (Table 10). This experiment demonstrates the loss of a subpopulation of O-acetylated sialic acids in colonic mucins as a result of the removal of fecal flow. This implicates a role for the bacterial flora in the regulation of expression of this glycotope and opens the way to examine the nature of this down-regulation and the signal(s) responsible for it.

6. CONCLUSIONS

This brief overview presents evidence in favor of a significant role for glycobiology in bacteria–host relationships. Current data and ongoing discussions have highlighted this as an important and developing area of interest and one where relatively little information is available. A closer look at the sweet coating of mucosal surfaces and the way bacteria cope with it are clearly on the agenda for the immediate future.

7. REFERENCES

Appelmelk, B.J., Simoons-Smit, I., Negrini, R., Moran, A.P., Aspinall, G.O., Forte, J.G., De Vries, T., Quan, H., Verboom, T., Maaskant, J.J., Ghiara, P., Kuipers, E.J., Bloemena, E., Tadema, T.M., Townsend, R.R., Tyagarajan, K., Crothers, J.M., Jr., Monteiro, M.A., Savio, A., and De Graaff, J., 1996, Potential role of molecular mimicry between *Helicobacter pylori* lipopolysaccharide and host Lewis blood group antigens in autoimmunity, *Infect Immun.* 64:2031–2040.

Basbaum, C., Lemjabbar, H., Longphre, M., Li, D., Gensch, E., and McNamara, N., 1999, Control of mucin transcription by diverse injury-induced signaling pathways, *Am J Respir Crit Care Med.* 160:S44–S48.

Boriello, S.P., 2002, The normal flora of the gastrointestinal tract, in: *Gut Ecology*. A.L. Hart, A.J. Stagg, H. Graffner, H. Glise, P.G. Falk, and M.A. Kamm, eds., Martin Dunitz, London, pp. 3–12.

Briselden, A.M., Moncla, B.J., Stevens, C.E., and Hillier, S.L., 1992, Sialidases (neuraminidases) in bacterial vaginosis and bacterial vaginosis-associated microflora, *J Clin Microbiol.* 30:663–666.

Brockhausen, I., 1999, Pathways of O-glycan biosynthesis in cancer cells, *Biochim Biophys Acta.* 1473:67–95.

Bry, L., Falk, P.G., and Gordon, J.I., 1996, Genetic engineering of carbohydrate biosynthetic pathways in transgenic mice demonstrates cell cycle-associated regulation of glycoconjugate production in small intestinal epithelial cells, *Proc Natl Acad Sci USA.* 93:1161–1166.

Campbell, F., Appleton, M.A.C., Fuller, C.E., Greef, M.P., Hallgrimsson, J., Katoh, R., Ng, O.L.I., Satir, A., Williams, G.T., and Williams, E.D., 1994, Racial variation in the O-acetylation phenotype of human colonic mucosa, *J Pathol.* 174:169–174.

Cook, G.M., 1994, Recognising the attraction of sugars at the cell surface, *Bioessays.* 16:287–295.

Corfield, A.P., Carroll, D., Myerscough, N., and Probert, C.S., 2001, Mucins in the gastrointestinal tract in health and disease, *Front Biosci.* 6:D1321–D1357.

Corfield, A.P., Longman, R., Sylvester, P., Arul, S., Myerscough, N., and Pigatelli, M., 2000, Mucins and mucosal protection in the gastrointestinal tract: new prospects for mucins in the pathology of gastrointestinal disease, *Gut.* 47:598–594.

Corfield, A.P., Myerscough, N., Gough, M., Brockhausen, I., Schauer, R., and Paraskeva, C., 1995, Glycosylation patterns of mucins in colonic disease, *Biochem Soc Trans.* 23:840–845.

Corfield, A.P., Myerscough, N., Warren, B.F., Durdey, P., Paraskeva, C., and Schauer, R., 1999, Reduction of sialic acid O-acetylation in human colonic mucins in the adenoma-carcinoma sequence, *Glycoconj J.* 16:307–317.

Corfield, T., 1992, Bacterial sialidases—roles in pathogenicity and nutrition, *Glycobiology.* 2:509–521.

Costerton, J.W., 1995, Overview of microbial biofilms, *J Indust Microbiol.* 15:137–140.

Dennis, J.W., Granovsky, M., and Warren, C.E., 1999, Protein glycosylation in development and disease, *Bioessays.* 21:412–421.

Deplancke, B., and Gaskins, H.R., 2001, Microbial modulation of innate defense: goblet cells and the intestinal mucus layer, *Am J Clin Nutr.* 73:1131S–1141S.

Drickamer, K. and Taylor, M.E., 1998, Evolving views of protein glycosylation, *TIBS.* 23:321–324.

Dwek, R.A., 1995, Glycobiology: more functions for oligosaccharides, *Science.* 269:1234–1235.

Edwards, C.M., Corfield, A.P., Jewell, D.P., Biddolph, S., and Warren, B.F., 2000, Mucin bound O-acetylated sialic acids and diversion colitis, *GUT.* 46:A81.

Filipe, I.M. and Ramachandra, S., 1995, The histochemistry of intestinal mucins: changes in disease, in: *Gastrointestinal and oesophageal pathology*. R. Whitehead, ed., Churchill Livingstone, Edinburgh, pp. 73–95.

Freitas, M., Axelsson, L.-G., Cayuela, C., Midvedt, T., and Trugnan, G., 2002, Microbial-host interactions specifically control the glycosylation pattern in intestinal mouse mucosa, *Histochem Cell Biol.* 118:149–161.

Freitas, M. and Cayuela, C., 2000, Microbial modulation of host intestinal glycosylation patterns, *Microb Ecol Health Dis.* Suppl. 2:165–178.

Gabius, H.J. and Gabius, S., 1997, *Glycosciences: status and perspectives*, Chapman and Hall, Weinheim, Germany.

Gagneux, P. and Varki, A., 1999, Evolutionary and considerations in relating oligosaccharide diversity to biological functions, *Glycobiology.* 9:744–755.

Hakomori, S., 2002, The glycosynapse, *Proc Natl Acad Sci USA*. 99:225–232.

Hooper, L.V., Bry, L., Falk, P.G., and Gordon, J.I., 1998, Host-microbial symbiosis in the mammalian intestine: exploring an internal ecosystem, *Bioessays*. 20:336–343.

Hooper, L.V., Falk, P.G., and Gordon, J.I., 2000, Analyzing the molecular foundations of commensalism in the mouse intestine, *Curr Opin Microbiol.* 3:79–85.

Hooper, L.V. and Gordon, J.I., 2001a, Commensal host-bacterial relationships in the gut, *Science*. 292:1115–1118.

Hooper, L.V. and Gordon, J.I., 2001b, Glycans as legislators of host microbial interactions: spanning the spectrum from symbiosis to pathogenicity, *Glycobiology*. 11:1R–10R.

Hooper, L.V., Midtvedt, T., and Gordon, J.I., 2002, How host-microbial interactions shape the nutrient environment of the Mammalian intestine, *Annu Rev Nutr.* 22:283–307.

Hooper, L.V., Wong, M.H., Thelin, A., Hansson, L., Falk, P.G. and Gordon, J.I., 2001, Molecular analysis of commensal host-microbial relationships in the intestine, *Science*. 291:881–884.

Hooper, L.V., Xu, J., Falk, P.G., Midtvedt, T., and Gordon, J.I., 1999, A molecular sensor that allows a gut commensal to control its nutrient foundation in a competitive ecosystem, *Proc Natl Acad Sci USA*. 96:9833–9838.

Hoskins, L.C., Agustines, M., McKee, W.B., Boulding, E.T., Kriaris, M., and Niedermeyer, G., 1985, Mucin degradation in human ecosystems. Evidence for the existence and role of bacterial subpopulations producing glycosidases as extracellular enzymes, *J Clin Invest.* 75:944–953.

Hoskins, L.C. and Boulding, E.T., 1976, Degradation of blood group antigens in human colon ecosystems. *In vitro* production of ABH blood group degrading enzymes by enteric bacteria, *J Clin Invest.* 57:63–73.

Hoskins, L.C., Boulding, E.T., Gerken, T.A., Harouny, V.R., and Kriaris, M., 1992, Mucin glycoprotein degradation by mucin oligosaccharide—degrading strains of human faecal bacteria. Characterization of saccharide cleavage products and their potential role in nutritional support of larger faecal bacterial populations, *Microb Ecol Health Dis.* 5:193–207.

Howe, L., Wiggins, R., Soothill, P.W., Millar, M.R., Horner, P.J., and Corfield, A.P., 1999, Mucinase and sialidase activity of the vaginal microflora: implications for the pathogenesis of preterm labour [In Process Citation], *Int J STD AIDS*. 10:442–447.

Hughes, N.R., Walls, R.S., Newland, R.C., and Payne, J.E., 1986, Antigen expression in normal and neoplastic colonic mucosa: three tissue specific antigens using monoclonal antibodies to isolated colonic glands, *Cancer Res.* 46:2164–2171.

Jass, J.R. and Roberton, A.M., 1994, Colorectal mucin histochemistry in health and disease: a critical review, *Pathology International*. 44:487–504.

Lemjabbar, H. and Basbaum, C., 2002, Platelet-activating factor receptor and ADAM10 mediate responses to *Staphylococcus aureus* in epithelial cells, *Nat Med.* 8:41–46.

McGregor, J.A., French, J.I., Jones, W., Milligan, K., McKinney, P.J., Patterson, E., and Parker, R., 1994, Bacterial vaginosis is associated with prematurity and vaginal fluid mucinase and sialidase: results of a controlled trial of topical clindamycin cream, *Am J Obstet Gynecol.* 170:1048–1059; discussion 1059–1060.

Milton, J.D., Eccleston, D., Parker, N., Raouf, A., Cubbin, C., Hoffman, J., Hart, C.A., and Rhodes, J.M., 1993, Distribution of O-acetylated sialomucin in the normal and diseased human gastrointestinal tract assessed using a newly characterized monoclonal antibody, *J Clin Path.* 45:211–217.

Montreuil, J., Vliegenthart, J.F.G., and Schachter, H., 1995, *Glycoproteins*, Elsevier Science BV., Amsterdam.

Mulvey, M.A., Schilling, J.D., Martinez, J.J., and Hultgren, S.J., 2000, Bad bugs and beleagured bladders: interplay between uropathogenic *Escherichia coli* and innate host defenses, *Proc Natl Acad Sci USA*. 97:8829–8835.

Philpott, D.J., Girardin, S.E., and Sansonetti, P.J., 2001, Innate immune responses of epithelial cells following infection with bacterial pathogens, *Cur Opin Immunol.* 13:410–416.

Ramphal, R. and Arora, S.K., 2001, Recognition of mucin components by *Pseudomonas aeruginosa*, *Glycoconj J.* 18:709–713.

Reid, P.E., Dunn, W.L., Ramey, C.W., Coret, E., Trueman, L., and Clay, M.G., 1984, Histochemical identification of side chain substituted O-acetylated sialic acids: The PAT-KOH-Bh-PAS and the PAPT-KOH-Bh-PAS procedures, *Histochem J.* 16:623–639.

Richman, P.I. and Bodmer, W.F., 1987, Monoclonal antibodies to human colorectal epithelium: Markers for differentation and tumour characterization, *Int J Cancer*. 39:317–328.

Roe, R., Corfield, A.P., and Williamson, R.C.N., 1989, Sialic acid in colonic mucin: an evaluation of modified PAS reactions in single and combination histochemical procedures, *Histochem J.* 21:216–222.

Roussel, P. and Lamblin, G., 1996, Human mucosal mucins in diseases, in: *Glycoproteins and disease*. J. Montreuil, J.F.G. Vliegenthart, and H. Schachter, eds., Elsevier Science BV, Amsterdam, pp. 351–393.

Rudd, P.M. and Dwek, R.A., 1997, Glycosylation: heterogeneity and the 3D structure of proteins, *Crit Rev Biochem Molec Biol.* 32: 1–100.

Rudd, P.M., Elliott, T., Cresswell, P., Wilson, I.A. and Dwek, R.A., 2001, Glycosylation and the immune system, *Science*, 291:2370–2376.

Scharfman, A., Degroote, S., Beau, J., Lamblin, G., Roussel, P., and Mazurier, J., 1999, *Pseudomonas aeruginosa* binds to neoglycoconjugates bearing mucin carbohydrate determinants and predominantly to sialyl-lewis x conjugates, *Glycobiology.* 9:757–764.

Schauer, R., deFreese, A., Gollub, M., Iwersen, M., Kelm, S., Reuter, G., Schlenzka, W., van Damme Feldhaus, V., and Shaw, L., 1997, Functional and biosynthetic aspects of sialic acid diversity, *Indian J Biochem Biophys.* 34:131–141.

Smithson, J.E., Campbell, A., Andrews, J.M., Milton, J.D., Pigott, R., and Jewell, D.P., 1997, Altered expression of mucins through the colon in ulcerative colitis, *Gut.* 40:234–240.

Tannock, G.W., 1994, The acquisition of the normal microflora of the gastrointestinal tract, in: *Human health: the contribution of microorganisms.* S.A.W. Gibson, ed., Springer Verlag, London, pp. 1–16.

Tannock, G.W., 1999, *Medical importance of the normal microflora*, Kluwer Academic Publishers, Dordrecht, Boston, London.

Taylor, M.E. and Drickamer, K., 2003, *Introduction to Glycobiology*, Oxford University Press, Oxford.

Varki, A., 1992, Diversity in the sialic acids, *Glycobiology.* 2:25–40.

Varki, A., 1993, Biological roles of oligosaccharides: all of the theories are correct, *Glycobiology.* 3:97–130.

Varki, A., 1997, Sialic acids as ligands in recognition phenomena (review), *FASEB J.* 11:248–255.

Varki, A., Cummings, R.D., Esko, J., Freeze, H., Hart, G., and Marth, J.D., 1999, *Essentials of Glycobiology*, Cold Spring Harbor Laboratory Press, Cold Spring Harbor, New York.

Veh, R.W., Meessen, D., Kuntz, D., and May, B., 1982, Histochemical demonstration of side-chain substituted sialic acids, in: *Colon carcinogenesis.* R.A. Malt, and R.C.N. Williamson, eds., MTP Press, Lancaster, UK, pp. 355–365.

Wiggins, R., Crowley, T., Horner, P.J., Soothill, P.W., Millar, M.R., and Corfield, A.P., 2000, Use of 5-bromo-4-chloro-3-indolyl-a-D-n-acetylneuraminic acid in a novel spot test to identify sialidase activity in vaginal swabs from women with bacterial vaginosis. *J Clin Microbiol.* 38:3096–3097

Wiggins, R., Hicks, S.J., Soothill, P.W., Millar, M.R., and Corfield, A.P., 2001, Mucinases and sialidases: their role in the pathogenesis of sexually transmitted infections in the female genital tract, *Sex Transm Infect.* 77:402–408.

Wiggins, R., Millar, M.R., Soothill, P.W., Hicks, S.J., and Corfield, A.P., 2002, Application of a novel human cervical mucin-based assay demonstrates the absence of increased mucinase activity in bacterial vaginosis, *Int J STD & AIDS.* 13:755–760.

2

THE GLYCOSYLATION OF AIRWAY MUCINS IN CYSTIC FIBROSIS AND ITS RELATIONSHIP WITH LUNG INFECTION BY *PSEUDOMONAS AERUGINOSA*

Philippe Roussel and Geneviève Lamblin

Département de Biochimie
Faculté de Médecine et Université de Lille 2
place de Verdun
59045 Lille, France

1. INTRODUCTION

Cystic fibrosis (CF) is the most common severe genetic disease among Caucasians (1/2500–1/3000 births). It affects the exocrine glands and, in its most typical form, the main symptoms are a chronic pulmonary disease, a pancreatic insufficiency with fat malabsorption, a meconium ileus at birth (in 10% CF neonates) and, later on, cirrhosis and male sterility. The diagnosis is based on the elevation of sweat electrolytes (sweat chloride: ≥70 mEq/L). In CF, there is a mucus hyper secretion as in chronic bronchitis. However, unlike chronic bronchitis, the CF lung infection is very peculiar and is characterized by infection due to *Staphylococcus aureus* in early life and, rapidly if not directly, by *Pseudomonas aeruginosa* which is almost impossible to eradicate and is responsible for most of the morbidity and mortality of the disease (Welsh *et al.*, 1995).

Cystic fibrosis is due to mutations of a gene localized on chromosome 7, *Cftr*, encoding for a *N*-glycosylated membrane glycoprotein of 1480 amino acids, CFTR (cystic fibrosis transmembrane conductance regulator) (Zielenski and Tsui, 1995). CFTR is a chloride channel of low conductance activated by protein kinase A, which influences other ion channels, such as the sodium channel ENac and the chloride channel ORCC (Devidas and Guggino, 1997), and which has probably additional unknown functions in cellular physiology which influence this susceptibility to infection.

Nearly 1000 mutations of the CF gene have been observed so far (Cystic Fibrosis mutation data base). However, in the American and Northern European populations, one mutation, the deletion of a phenylalanine residue (ΔF508), is found in about 70% of

Glycobiology and Medicine, edited by John S. Axford
Kluwer Academic / Plenum Publishers, New York, 2003

the CF chromosomes and more than 90% of the CF patients have at least one ΔF508 allele (Zielenski and Tsui, 1995). This ΔF508 mutation generates an endoplasmic reticulum storage disease since most of the mutated CFTR fails to be processed past its immature high-mannose state and is subsequently degraded (Riordan, 1999).

In spite of the discovery of the CF gene, in 1989, the pathophysiology of the lung infection is still mysterious and the treatment of the disease, although improving, is largely empirical. The airway mucosa is normally protected by the muco-ciliary system, which is made of a layer of mucus mobilized by cilia beating in the fluid film covering the surface of the airway epithelium. This system acts as an escalator trapping inhaled particles and microorganisms, which are moved up to the pharynx where they are normally swallowed. It normally maintains the bronchial tree in a sterile state.

In CF, the abnormalities of the fluid phase are still a matter of debate. A decrease in the volume of the fluid phase remaining isotonic is the most favored hypothesis (Matsui *et al.*, 2000). This affects the functioning of the muco-ciliary system. However, the question of water and solute modifications as the predominant factor in the genesis of the lung bacterial colonization is also a matter of controversy.

A major problem in the pathophysiology of CF is to relate these abnormalities to lung infection by *P. aeruginosa*, the main pathogen encountered in this disease. In the typical forms of CF, the airways are inflamed and infected. They have a tendency to be filled with mucus plugs that have entrapped bacteria and leukocytes and which are difficult to eliminate by normal muco-ciliary activity or coughing (*mucoviscidosis*).

In CF patients, airway inflammation, characterized by an influx of neutrophils and high levels of proinflammatory cytokines, has been described as early, severe and sustained (Cantin, 1995). It has been suspected that lung inflammation in CF patients might occur before bacterial colonization (Bonfield *et al.*, 1995; Chmiel *et al.*, 2002; Konstan *et al.*, 1994). However recent clinical data suggest that while a significant relationship between infection and inflammation was observed, the possibility of intrinsic inflammation could not be excluded (Dakin *et al.*, 2002).

Long before the discovery of the CF gene, investigators have been searching for glycosylation abnormalities of the airways that might pave the way for lung colonization. Alterations of glycoproteins and glycolipids from CF cells have been described but no unique alteration has been found.

The purpose of the present review is to focus on the glycosylation and sulfation of CF airway mucins which have a high affinity for *P. aeruginosa*, to compare their modifications with the alterations of membrane glycoconjugates observed in CF airway cells in culture, and to suggest pathophysiological explanations for the observed differences.

2. GLYCOSYLATION ALTERATIONS OF CF CELLS

Various abnormalities of membrane-bound glycoproteins or glycolipids from CF cells have been described.

An increased expression of asialo-GM1 has been observed on the surface of respiratory epithelial cells in primary culture and more asialo-GM1 residues were observed after exposure to *P. aeruginosa* exoproducts (Saiman and Prince, 1993). The *pilus* adhesins of *P. aeruginosa* PAK and PAO bind to the gangliosides asialo-GM1 and asialo-GM2

(Krivan, 1988) and the carbohydrate sequence β-D-GalNAc(1-> 4) β-D-Gal is the minimal carbohydrate receptor sequence of asialo-GM1 and asialo-GM2 (Sheth *et al.*, 1994).

Similarly, a decreased sialylation of glycoproteins from CF cells has been reported in various CF cells derived from the respiratory mucosa from CF patients (Barasch *et al.*, 1991; Dosanijh *et al.*, 1994). An increased fucosylation of glycoconjugates has been observed in CF cells in culture, that was reversible when these cells were transfected with wild-type *Cftr* (Glick *et al.*, 2001; Rhim *et al.*, 2001). The undersialylation was also reversible when these cells were transfected with wild-type *Cftr*.

As far as sulfation is concerned, the glycoconjugates from CF cells in culture are hypersulfated (Cheng *et al.*, 1989; Frates *et al.*, 1983; Mohapatra *et al.*, 1995). Several hypotheses have been postulated to explain these abnormalities. It has been suggested that, in CF cells, defective acidification of the *trans*-Golgi network may modify the activity (Barasch and Al-Awqati, 1993) or the cellular localization (Glick *et al.*, 2001) of several transferases. However this issue of defective acidification of the *trans*-Golgi network is controversial (Seksek *et al.*, 1996) and it has been recently suggested that there was no defect in acidification but rather a hyperacidification of endosomal organelles in CF lung epithelial cells (Poschet *et al.*, 2001).

It has also been suggested that the concentration of PAPS, the sulfate donor, is regulated in part by CFTR (Pasyk and Foskett, 1997), and therefore may influence the sulfation process. The wild-type *Cftr* would tend to lower the PAPS concentration in the Golgi lumen by letting PAPS leak out of the Golgi, whereas the lack of normal CFTR in CF would increase PAPS concentration in the Golgi and therefore favor sulfation reactions.

In CF patients, the vast majority of the *Pseudomonas* cells are mixed with mucins in the airway lumen (Jeffery and Brain, 1988; Simel *et al.*, 1984). They bind poorly to intact tracheal cells (Ramphal *et al.*, 1980). Therefore the role of the bacteria adhering to airway epithelial cells may not be the main factor for the persistent colonization of the CF airways by *P. aeruginosa*.

3. AIRWAY MUCINS

Human airway mucins represent a very large family of polydisperse high molecular mass glycoproteins, which are part of the airway innate immunity. In electron microscopy, mucins purified from human bronchial secretion appear as long, polydisperse, linear, and apparently flexible threads (Lamblin *et al.*, 2001). Their size varies from a few hundred nanometers up to more than five μm, and decreases after reduction of disulfide bridges. They can be schematized as bottlebrush with a peptidic axis covered by hundreds of carbohydrate chains (Figure 1). Most mucins are secreted in the airway lumen in contrast to membrane-bound mucins, which are attached to the apical part of airway epithelial cells.

Apomucins, which correspond to their peptide part, are encoded by at least 6 different mucin genes (Debailleul *et al.*, 1998; Moniaux *et al.*, 2001). MUC1 and MUC4 correspond to membrane-bound mucins, whereas MUC2, MUC5B, MUC5AC and MUC7 are secreted in the airway lumen. MUC2, MUC5AC and MUC5B are very large mucins secreted by goblet cells and/or mucus cells of the tracheobronchial glands. They are encoded by *MUC* genes, which have large similarities, and which belong to a cluster of genes on chromosome 11. Some *MUC* genes are polymorphic and have variable numbers of tandem repeats (VNTR). These large mucins can dimerize and multimerize.

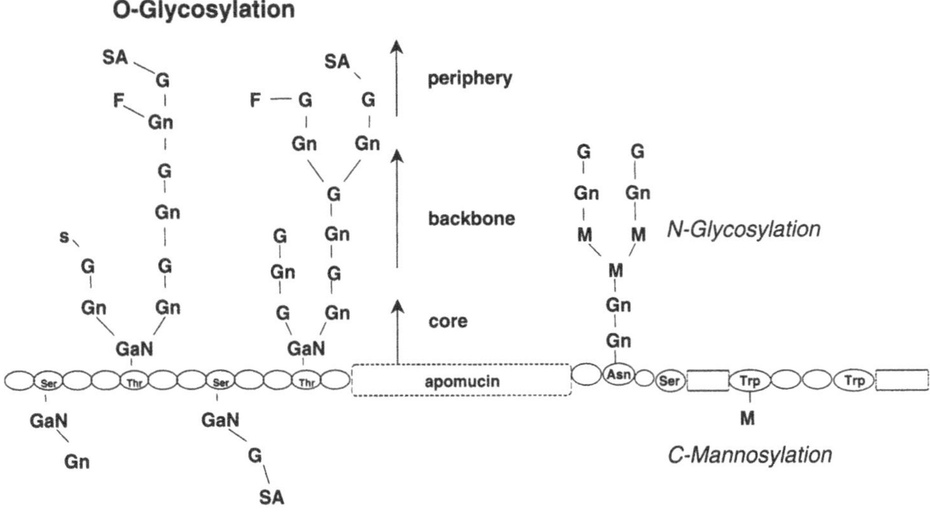

Figure 1. Schematic representation of mucin carbohydrate chains. Most chains are O-glycans, that is carbohydrate chains O-glycosidically linked to apomucin by linkages involving *N*-acetylgalactosamine (GaN) and a hydroxy-amino acid (serine or threonine). Each O-glycan can be described with a core, a backbone and a periphery. There are also a few N-glycans having the typical central pentasaccharide which are N-linked to the asparagine residue of a sequon (Asn-X-Ser [or Thr]) and possibly C-mannosyl residues attached to the tryptophan residue of a sequence (Trp-X-X-Trp) (Perez-Vilar *et al.*, 2002). F = fucose; G = galactose; Gn = *N*-acetylglucosamine; GaN = *N*-acetylgalactosamine; M = mannose; SA = sialic acid; s = sulfate.

Table 1. *MUC* genes expressed in human airways.

Genes	Chromosome	Mucin localization	aa (n)	Repeat	VNTR	Regulation
MUC1	1q21–24	Membrane-bound epithelial cell		20	+	
MUC4	3q29	Membrane-bound epithelial cell	4500–8500	16	+	
MUC2	11p15.5	Goblet cells & mucous cells	5179	23	+	TNFα, IL9, IL4, EGF, TGFα
MUC5AC	11p15.5	Goblet cells	5225	8	+	TNFα, IL9, IL4, EGF, TGFα
MUC5B	11p15.5	Mucous cells	5662	29		
MUC7	4q13-q21	Serous cells	377	23		

MUC7 corresponds to a rather small mucin secreted by the glandular serous cells (Table 1). The expression of some of these genes (at least *MUC2* and *MUC5AC*) is induced by bacterial products (lipopolysaccharide and flagellin of *Pseudomonas aeruginosa*, lipoteichoic acid of Gram-positive bacteria) (McNamara and Basbaum, 2001), as well as by tobacco smoke and different mediators: TNFα (Levine *et al.*, 1995), PGE2 (Borchers *et al.*, 1999), IL-9 (Louahed *et al.*, 2000; Longphre *et al.*, 1999), IL-4 (Dabbagh *et al.*, 1999), EGF and TGFα (Perrais *et al.*, 2002).

Human airway mucins are highly glycosylated (70–80% per weight). They contain from a single to several hundred carbohydrate chains. The carbohydrate chains that cover the apomucins are extremely diverse, adding to the complexity of these molecules. Structural information is available for more than 150 different *O*-glycan chains corresponding to the shortest chains (less than 12 sugars) (Lamblin *et al.*, 2001).

The biosynthesis of these carbohydrate chains is a stepwise process involving many glycosyl- or sulfo-transferases (Lamblin *et al.*, 2001). The only structural element shared by all mucin *O*-glycan chains is a GalNAc residue linked to a serine or threonine residue of the apomucin. There is growing evidence that the apomucin sequences influence the first glycosylation reactions (initiation of the chains) (reviewed in Lamblin *et al.*, 2001). The elongation of the chains leads to various linear or branched extensions organized as building blocks made of a disaccharide Galβ1-4GlcNAc- (some chains correspond to lactosaminoglycans) (Figure 1). Their nonreducing end, which corresponds to the termination of the chains, may bear different carbohydrate structures, such as histo-blood groups A or B determinants, H and sulfated H determinants, Lewis a, Lewis b or Lewis y epitopes, as well as sialyl- or sulfo- (sometimes sialyl- and sulfo-) Lewis a or Lewis x determinants.

The synthesis of these different terminal determinants involves different pathways with a whole set of glycosyl- and sulfo-transferases. Figure 2 represents biosynthetic pathways leading to some very common glycotopes such as the H determinant, which is a signature of the secretor status of an individual, the sulfated or sialylated derivatives of a terminal disaccharide Galβ1-4GlcNAc, or the various sialylated or/and sulfated derivatives of the Lewis x epitope.

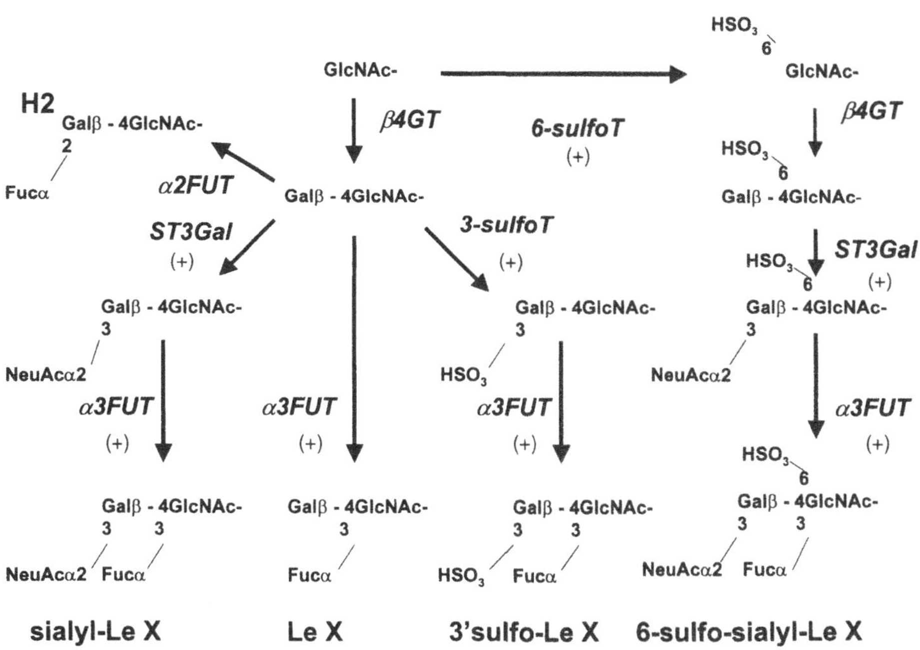

Figure 2. Transferases involved in the biosynthesis of common non-reducing ends of airway mucin O-glycans. (+) indicates the effect of TNFα on transferases.

Due to their wide structural diversity forming a combinatory of carbohydrate determinants as well as to their location at the surface of the airways, mucins are involved in multiple interactions with microorganisms and are very important in the protection of the underlying airway mucosa (Lamblin *et al.*, 2001).

4. SULFATION AND GLYCOSYLATION ALTERATIONS OF CF AIRWAY MUCINS

In addition to its role on the glycosylation of membrane glycoconjuates in epithelial cells, altered CFTR seems also to affect, in some way, the biosynthesis of glycoproteins secreted by the mucin-producing cells, and different abnormalities of mucins have been described in CF. Surprisingly, some glycosylation modifications of CF airway mucins are completely different from those observed in the glycoconjugates of CF airway cells in culture.

Davril *et al.* (1999) have purified airway mucins from 39 patients suffering either from cystic fibrosis or from chronic bronchitis. The bulk of mucins of each patient was purified using two steps of density gradient ultracentrifugation. For each patient, the importance of infection in the mucus secretion was estimated by measuring an index corresponding to the ratio of DNA present in the secretion to the carbohydrate content corresponding to the mucins. Measuring this index allowed to separate four groups of patients: patients suffering from cystic fibrosis and patients suffering from chronic bronchitis, with or without severe infection. Then, the 39 airway mucins were compared for their sialic acid and sulfate contents, as well as for the sialyl-Lewis x expression (Tables 2 and 3).

When comparing a group of mucins from CF patients with mucins from other patients, there was a significant increase in the sulfate content of the CF mucins (Table 2). This confirmed previous works showing a hypersulfation of glycoproteins or mucins secreted by CF patients (Chace *et al.*, 1985; Frates *et al.*, 1983; Lamblin *et al.*, 1977; Roussel *et al.*, 1975; Zhang *et al.*, 1995). However, when comparing the mucins from infected patients (CF and non-CF) with those from noninfected patients, the difference was even more significant raising the question of a possible influence of severe inflammation on the sulfation process of airway mucins (Table 2).

Surprisingly the sialic acid content of airway mucins from CF patients and from chronic bronchitis patients with severe infection was higher than that from non infected

Table 2. Comparison of the sulfate content of airway mucins secreted by different groups of patients (Davril *et al.*, 1999).

	Cystic fibrosis (CF) ($n = 14$)	Chronic bronchitis (CB) ($n = 24$)	p
Sulfate[a]	3.31 ± 0.25	2.67 ± 0.56	0.024
	Infected patients (CF + CB) ($n = 20$)	Non-infected patients (CF + CB) ($n = 18$)	
Sulfate	3.16 ± 0.18	2.62 ± 0.15	0.014

[a]Sulfate is expressed in % by weight.

Table 3. Sialic acid content and sialyl-Lewis x reactivity of airway mucins secreted by different groups of patients (Davril *et al.*, 1999).

	Infected CF ($n = 13$)	Infected CB ($n = 7$)	p
Sialic acid[a]	7.66 ± 0.47	6.30 ± 0.97	NS
Sialyl-Lewis x[b]	1.19 ± 0.12	1.11 ± 0.09	NS
	Infected patients (CF + CB) ($n = 20$)	Noninfected patients (CF + CB) ($n = 18$)	
Sialic acid	7.18 ± 0.46	3.92 ± 0.41	0.0001
Sialyl-Lewis x	1.16 ± 0.08	0.80 ± 0.11	0.035

[a]Sialic acid is expressed in N-acetylneuraminic acid % by weight.
[b]The reactivity of the mucin samples with an anti-sialyl-Lewis x antibody was determined in ELISA assay by measuring the absorbance at 490 nm (Davril *et al.*, 1999).
NS = Not significant.

patients (Table 3). This was also true for the sialyl-Lewis x reactivity of these different groups of mucins (Table 3).

Similar modifications of the sialic acid content have been observed in the bulk of salivary mucins secreted by patients suffering from CF. Carnoy *et al.* (1993) have shown an increased sialylation of CF salivary mucins as compared to the mucins from controls, and more recently, Shori *et al.* (2001) have shown an increased expression of sialyl-Lewis x determinants in CF salivary mucins. One should also mention that these data are in agreement with structural studies carried out on the acidic carbohydrate chains of the mucins from a patient suffering from CF severely infected by *P. aeruginosa* (Lo-Guidice *et al.*, 1994) and from two adult patients suffering from chronic bronchitis without severe infection by *P. aeruginosa* (Lo-Guidice *et al.*, 1997; Degroote *et al.*, 2003). Many oligosaccharide bearing sialyl-Lewis x epitopes were observed in the mucins from the CF patient, in contrast to the mucins from the patients suffering from chronic bronchitis without severe infection. Moreover a recent study of the polylactosaminoglycan chains of the mucins from patients suffering from CF and from chronic bronchitis with or without severe infection has shown that mucins from severely infected patients contain more polylactosaminoglycan chains terminated with a sialyl-N-acetyllactosamine group or with a sialyl-Lewis x epitope than the mucins from non severely infected patients (Morelle *et al.*, 2001).

5. INFLAMMATION AND AIRWAY MUCINS IN CF

Inflammation as such may modify the synthesis and glycosylation of certain glycoproteins. The increased secretion of acute phase glycoproteins synthesized in the liver in relation to inflammation is well known, but several recent reports have indicated possible glycosylation modifications of acute phase glycoproteins such as an increased expression of sialyl-Lewis x epitopes due to the secretion of cytokines (De Graaf *et al.*, 1993).

The different modifications observed in the mucins of severely infected patients suffering from CF or from chronic bronchitis strongly suggest that inflammation affects the glycosylation and sulfation processes in the airway mucosa of these patients.

TNFα is an important factor of airway mucosa inflammation, acting as an initial inflammatory cytokine that subsequently regulates both early neutrophil infiltration and eosinophil recruitment into the lung and airspace (Lukacs *et al.*, 1995). TNFα, as other cytokines, is found in the airways of patients suffering from bronchial diseases such as chronic bronchitis or CF (Karpati *et al.*, 2000; Osika *et al.*, 1999). In order to investigate the role of cytokines on mucin sulfation and glycosylation, explants of human airway mucosa have been exposed to TNFα (Delmotte *et al.*, 2002). The activities of galactosyl 3-*O*- and N-acetylglucosaminyl-6-*O*-sulfotransferases, of α1-3 fucosyltransferases and α2-3 sialyltransferases were increased in contrast to that of α1-2 fucosyltransferase, which was not modified (Tables 4 and 5).

TNFα also increases messenger expression of α2,3-sialyltransferases *ST3-GalIII* and *ST3-GalIV*, and of α1,3-fucosyltransferases *FUT3* and *FUT4* (Delmotte *et al.*, 2002). Therefore by acting on these different enzymes, TNFα is involved in the biosynthesis of Lewis x, sialyl-Lewis x, sulfo-Lewis x and sulfo-sialyl-Lewis x determinants by the human bronchial mucosa (Figure 2). It is possible that other cytokines, which are abundant in CF mucins and in mucins from severely infected patients, might be involved in the biosynthesis of these different epitopes.

Table 4. Effects of TNFα (20 ng/ml) on galactosyl 3-*O*- and N-acetylglucosaminyl-6-*O*-sulfotransferases activities of human airway explants (Delmotte *et al.*, 2002).

Transferases	Incubation time (h)	*n*	Control Range	Control Median	+ TNFα Range	+ TNFα Median	P^b
Gal-3-*O*-sulfotransferase[a]	4	5	6.82–11.6	8.91	23.7–29.4	25.1	*0.0431*
GlcNAc-6-*O*-sulfotransferase[a]	4	5	8.37–20.6	15.1	19.2–39.3	33.4	*0.0431*

[a]Activities was expressed as picomoles/mg protein/min.
[b]*P* was calculated using the Wilcoxon signed-rank test.

Table 5. Effects of TNFα (20 ng/ml) on fucosyltransferases and α2,3-sialyltransferases of human airway explants (Delmotte *et al.*, 2002).

Transferases	Incubation time (h)	*n*	Control Range	Control Median	+TNFα Range	+TNFα Median	P^c
α2-FUT[a]	4	5	121–193.1	158.5	130.8–191.3	166	>0.05
α3-FUT[a]	4	5	45.3–111	98	86.1–177	151	0.0431
	16	5	68.9–116	84.5	205–588	236.2	0.0277
α2-3SiaT[b]	4	5	24.6–55	38.6	25.6–130	63.1	0.0431
	16	5	85.6–122	105.7	390.6–501	453.4	0.0431

[a]FUT activities were expressed as femtomoles/mg protein/min.
[b]SiaT activities was expressed as picomoles/mg protein/min.
[c]*P* was calculated using the Wilcoxon signed-rank test.

6. HOW TO EXPLAIN THE DISCREPANCIES BETWEEN THE DATA OBSERVED WITH MUCINS SECRETED BY CF PATIENTS AND THOSE WITH AIRWAY CELLS IN CULTURE?

There are apparent discrepancies between the data obtained for membrane glycoconjugates of CF airway cells in culture and the results obtained with airway mucins from CF patients: undersialylation in one case and hypersialylation in the other. In order to explain these differences, several comments have to be made.

(i) CFTR is highly expressed in non-ciliated epithelial cells, in some duct cells and in serous cells of the tubular glands (Engelhardt *et al.*, 1992, 1994). In contrast, the expression of CFTR in cells synthesizing mucins (goblet cells and mucous glands of the acinar cells) is very low, if any (Jacquot *et al.*, 1993). Therefore the effect of the CFTR defect on mucin glycosylation may not be primary but secondary.

(ii) *In vivo*, there are differences in the sialylation pattern of the different cells of the bronchial epithelium: the sialylation of the terminal galactose residues of mucins contained in goblet cells occurs mostly on the 3-OH of these residues, whereas the sialylation of membrane-bound glycoproteins of the other bronchial cells occurs predominantly on the 6-OH of the terminal galactose residues (Couceiro *et al.*, 1993). This last observation is also in agreement with all the structural works performed on carbohydrate chains of secreted human bronchial mucins, showing very little sialylation of bronchial mucins on the 6-OH of a terminal galactose (Lamblin *et al.*, 2001).

(iii) The phenotype of airway cell lines that express CFTR is not always well characterized. Some cell lines may have a serous or sero-mucous phenotype: they may express some mucin genes but not synthesize high molecular weight mucins (Lo-Guidice *et al.*, 1997), and the expression of some glycosyltransferases may vary according to culture conditions (Delmotte *et al.*, 2001).

Therefore a possible explanation of the discrepancies observed in the sialylation of membrane glycoconjugates of CF cells in culture and airway mucins secreted by CF patients might be as follows:

(i) The alterations of membrane glycoproteins from airway cells expressing a mutated *Cftr* (increased fucosylation and decreased sialylation) (Rhim *et al.*, 2001) do correspond to a direct effect of CFTR on the glycosyltransferases of these airway epithelial cells, by a mechanism that is still discussed (modification of the pH in the Golgi, mislocalization of glycosylransferases?). The previous data concerning the sialylation defect of gangliosides (Saiman and Prince, 1993) might have a similar interpretation. The exact phenotype of these airway cells is unknown but, since they express CFTR, they should probably correspond to cells, which do not synthesize secreted mucins, such as MUC2, MUC5AC or MUC5B.

(ii) The modifications observed in the secreted airway mucins from patients suffering from CF or from severe chronic bronchitis, infected by *P. aeruginosa* (increased sialylation and sulfation as well as increased expression of the sialyl-Lewis x determinants) are most probably related to inflammation acting on goblet cells and mucous cells, which have no, or weak, expression of *Cftr*. Its mechanism may be secondary to bacterial infection, or primary as suggested by the hyperreactivity of the airways of the CF mice, or by several clinical studies indicating that inflammation in CF airways precedes bacterial infection. Different mediators, such as TNFα, might be involved in such an effect.

7. THE AFFINITY OF AIRWAY MUCINS FOR *PSEUDOMONAS AERUGINOSA*

Pseudomonas aeruginosa is an opportunistic bacteria with a strong tropism for airways of patients suffering from a defect in host defense mechanisms. It is responsible for ventilator-associated pneumonia, for the early colonization of most patients suffering from CF and of a few cases of chronic bronchial diseases, but after a much longer evolution. It can be trapped by mucus of patients suffering from CF and chronic bronchitis, and its niche is constituted by the mucus layer covering the airways.

Pseudomonas aeruginosa recognizes mucins, the main component of mucus and early studies have demonstrated that sialic acid and N-acetylglucosamine were involved in the interactions between *P. aeruginosa* and mucins (reviewed in Ramphal and Arora, 2001). Airway mucins and salivary mucins from CF patients have been found to have an increased affinity for *P. aeruginosa* (Carnoy *et al.*, 1993; Devaraj *et al.*, 1994), but all airway mucins studied so far bind, more or less, this microorganism. It was therefore suggested that *P. aeruginosa* had one or several adhesins with affinity for some of the diverse carbohydrate chains covering the airway mucins (Carnoy *et al.*, 1994).

Pseudomonas aeruginosa is a piliated microorganism and pilin has been found to bind to the Galβ1-4GalNAc sequence observed in asialo-GM1. This disaccharide sequence, which is characteristic of gangliosides, has never been found in mucins. Moreover the preparation of non-piliated strains of *P. aeruginosa* showed that the bacteria still adhere to mucins, even more strongly (Ramphal *et al.*, 1991b), as well as to different glycopeptides obtained by proteolysis of airway mucins (Ramphal *et al.*, 1989). Non-pilus adhesins have been identified using mutated strains of *P. aeruginosa* (PAK and PAO1) (Carnoy *et al.*, 1994). The observation that nonmotile mutants had a low affinity for airway mucins, indicated that proteins of the flagella might be involved in the binding to mucins and the flagellar cap protein (Fli D) of strain PAK has been identified as a major factor of the binding to mucins (Arora *et al.*, 1998). More recently the flagellar cap protein (Fli D) and the flagellin (Fli C), of strain PAO1 have been found to be involved in the binding to mucins (Scharfman *et al.*, 2001). In order to identify some of the carbohydrate ligands involved in the binding to *P. aeruginosa*, various approaches have been used. They were mainly based on the use of a single type of oligosaccharide representing the non-reducing end of one of the multiple chains covering the airway mucins. These oligosaccharides were either substituted with an hydrophobic tail in order to allow their binding to plastic, or they were hooked to a polyacrylamide to generate a neoglycoconjugate with several identical chains, which, moreover, could be labeled with fluorescein. The binding of these different neoglycoconjugates to various strains of bacteria has been studied using either a microtiter plate adhesion assay or flow cytometry, and various carbohydrate ligands have been defined (Table 6). The best ligands are the sialyl-Lewis x and the 3-sialyl-6-sulfo-Lewis x determinants (Scharfman *et al.*, 1999, 2000). Moreover, in the case of the PAO1 strain, specific ligands of the flagellar proteins have been identified, sialyl-Lewis x and Lewis x for the flagellar cap protein, and Lewis x for the flagellin (Scharfman *et al.*, 2001). These ligands were not the specific ligands of the corresponding proteins of the PAK strain.

Table 6. Carbohydrate ligands of *P. aeruginosa* strains.

Carbohydrate ligands	*P. aeruginosa* strains	Reference
Galβ1,4GlcNAcβ1,3Galβ1,4Glc-	M35 (mucoid) 1244 (piliated and non-piliated)	Ramphal *et al.*, 1991a; Rosenstein *et al.*, 1992
	PAK (piliated and non-piliated)	Ramphal *et al.*, 1991a
Galβ1,3GlcNAcβ1,3Galβ1,4Glc-	M35 (mucoid) 1244 (piliated and non-piliated)	Ramphal *et al.*, 1991a; Rosenstein *et al.*, 1992
		Ramphal *et al.*, 1991a
NeuAcα2,3Galβ1,3GlcNAcβ1,3Gal-	1244 (piliated)	Ramphal *et al.*, 1991a
Galβ1,3[Fucα1,4]GlcNAc- *(Lewis a)*	1244 (non-piliated) clinical strains	Scharfman *et al.*, 1999
Galβ1,4[Fucα1,3]GlcNAc- *(Lewis x)*	1244 (non-piliated) clinical strains	Scharfman *et al.*, 1999
Fucα1,2Galβ1,4[Fucα1,3]GlcNAc- *(Lewis y)*	1244 (non-piliated) clinical strains	Scharfman *et al.*, 1999
NeuAcα2,3Galβ1,4[Fucα1,3]GlcNAc- *(sialyl-Lewis x)*	1244 (non-piliated) clinical strains	Scharfman *et al.*, 1999
3-sulfo-Galβ1,4[Fucα1,3]GlcNAc- *(3-sulfo-Lewis x)*	1244 (non-piliated) clinical strains	Scharfman *et al.*, 1999
NeuAcα2,3Galβ1,4[6-sulfo][Fucα1,3]GlcNAc- *(3-sialyl-6-sulfo-Lewis x)*	1244 (non-piliated) clinical strains	Scharfman *et al.*, 2000

Therefore, neoglycoconjugates are very useful to analyze the affinity of a given type of ligand, but one should mention: (i) that they may not reproduce the environment of that ligand in the airway mucin molecule and (ii) that the affinity of a bacteria for a specific ligand on a mucin molecule or on a neoglycoconjugate may not be necessarily identical.

In summary, the binding of *P. aeruginosa* to mucins probably involves different types of carbohydrate, especially the acidic derivatives of the Lewis x determinants induced by inflammation.

8. A POSSIBLE MODEL TO EXPLAIN THE AIRWAY MUCUS COLONIZATION IN CF

The airway colonization by *P. aeruginosa* in CF is most probably a multifactorial phenomenon involving at least three parameters that influence the muco-ciliary escalator.

(i) *The first parameter is the influence of the hydration of the film, which covers the airway epithelium.* Although it is still a matter of debate, the volume of the liquid layer is probably reduced. This may affect the mechanical efficiency of the muco-ciliary system. However this abnormality may not be the only reason for the specific colonization by *P. aeruginosa*, since patients suffering from immotile cilia syndromes get colonized at a much older age than patients with CF, if any.

(ii) *The second parameter is the hyperreactivity of the airway epithelial cells.* The increased inflammation in CF corresponds, at least, to an exaggerated and prolonged inflammatory response to a given bacterial stimulus. The data obtained with CF mice have been disappointing because, in contrast to the human disease, there was no spontaneous infection of the airways by *P. aeruginosa*. However, van Heeckeren *et al.* (1997) have reported elevated levels of inflammatory cytokines corresponding to an excessive inflammatory response of the airways of these mice, when they were challenged with *P. aeruginosa*. Sajjan *et al.* (2001) have also shown that $Cftr^{-/-}$ knockout mice exposed to repeated instillation of *Burkholderia cepacia* demonstrated an enhanced inflammatory response but apparently less effective than the $Cftr^{+/+}$. More recently, Oceandy *et al.* (2002) found that gene complementation of the airway epithelium in the CF mouse corrects the inflammatory abnormalities. Several reports have shown that airway epithelial cells expressing mutated *Cftr* have increased secretion of pro-inflammatory cytokines and exaggerated activation of NF-κB (Blackwell *et al.*, 2001). In addition, CF airways cells in culture bind more bacteria. When stimulated by *P. aeruginosa* or *Staphylococcus aureus*, they secrete more G-CSF and GM-CSF and increase recruitment and survival of PMN (Saba *et al.*, 2002). In cell cultures these effects involve asialo-glycolipid receptors, which are more abundant on CF airway cells (Saiman and Prince, 1993). However the question of the primary role of this receptor is still open since, in explants, these receptors seem to be only expressed on regenerating epithelial cells and not on the other cells (de Bentzman *et al.*, 1996).

(iii) *Last but not the least, the third parameter corresponds to the hypersecretion and biosynthesis modifications of CF airway mucins, produced by goblet and mucous cells, which react to various stimuli, especially cytokines, related to inflammation and infection.* This increased secretion of mucins is responsible for the airway mucus plugs (*mucoviscodosis*) embedding most of the *P. aeruginosa* cells present in the airways. These modifications are not primary since the mucin-secreting cells do not seem to express *Cftr*. They most probably result from two different types of stimulation, one acting on the expression of mucin genes and the other one on the glycosylation and sulfation processes of these mucins. *P. aeruginosa*, flagellin, and LPS up-regulate MUC2 transcription through ATP release (McNamara and Basbaum, 2001). *Staphylococcus aureus* and lipoteichoic acid also stimulate MUC2 induction (McNamara and Basbaum, 2001). Various cytokines (TNFα, IL9 and IL4) also induce the expression of several mucin genes. Similarly inflammation increases the expression of different transferases involved in the biosynthesis of various Lewis x derivatives, such as the sialyl-Lewis x or the sulfo-sialyl-Lewis x (Delmotte *et al.*, 2002), which are the best ligands so far identified for *Pseudomonas* adhesins (Scharfman *et al.*, 2000). Finally, the bacteria embedded in mucins may escape the process of opsonophagocytosis (Ramphal and Vishwanath, 1987).

In conclusion, in CF, abnormal hyperreactivity of airway epithelial cells generate inflammatory mediators able to modify the mucin biosynthesis in goblet and mucous cells, especially by increasing expression of the acidic derivatives of the Lewis x determinants, which probably play a major role in the binding to *P. aeruginosa* and in the chronic colonization of the CF airways by this bacteria.

9. ACKNOWLEDGMENTS

We would like also to acknowledge the long support of the association Vaincre la Mucoviscidose.

10. REFERENCES

Arora, S.K., Ritchings, B.W., Almira, E.C., Lory, S., and Ramphal, R., 1998, The *Pseudomonas aeruginosa* flagellar cap protein, FliD, is responsible for mucin adhesion, *Infect Immun.* 66:1000–1007.

Barasch, J., and Al-Awqati, Q., 1993, Defective acidification of the biosynthetic pathway in cystic fibrosis, *J Cell Sci Suppl.* 17:229–233.

Barasch, J., Kiss, B., Prince, A., Saiman, L., Gruenert, D., and Al-Awqati, Q., 1991, Defective acidification of intracellular organelles in cystic fibrosis, *Nature.* 352:70–73.

Blackwell, T.S., Stecenko, A.A., and Christman, J.W., 2001, Dysregulated NF-κB activation in cystic fibrosis: evidence for a primary inflammatory disorder, *Am J Physiol Lung Cell Mol Physiol.* 281:L69–L70.

Bonfield, T.L., Panuska, J.R., Konstan, M.W., Hilliard, K.A., Hilliard, J.B., Ghnaim, H., and Berger, M., 1995, Inflammatory cytokines in cystic fibrosis lungs, *Am J Respir Crit Care Med.* 152:2111–2118.

Borchers, M.T., Carty, M.P., and Leikauf, G.D., 1999, Regulation of human airway mucins by acrolein and inflammatory mediators, *Am J Physiol.* 276:L549–L555.

Cantin, A., 1995, Cystic fibrosis lung inflammation: early, sustained, and severe, *Am J Respir Crit Care Med.* 151:939–941.

Carnoy, C., Ramphal, R., Scharfman, A., Houdret, N., Lo-Guidice, J.-M., Klein, A., Galabert, C., Lamblin, G., and Roussel, P., 1993, Altered carbohydrate composition of salivary mucins from patients with cystic fibrosis and the adhesion of *Pseudomonas aeruginosa, Am J Respir Cell Mol Biol.* 9:323–334.

Carnoy, C., Scharfman, A., Van Brussel, E., Lamblin, G., Ramphal, R., and Roussel, P., 1994, *Pseudomonas aeruginosa* outer membrane adhesins for human respiratory mucus glycoproteins, *Infect Immun.* 62:1896–1900.

Chace, K.V., Flux, M., and Sachdev, G.P., 1985, Comparison of physicochemical properties of purified mucus glycoproteins isolated from respiratory secretions of cystic fibrosis and asthmatic patients, *Biochemistry.* 24:7334–7341.

Cheng, P.W., Boat, T.F., Cranfill, K., Yankaskas, J.R., and Boucher, R.C., 1989, Increased sulfation of glycoconjugates by cultured nasal epithelial cells from patients with cystic fibrosis, *J Clin Invest.* 84:68–72.

Chmiel, J.F., Berger, M., and Konstan, M.W., 2002, The role of inflammation in the pathophysiology of CF lung disease, *Clin Rev Allergy Immunol.* 23:5–27.

Couceiro, J.N., Paulson, J.C., and Baum, L.G., 1993, Influenza virus strains selectively recognize sialyloligosaccharides on human respiratory epithelium: the role of the host cell in selection of hemagglutinin receptor specificity, *Virus Res.* 29:155–165.

Cystic Fibrosis Mutation Data Base—http://www.genet.sickkids.on.ca/cftr/201.

Dabbagh, K., Takeyama, K., Lee, H.M., Ueki, I.F., Lausier, J.A., and Nadel, J.A., 1999, IL-4 induces mucin gene expression and goblet cell metaplasia *in vitro* and *in vivo, J Immunol.* 162:6233–6237.

Dakin, C.J., Numa, A.H., Wang, H., Morton, J.R., Vertzyas, C.C., and Henry, R.L., 2002, Inflammation, infection, and pulmonary function in infants and young children with cystic fibrosis, *Am J Respir Crit Care Med.* 165:904–10.904–10.

Davril, M., Degroote, S., Humbert, P., Galabert, C., Dumur, V., Lafitte, J.-J., Lamblin, G., and Roussel, P., 1999, The sialylation of bronchial mucins secreted by patients suffering from cystic fibrosis or from chronic bronchitis is related to the severity of airway infection, *Glycobiology.* 9:311–321.

Debailleul, V., Laine, A., Huet, G., Mathon, P., d'Hooghe, M.C., Aubert, J.-P., and Porchet, N., 1998, Human mucin genes *MUC2, MUC3, MUC4, MUC5AC, MUC5B*, and *MUC6* express stable and extremely large mRNAs and exhibit a variable length polymorphism—An improved method to analyze large mRNAs, *J Biol Chem.* 273:881–890.

de Bentzmann, S., Roger, P., Dupuit, F., Bajolet-Laudinat, O., Fuchey, C., Plotkowski, M.C., and Puchelle, E., 1996, Asialo GM1 is a receptor for *Pseudomonas aeruginosa* adherence to regenerating respiratory epithelial cells, *Infect Immun.* 64:1582–1588.

De Graaf, T.W., Van der Stelt, M.E., Anbergen, M.G., and van Dijk, W., 1993, Inflammation-induced expression of sialyl-Lewis X-containing glycan structures on α1-acid glycoprotein (orosomucoid) in human sera, *J Exp Med.* 177:657–666.

Degroote, S., Maes, E., Humbert, P., Delmotte, P., Lamblin, G., and Roussel, P., 2003, Sulfated oligosaccharides isolated from the respiratory mucins of a secretor patient suffering from chronic bronchitis, *Biochimie.* 85:369–379

Delmotte, P., Degroote, S., Lafitte, J.-J., Lamblin, G., Perini, J.-M., and Roussel, P., 2002, Tumor necrosis factor alpha increases the expression of glycosyltransferases and sulfotransferases responsible for the biosynthesis of sialylated and/or sulfated Lewis x epitopes in the human bronchial mucosa, *J Biol Chem.* 277: 424–431.

Delmotte, P., Degroote, S., Merten, M., Bernigaud, A., Van Seuningen, I., Figarella, C., Roussel, P., and Perini, J.-M., 2001, Influence of culture conditions on the alpha 1,2-fucosyltransferase and MUC gene expression of a transformed cell line MM-39 derived from human tracheal gland cells, *Biochimie.* 83:749–755.

Devaraj, N., Sheykhnazari, M., Warren, W.S., Bhavanandan, V.P., 1994, Differential binding of *Pseudomonas aeruginosa* to normal and cystic fibrosis tracheobronchial mucins, *Glycobiology*. 4:307–316.

Devidas, S. and Guggino, W.B., 1997, CFTR: domains, structure, and function, *Bioenerg Biomembr.* 29:443–451.

Dosanjh, A., Lencer, W., Brown, D., Ausiello, D.A., and Stow, J.L., 1994, Heterologous expression of ΔF508 CFTR results in decreased sialylation of membrane glycoconjugates, *Am J Physiol.* 266:C360–C366.

Engelhardt, J.F., Yankaskas, J.R., Ernst, S.A., Yang, Y., Marino, C.R., Boucher, R.C., Cohn, J.A., and Wilson, J.M., 1992, Submucosal glands are the predominant site of CFTR expression in the human bronchus, *Nat Genet.* 3:240–248.

Engelhardt, J.F., Zepeda, M., Cohn, J.A., Yankaskas, J.R., Wilson, J.M., 1994, Expression of the cystic fibrosis gene in adult human lung, *J Clin Invest.* 93:737–749.

Frates, R.C. Jr, Kaizu, T.T., and Last, J.A., 1983, Mucus glycoproteins secreted by respiratory epithelial tissue from cystic fibrosis patients, *Pediatr Res.* 17:30–34.

Glick, M.C., Kothari, V.A., Liu, A., Stoykova, L.I., and Scanlin, T.F., 2001, Activity of fucosyltransferases and altered glycosylation in cystic fibrosis airway epithelial cells, *Biochimie.* 83:743–747.

Jacquot, J., Puchelle, E., Hinnrasky, J., Fuchey, C., Bettinger, C., Spilmont, C., Bonnet, N., Dieterle, A., Dreyer, D., Pavirani, A. *et al.*, 1993, Localization of the cystic fibrosis transmembrane conductance regulator in airway secretory glands, *Eur Respir J.* 6:169–176.

Jeffery, P.K. and Brain, A.P.R., 1988, Surface morphology of human airway mucosa: normal, carcinoma or cystic fibrosis, *Scan Microsc.* 2:345–351.

Karpati, F., Hjelte, F.L., and Wretlind, B., 2000, TNF-alpha and IL-8 in consecutive sputum samples from cystic fibrosis patients during antibiotic treatment, *Scand J Infect Dis.* 32:75–79.

Konstan, M.W., Hilliard, K.A., Norvell, T.M., and Berger, M., 1994, Bronchoalveolar lavage findings in cystic fibrosis patients with stable, clinically mild lung diseases suggest ongoing infection and inflammation, *Am J Respir Crit Care Med.* 150:448–454.

Krivan, H.C., Ginsburg, V., Roberts, D.D., 1988, *Pseudomonas aeruginosa* and *Pseudomonas cepacia* isolated from cystic fibrosis patients bind specifically to gangliotetraosylceramide (asialo GM1) and gangliotriaosylceramide (asialo GM2), *Arch Biochem Biophys.* 260:493–496.

Lamblin, G., Degroote, S., Perini, J.-M., Delmotte, P., Scharfman, A., Davril, M., Lo-Guidice J.-M., Houdret, N., Dumur, V., Klein, A., and Roussel, P., 2001, Human airway mucin glycosylation: a combinatory of carbohydrate determinants which vary in cystic fibrosis, *Glycoconj J.* 18:661–684.

Lamblin, G., Lafitte, J.-J., Lhermitte, M., Degand, P., and Roussel, P., 1977, Mucins from cystic fibrosis sputum, *Mod Probl Paediat.* 19:153–164.

Levine, S.J., Larivee, P., Logun, C., Angus, C.W., Ognibene, F.P., and Shelhamer, J.H., 1995, Tumor necrosis factor-α induces mucin hypersecretion and *MUC-2* gene expression by human airway epithelial cells, *Am J Respir Cell Mol Biol.* 12:196–204.

Lo-Guidice, J.-M., Herz, H., Lamblin, G., Plancke, Y., Roussel, P., and Lhermitte, M., 1997, Structures of sulfated oligosaccharides isolated from the respiratory mucins of a non-secretor (O, $Le^{a+\ b-}$) patient suffering from chronic bronchitis, *Glycoconj J.* 14:113–125.

Lo-Guidice, J.-M., Merten, M.D., Lamblin, G., Porchet, N., Houvenaghel, M.-C., Figarella, C., Roussel, P., and Perini, J.-M., 1997, Mucins secreted by a transformed cell line derived from human tracheal gland cells, *Biochem J.* 326:431–437.

Lo-Guidice, J.-M., Wieruszewski, J.-M., Lemoine, J., Verbert, A., Roussel, P., and Lamblin, G., 1994, Sialylation and sulfation of the carbohydrate chains in respiratory mucins from a patient with cystic fibrosis, *J Biol Chem.* 269:18794–18813.

Longphre, M., Li, D., Gallup, M., Drori, L., Ordonez, C.L., Redman, T., Wenzel, S., Bice, D.E., Fahy, J.V., and Basbaum, C., 1999, Allergen-induced IL-9 directly stimulates mucin transcription in epithelial cells, *J Clin Invest.* 104:1375–1382.

Louahed, J., Toda, M., Jen, J., Hamid, Q., Renauld, J.C., Levitt, R.C., and Nicolaides, N.C., 2000, Interleukin-9 upregulates mucus expression in the airways, *Am J Respir Cell Mol Biol.* 22:649–656.

Lukacs, N.W., Strieter, R.M., Chensue, S.W., Widmer, M., and Kunkel, S.L., 1995, TNF-alpha mediates recruitment of neutrophils and eosinophils during airway inflammation, *J Immunol.* 154:5411–5417.

McNamara, N. and Basbaum, C., 2001, Signaling networks controlling mucin production in response to Gram-positive and Gram-negative bacteria, *Glycoconj J.* 18:715–722.

Matsui, H., Davis, C.W., Tarran, R., and Boucher, R.C., 2000, Osmotic water permeabilities of cultured, well-differentiated normal and cystic fibrosis airway epithelia, *J Clin Invest.* 105:1418–1427.

Mohapatra, N.K., Cheng, P.W., Parker, J.C., Paradiso, A.M., Yankaskas, J.R., Boucher, R.C., and Boat, T.F., 1995, Alteration of sulfation of glycoconjugates, but not sulfate transport and intracellular inorganic sulfate content in cystic fibrosis airway epithelial cells, *Pediatr Res.* 38:42–48.

Moniaux, N., Escande, F., Porchet, N., Aubert, J.-P., and Batra, S.K., 2001, Structural organization and classification of the mucin genes, *Front Biosci.* 6:d1192–1206.

Morelle, W., Sutton-Smith, M., Morris, H.R., Davril, M., Roussel, P., and Dell, A., 2001, FAB-MS characterization of sialyl Lewis^x determinants on polylactosamine chains of human airway mucins secreted by patients suffering from cystic fibrosis or chronic bronchitis, *Glycoconj J.* 18:699–708.

Oceandy, D., McMorran, B.J., Smith, S.N., Schreiber, R., Kunzelmann, K., Alton, E.W., Hume, D.A., and Wainwright, B.J., 2002, Gene complementation of airway epithelium in the cystic fibrosis mouse is necessary and sufficient to correct the pathogen clearance and inflammatory abnormalities, *Hum Mol Genet.* 11:1059–1067.

Osika, E., Cavaillon, J.-M., Chadelat, K., Boule, M., Fitting, C., Tournier, G., and Clement, A., 1999, Distinct sputum cytokine profiles in cystic fibrosis and other chronic inflammatory airway disease, *Eur Respir J.* 14:339–346.

Pasyk, E.A. and Foskett, J.K., 1997, Cystic fibrosis transmembrane conductance regulator-associated ATP and adenosine 3′-phosphate 5′-phosphosulfate channels in endoplasmic reticulum and plasma membranes, *J Biol Chem.* 272:7746–7751.

Perez-Vilar, J., Randell, S.H., and Boucher, R., 2002, The cys subdomains of human gel-forming mucins are C-mannosylated domains involved in weak protein-protein interactions, *Pediatric Pulmonol.* Suppl 24:190.

Perrais, M., Pigny, P., Copin, M.-C., Aubert, J.-P., and Van Seuningen, I., 2002, Induction of MUC2 and MUC5AC mucins by factors of the epidermal growth factor (EGF) family is mediated by EGF receptor/Ras/Raf/extracellular signal-regulated kinase cascade and Sp1, *J Biol Chem.* 277:32258–32267.

Poschet, J.F., Boucher, J.C., Tatterson, L., Skidmore, J., Van Dyke, R.W., and Deretic, V., 2001, Molecular basis for defective glycosylation and *Pseudomonas* pathogenesis in cystic fibrosis lung, *Proc Natl Acad Sci USA.* 98:13972–13977.

Ramphal, R. and Arora, S.K., 2001, Recognition of mucin components by *Pseudomonas aeruginosa, Glycoconj J.* 18:709–713.

Ramphal, R., Carnoy, C., Fievre, S., Michalski, J.-C., Houdret, N., Lamblin, G., Strecker, G., and Roussel, P., 1991a, *Pseudomonas aeruginosa* recognizes carbohydrate chains containing type 1 (Galβ1-3GlcNAc) or type 2 (Galβ1-4GlcNAc) disaccharide units, *Infect Immun.* 59:700–704.

Ramphal, R., Houdret, N., Koo, L., Lamblin, G., and Roussel, P., 1989, Differences in adhesion of *Pseudomonas aeruginosa* to mucin glycopeptides from sputa of patients with cystic fibrosis and chronic bronchitis, *Infect Immun.* 57:3066–30671.

Ramphal, R., Koo, L., Ishimoto, K.S., Totten, P.A., Lara, J.C., and Lory, S., 1991b, Adhesion of *Pseudomonas aeruginosa* pilin-deficient mutants to mucin, *Infect Immun.* 59:1307–1311.

Ramphal, R., Small, P.M., Shands, J.W. Jr, Fischlschweiger, W., and Small, P.A. Jr, 1980, Adherence of *Pseudomonas aeruginosa* to tracheal cells injured by influenza infection or by endotracheal intubation. *Infect Immun.* 27:614–619.

Ramphal, R. and Vishwanath, S., 1987, Why is *Pseudomonas* the colonizer and why does it persist, *Infection.* 15:281–287.

Rhim, A.D., Stoykova, L., Glick, M.C., and Scanlin, T.F., 2001, Terminal glycosylation in cystic fibrosis (CF): a review emphasizing the airway epithelial cell, *Glycoconj J.* 8:649–659

Riordan, J.R., 1999, Cystic fibrosis as a disease of misprocessing of the cystic fibrosis transmembrane conductance regulator glycoprotein, *Am J Hum Genet.* 64:1499–1504.

Rosenstein, I.J., Yuen, C.T., Stoll, M.S., and Feizi, T., 1992, Differences in the binding specificities of *Pseudomonas aeruginosa* M35 and *Escherichia coli* C600 for lipid-linked oligosaccharides with lactose-related core regions, *Infect Immun.* 60:5078–5084.

Roussel, P., Lamblin, G., Degand, P., Walker-Nasir, E., and Jeanloz, R.W., 1975, Heterogeneity of the carbohydrate chains of sulfated bronchial glycoproteins isolated from a patient suffering from cystic fibrosis, *J Biol Chem.* 250:2114–2122.

Saba, S., Soong, G., Greenberg, S., and Prince, A., 2002, Bacterial stimulation of epithelial G-CSF and GM-CSF expression promotes PMN survival in CF airways, *Am J Respir Cell Mol Biol.* 27:561–567.

Saiman, L. and Prince, A., 1993, *Pseudomonas aeruginosa* pili binds asialo-GM1 which is increased at the surface of cystic fibrosis epithelial cells, *J Clin Invest.* 92:1875–1880.

Sajjan, U., Thanassoulis, G., Cherapanov, V., Lu, A., Sjolin, C., Steer, B., Wu, Y.J., Rotstein, O.D., Kent, G., McKerlie, C., Forstner, J., and Downey, P., 2001, Enhanced susceptibility to pulmonary infection with *Burkholderia cepacia* in *Cftr*$^{-/-}$ mice, *Infect Immun.* 69:5138–5150.

Scharfman, A., Arora, S.K., Delmotte, P., Van Brussel, E., Mazurier, J., Ramphal, R., and Roussel, P., 2001, Recognition of Lewis x derivatives present on mucins by flagellar components of *Pseudomonas aeruginosa, Infect Immun.* 69:5243–5248.

Scharfman, A., Degroote, S., Beau, J., Lamblin, G., Roussel, P., and Mazurier, J., 1999, *Pseudomonas aeruginosa* binds to neoglycoconjugates bearing mucin carbohydrate determinants and predominantly to sialyl-Lewis x conjugates. *Glycobiology.* 9:757–764.

Scharfman, A., Delmotte, P., Beau, J., Lamblin, G., Roussel, P., and Mazurier, J., 2000, Sialyl-Le(x) and sulfo-sialyl-Le(x) determinants are receptors for *P. aeruginosa. Glycoconj J.* 10: 735–740.

Seksek, O., Biwersi, J., and Verkman, A.S., 1996, Evidence against defective *trans*-Golgi acidification in cystic fibrosis, *J Biol Chem.* 271:15542–15548.

Sheth, H.B., Lee, K.K., Wong, W.Y., Srivastava, G., Hindsgaul, O., Hodges, R.S., Paranchych, W., and Irvin, R.T., 1994, The pili of *Pseudomonas aeruginosa* strains PAK and PAO bind specifically to the carbohydrate sequence beta GalNAc(1-4)betaGal found in glycosphingolipids asialo-GM1 and asialo-GM2, *Mol Microbiol.* 11:715–23202.

Shori, D.K., Genter, T., Hansen, J., Koch, C., Wyatt, H., Kariyawasam, H.H., Knight, R.A., Hodson, M.E., Kalogeridis, A., and Tsanakas, I., 2001, Altered sialyl- and fucosyl-linkage on mucins in cystic fibrosis patients promotes formation of the sialyl-Lewis X determinant on salivary MUC-5B and MUC-7, *Pflugers Arch.* 443 Suppl:S55–S61.

Simel, D.L., Mastin, J.P., Pratt, P.C., Wisseman, C.L., Shelburne, J.D., Spock, A., Ingram, P., 1984, Scanning electron microscopic study of the airways in normal children and in patients with cystic fibrosis and other lung diseases, *Pediatr Pathol.* 2:47–64.

van Heeckeren, A., Walenga, R., Konstan, M.W., Bonfield, T., Davis, P.B., and Ferkol, T., 1997, Excessive inflammatory response of cystic fibrosis mice to bronchopulmonary infection with *Pseudomonas aeruginosa, J Clin Invest.* 100:2810–2815.

Welsh, M.J., Tsui, L.-C., Boat, T.F., and Beaudet, A.L., 1995, Cystic fibrosis, in: *The metabolic and molecular bases of inherited disease*. C.R. Scriver, A.L. Beaudet, W.S. Sly, and D. Valle, eds., (McGraw-Hill Inc.) pp. 3799–3876.

Zhang, Y., Doranz, B., Yankaskas, J.R., and Engelhardt, J.F., 1995, Genotypic analysis of respiratory mucous sulfation defects in cystic fibrosis, *J Clin Invest.* 96:2997–3004.

Zielenski, J. and Tsui, L.-C., 1995, Cystic fibrosis: genotypic and phenotypic variations, *Annu Rev Genet.* 29:777–807.

3

STRUCTURAL BASIS FOR BACTERIAL ADHESION IN THE URINARY TRACT

Jenny Berglund and Stefan D. Knight

Department of Molecular Biosciences/Structural Biology
Uppsala Biomedical Center
Swedish University of Agricultural Sciences
Box 590, SE-753 24 Uppsala, Sweden

1. INTRODUCTION

Most bacterial infections occur in the respiratory, gastrointestinal, or urinary tract. These spaces offer attractive advantages for bacteria in the form of nutrient availability, and are readily accessible from the outside world. In many cases bacterial habitation is not compatible with well being of the host, and mammals have developed powerful counter-measures and clearance mechanisms in order to limit bacterial colonization. Pathogenic bacteria have in turn developed solutions to overcome these challenges. One of the first obstacles facing bacteria striving to colonize one of the epithelial tracts is the cleansing action exerted by the flow of, for example, saliva, mucus, or urine. It is therefore no surprise that bacterial pathogenesis frequently involves adhesion of the pathogen to host epithelial tissues, and that adhesion in many cases is a first crucial event in establishing colonization and infection.

Urinary tract infections (UTIs) are among the most common bacterial infections, estimated to affect at least 50% of women over life at a yearly cost of ~$2 billion in the US alone (Foxman, 2002). The most common cause of UTI is infection by uropathogenic *Escherichia coli* (UPEC), which accounts for about 80% of reported cases (Ronald, 2002). UTI can in most cases be effectively treated with antibiotics, but recurrence is a problem as is the emergence of resistant strains (Gupta *et al.*, 2001; Johnson *et al.*, 2002; Nicolle, 2002; Ronald, 2002). Currently no approved UPEC vaccine is available.

A number of UPEC adhesive organelles (Table 1) have been identified as critical UTI virulence factors (Johnson, 1991; Foxman *et al.*, 1995; Mulvey, 2002; Wang *et al.*, 2002) and are attractive targets for novel therapeutics and vaccines. Type 1 pili recognize mannose-containing receptors present on the luminal surface of the bladder epithelium and are critical for the establishment of cystitis (Bahrani-Mougeot *et al.*, 2002; Connell *et al.*, 1996;

Glycobiology and Medicine, edited by John S. Axford
Kluwer Academic / Plenum Publishers, New York, 2003

Table 1. UTI-associated adhesion organelles.

Organelle	UTI	Receptor(s)	Adhesin
Type 1 pilus	Cystitis	Mannose	FimH
P pilus	Pyelonephritis	Galabiose	PapG
F1C pilus	Ascending UTI	Galactosylceramide, globotriaosylceramide	FocH
S pilus	Ascending UTI	Sialic acid	SfaS
		Galactosylceramide	SfaH?
Dr adhesins (non-pilus)	Cystitis, Pyelonephritis	DAF, type IV collagen	DraE

Langermann *et al.*, 1997; Mulvey *et al.*, 1998; Thankavel *et al.*, 1997). P pili mediate binding to Galα(1-4)Gal-containing receptors in the upper urinary tract to cause pyelonephritis (Roberts *et al.*, 1994). The closely related F1C and S pili bind to galactosyl ceramide and globotriaosyl ceramide receptors (Bäckhed *et al.*, 2002; Khan *et al.*, 2000) and to sialic-acid-containing receptors (Hanisch *et al.*, 1993; Korhonen *et al.*, 1984) respectively and are implicated in ascending UTIs. Adhesive organelles belonging to the Dr family are frequently found on UPEC strains isolated from patients with cystitis or pyelonephritis (Nowicki *et al.*, 2001). These organelles do not appear to recognize carbohydrate structures but instead interact with the Dr(a+) antigen present on the decay-accelerating factor (DAF; CD55). At least one member of the Dr family also recognises type IV collagen.

In this short review we will summarize recent structural advances in understanding the biogenesis and binding properties of the adhesion organelles involved in colonization of the urinary tract and establishment of UTI.

2. ULTRASTRUCTURE AND BIOGENESIS OF UPEC ADHESIVE ORGANELLES

All of the UPEC adhesive organelles discussed in this review are large filamentous protein polymers. Type 1, P, F1C, and S pili are examples of the hair-like adhesive organelles referred to as pili or fimbriae. These ~7 nm wide and 1–2 μm long heteropolymeric appendages are constructed from a number of weakly homologous protein subunits (pilins) with a molecular weight of ~15 kDa. Typically, several hundred adhesive pili are evenly distributed over the bacterial cell surface. High-resolution electron microscopy of P, type 1, and S pili has shown them to be composite fibers with a thin flexible tip fibrillum at the distal end of a more rigid helical rod (Jones *et al.*, 1995, 1996; Kuehn *et al.*, 1992). F1C pili are genetically very similar to S pili and are expected to have a similar structure. The rod consists of several thousand copies of a major pilin arranged as a tight right-handed helix forming a hollow cylinder with an outer diameter of about ~7 nm and an inner diameter of 2–3 nm (Brinton, 1965; Bullitt and Makowski, 1995; Gong and Makowski, 1992; Hahn *et al.*, 2002). The tip fibrillum is an open helical structure with a diameter of about 2 nm (Jones *et al.*, 1995; Kuehn *et al.*, 1992). The P and S tip fibrillum is relatively long and flexible whereas it is short and stubby in type 1 pili (Jones *et al.*, 1996). Receptor binding is typically mediated by a specialized carbohydrate-binding adhesin (Table 1) that consists of an N-terminal sugar-binding domain fused to a C-terminal pilin domain and that is located at the distal end of the tip fibrillum (Jones *et al.*, 1995; Kuehn *et al.*, 1992). Specific adapter subunits link the adhesin to the tip fibrillum, and the tip fibrillum to the rod (Jones *et al.*, 1995; Kuehn *et al.*, 1992). In contrast to the rod-like pili, the Dr family of adhesive organelles represent a class of non-pilus structures that appear as very thin flexible fibrillae or have no discernible ultrastructure by electron microscopy. They are built up from a single type of pilin subunit, and lack a specialized two-domain carbohydrate-binding adhesin protein.

Despite the differences in appearance and binding specificity, the genes encoding all of these UPEC organelles are arranged in similar gene clusters and are assembled using a similar assembly machinery called the chaperone/usher pathway (Knight *et al.*, 2000; Sauer *et al.*, 2000a; Sauer *et al.*, 2000b; Thanassi and Hultgren, 2000; Thanassi *et al.*, 1998a) (Figure 1). This pathway is commonly used to assemble adhesive surface organelles in Gram negative bacteria. Chaperone/usher systems consist of a periplasmic chaperone

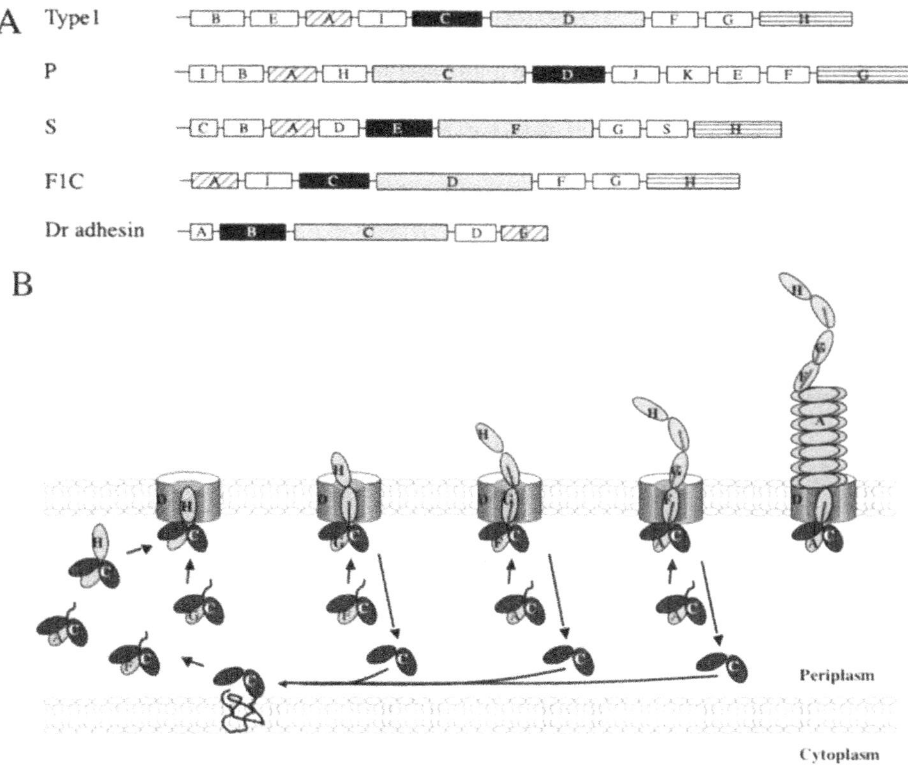

Figure 1. (A) Gene clusters coding for the biogenesis of type 1 (Klemm and Christiansen, 1987), P (Lindberg *et al.*, 1987), F1C (Riegman *et al.*, 1990), S (Schmoll *et al.*, 1990) and Dr family adhesins (Nowicki *et al.*, 1989; Swanson *et al.*, 1991). Chaperone subunits are colored black, ushers gray and major subunits are diagonally striped. Lectin-like two-domain subunits are horizontally striped. (B) Schematic of type 1 pilus assembly, illustrating organelle assembly *via* the chaperone/usher pathway. The periplasmic chaperone, FimC, assists in folding of the subunits (FimH, FimG, FimF, and FimA) as they emerge into the periplasm. FimC caps the assembly surface of the subunits, and targets them to the usher, FimD. The usher serves as an assembly platform where subunits are joined non-covalently into linear fibers and transported through the usher pore to the outside of the bacteria.

and an outer membrane usher that together mediate the assembly of a specific organelle. Organelle subunits are synthesized as pre-proteins in the cytoplasm and transported across the cytoplasmic membrane *via* the Sec machinery. Interaction with the periplasmic chaperone is necessary for efficient release from the inner membrane (Jones *et al.*, 1997). Organelle subunits are thought to fold onto the chaperone where they remain in a tightly bound soluble complex (Bullitt *et al.*, 1996; Jones *et al.*, 1997; Kuehn *et al.*, 1991, 1993; Soto *et al.*, 1998). In the absence of chaperone, subunits are unstable, aggregate, and are digested by periplasmic proteases. Usher molecules are large pore-forming proteins that mediate export of organelle subunits and that also function as platforms for assembly of the adhesive organelle (Dodson *et al.*, 1993; Klemm and Christiansen, 1990; Saulino *et al.*, 1998, 2000; Thanassi *et al.*, 1998b, 2002). The chaperone targets subunits to the usher where they are dissociated from the chaperone and assembled into adhesive organelles. The diameter of the usher pore is about 2 nm (Thanassi *et al.*, 1998b), which is too narrow to allow passage of a pilus rod. Pilus subunits are believed to be joined head-to-tail just before or during export through the usher, exported as a linear fiber, and packaged into a helical rod on the outside of the bacterial cell surface (Saulino *et al.*, 2000; Thanassi *et al.*, 1998b).

Periplasmic chaperones constitute a growing family of homologous proteins. Three-dimensional structures are available for the P pilus chaperone PapD (Holmgren and Branden, 1989; Kuehn *et al.*, 1993; Sauer *et al.*, 1999, 2002; Soto *et al.*, 1998), the type 1 pilus chaperone FimC (Choudhury *et al.*, 1999; Pellecchia *et al.*, 1998, 1999), and the S pilus chaperone SfaE (Knight *et al.*, 2002). The first structure for a non-pilus chaperone, Caf1M, was recently solved in our laboratory (Zavialov *et al.*, 2003). This chaperone directs the assembly of the F1 capsular antigen in *Yersinia pestis* (Chapman *et al.*, 1999; MacIntyre *et al.*, 2001; Zavialov *et al.*, 2002). All of the chaperones have very similar structures with two immunoglobulin (Ig)-like domains oriented at right angles and with a deep subunit-binding cleft between the domains (Figure 2). The bottom of the cleft contains a pair of positively charged basic residues (Arg 8 and Lys 112 in PapD) that are invariant throughout the family,

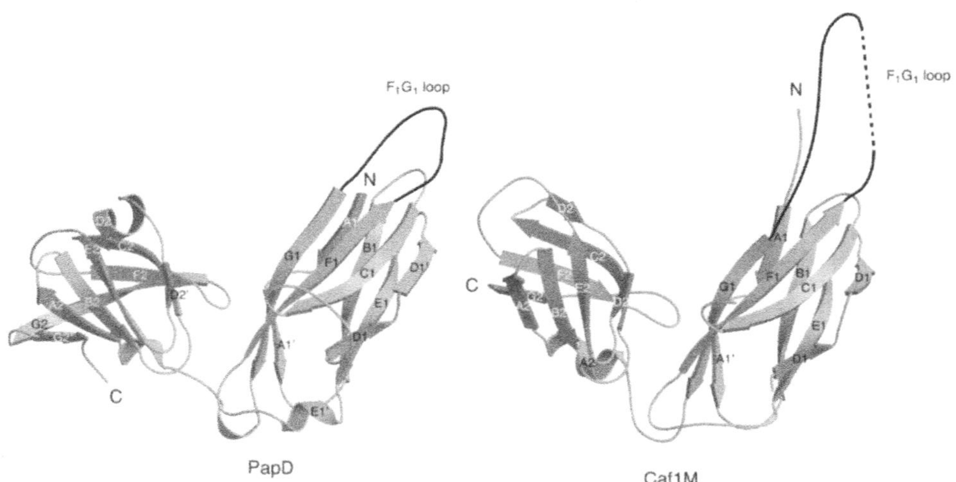

Figure 2. Ribbon diagrams of the PapD (FGS) and Caf1M (FGL) chaperones.

and that are indispensable for subunit binding and chaperone function (Hung *et al.*, 1996; Kuehn *et al.*, 1993; Slonim *et al.*, 1992). A conserved interdomain salt bridge maintains the relative orientation of the two domains, and might also be directly involved in chaperone function during organelle assembly (Hung *et al.*, 1996, 1999; Knight *et al.*, 2002). The 6th (F_1) and 7th (G_1) β-strands of the N-terminal domain are connected by a long and relatively flexible loop. The end of this loop and the beginning of the G_1 strand contains a conserved motif of alternating hydrophobic residues that is crucial for chaperone function (Chapman *et al.*, 1999; Hung *et al.*, 1996; MacIntyre *et al.*, 2001).

The length of the F_1-G_1 loop and the number of hydrophobic residues in the motif varies depending on the type of structure that is assembled (Chapman, 1999; Hung *et al.*, 1996; MacIntyre, 2001). Chaperones involved in pilus assembly (FGS chaperones) have a relatively short F_1-G_1 loop and three hydrophobic residues in this motif. Chaperones involved in assembly of non-pilus structures (FGL chaperones) have a longer F_1-G_1 loop and a longer motif (Figure 2). FGL chaperones are further characterized by an extended N-terminal sequence and by a disulfide bond linking the F_1 and G_1 β-strands (Chapman *et al.*, 1999; MacIntyre *et al.*, 2001).

The structural basis for chaperone action and pilus biogenesis was revealed by the X-ray structures for the FimC : FimH (Choudhury *et al.*, 1999) (Figure 3A) and PapD : PapK (Sauer *et al.*, 1999) chaperone:subunit complexes. FimH is the type 1-pilus mannose-binding adhesin; the PapK adapter subunit links the tip fibrillum to the rod in P pili. FimH consists of two domains separated by a short linker: the N-terminal lectin domain (residues 1–156) binds carbohydrate, the C-terminal pilin domain (residues 160–279) anchors FimH to the pilus tip fibrillum. PapK and the pilin-domain of FimH have similar Ig-like folds, except that the final (G) strand of the fold is missing. The missing strand leaves a large cleft between the pilin A and F strands, lined by hydrophobic residues from the core of the pilin. The chaperones bind the subunits by inserting their G_1 β-strand into this cleft in a process called donor strand complementation (DSC) (Figure 3B). The G_1 strand is lodged between the pilin A and F strands with extensive main-chain-to-main-chain hydrogen bonding between the pilin and the chaperone. The hydrophobic side chains from the conserved G_1 motif are inserted into the cleft and become an integral part of the pilin hydrophobic core. The incomplete fold of the pilins explain why they cannot fold efficiently and are unstable in the absence of chaperone. Experiments with FimH show that chaperone-independent folding can be achieved by genetically engineering a "self-complemented" subunit containing an extra sequence at the C-terminus corresponding to the missing G strand (Barnhart *et al.*, 2000).

Residues in the acceptor cleft between the pilin A and F strands have been shown to be part of an essential pilus assembly surface (Soto *et al.*, 1998). The N-terminus of pilus subunits, which is disordered in chaperone:subunit complexes, harbors a conserved β-strand motif similar to the chaperone G_1 motif. Mutations in this motif also block assembly (Soto *et al.*, 1998). Pilus assembly was therefore suggested to proceed by a donor strand exchange (DSE) mechanism in which the chaperone G_1 donor strand is replaced by the N-terminus of a second subunit, thereby joining subunits into a fiber (Figure 3C). We recently obtained a structure for the *Y. pestis* F1 capsular antigen consisting of two organelle subunits capped by a single Caf1M chaperone molecule (Zavialov *et al.*, 2003). The structure reveals two pilin-like subunits linked by DSC and provides the first direct evidence for the proposed DSE model for organelle assembly.

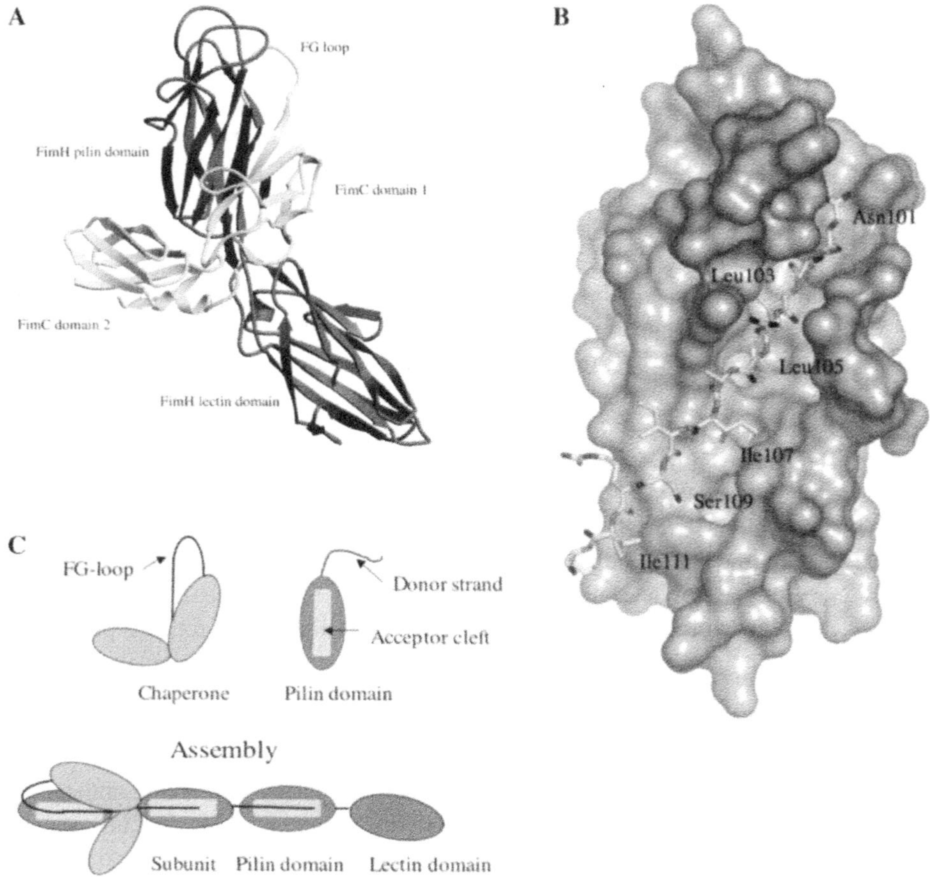

Figure 3. (A) Ribbon diagram of FimC:FimH complex. FimC chaperone in light gray, FimH in dark gray, G_1 donor strand of FimC colored black. (B) FimC G_1 donor-strand in FimH acceptor cleft. Donated hydrophobic residues are indicated. (C) Schematic of DSC. Pilin domains can be thought of as rearranged Ig domains where the C-terminal G strand has been moved to the N-terminus, creating a hydrophobic acceptor cleft on the surface of the Ig barrel and an N-terminal donor strand. In a chaperone:subunit complex, the acceptor cleft is occupied by the chaperone G_1 strand. In assembled organelles, the donor strand of one subunit binds in the acceptor cleft of the previous, thereby joining subunits together in a linear fiber.

3. RECEPTOR BINDING

3.1. Type 1 Pili

Attachment of UPEC to the bladder uroepithelium is mediated by binding of the type-1 pilus adhesin FimH to the TM4-family glycoprotein uroplakin Ia (Malagolini *et al.*, 2000; Wu and Sun, 1993; Zhou *et al.*, 2001). Binding to the single N-linked high-mannose carbohydrate on uroplakin Ia initiates a complex series of events including exfoliation of bladder epithelial cells to clear the infection, and bacterial invasion of newly exposed epithelial cells leading to persistence of infection (Martinez *et al.*, 2000; Mulvey *et al.*, 1998; Mulvey *et al.*, 2001).

In addition to uroplakin Ia, several other FimH receptors, both in and outside the urinary tract, have been reported. For example, binding of type 1-piliated UPEC to high-mannose moieties on the urinary Tamm-Horsfall protein inhibits uroplakin binding and has been suggested to be an important host defence mechanism (Pak *et al.*, 2001). Binding to the cell adhesion protein CD48 on macrophages leads to uptake of bacteria which then survive intracellularly (Baorto *et al.*, 1997), whereas binding to CD48 on mast cells leads to release of the proinflammatory mediator TNFα (Malaviya *et al.*, 1999). FimH-mediated binding to laminin (Kukkonen *et al.*, 1993), collagen (Pouttu *et al.*, 1999), fibronectin (Schembri *et al.*, 2000; Sokurenko *et al.*, 1992), as well as to abiotic surfaces (Pratt and Kolter, 1998; Schembri and Klemm, 2001) has also been reported.

Common to most glycoproteins recognized by FimH is the presence of one or more N-linked high-mannose structures but FimH can also bind to yeast mannans and mediate agglutination of yeast cells. FimH-mediated binding can be inhibited by *D*-mannose and a variety of natural and synthetic oligosaccharides containing terminal mannose residues (Firon *et al.*, 1982; Firon *et al.*, 1983; Firon *et al.*, 1984; Lindhorst *et al.*, 1998; Neeser *et al.*, 1986). Binding studies using lipid-conjugated oligosaccharides separated by thin-layer chromatography implicated $Man_5GlcNAc_2$ as the optimal FimH-receptor (Rosenstein *et al.*, 1988). Several aromatically substituted mannosides are potent inhibitors of FimH binding (Firon *et al.*, 1987). Based on inhibition studies, the carbohydrate binding site of FimH has been suggested to be an extended pocket with optimal fit for the high-mannose core trisaccharide α-*D*-Man(1-3)-β-*D*-Man(1-4)-*D*-GlcNAc ($Man_2GlcNAc$), and with a hydrophobic region in or close to the binding site (Firon *et al.*, 1983; Firon *et al.*, 1987; Sharon, 1987).

FimH-mediated binding requires terminal mannose residues. The reason for this became evident when the structure of the FimC:FimH complex (Figure 3A) was solved (Choudhury *et al.*, 1999). The ellipsoid-shaped FimH lectin domain consists of eleven β-strands in three β-sheets arranged as a pseudo-twofold β-barrel with a jelly roll-like topology. The tip of the domain has a deep pocket large enough to accommodate and completely bury a single mannose residue (Figure 4A). In the FimC:FimH structure, a carbohydrate analogue (C-HEGA, cyclohexylbutanoyl-*N*-hydroxyethyl-*D*-glucamide) occupied this pocket. Mutagenesis of residues in the pocket (Choudhury *et al.*, 1999; Hung *et al.*, 2002; Knight *et al.*, 2000; Schembri *et al.*, 2000), as well as co-crystal structures of FimC : FimH (Hung *et al.*, 2002) and of the lectin domain of FimH (J. Berglund & S. D. Knight, unpublished data) with *D*-mannose confirm that this pocket represents the (major) mannose-binding site in FimH. The cavity is lined with a number of polar and charged residues that form hydrogen bonds to the sugar (Choudhury *et al.*, 1999; Hung *et al.*, 2002). A hydrophobic rim surrounds the pocket, which might account for the increased affinity for aromatically substituted mannosides. Computational docking studies (D. Choudhury, J. Berglund, & S. D. Knight, unpublished data) using AutoDock3 (Goodsell and Olson, 1990; Morris *et al.*, 1998) suggest that trisaccharides bind to FimH with a terminal mannose in the C-HEGA pocket and the other end of the sugar between two tyrosine side chains forming a "tyrosine gate" (Figure 4B). The docking studies further suggest that aromatically substituted mannosides bind with the aromatic group in the tyrosine gate.

FimH inhibition studies with multiantennary mannosyl clusters (Lindhorst, 2002) have shown that linking two or three mannosyl residues in the same glycocluster can provide additional inhibitory potency far beyond what is explained by simple valency effects.

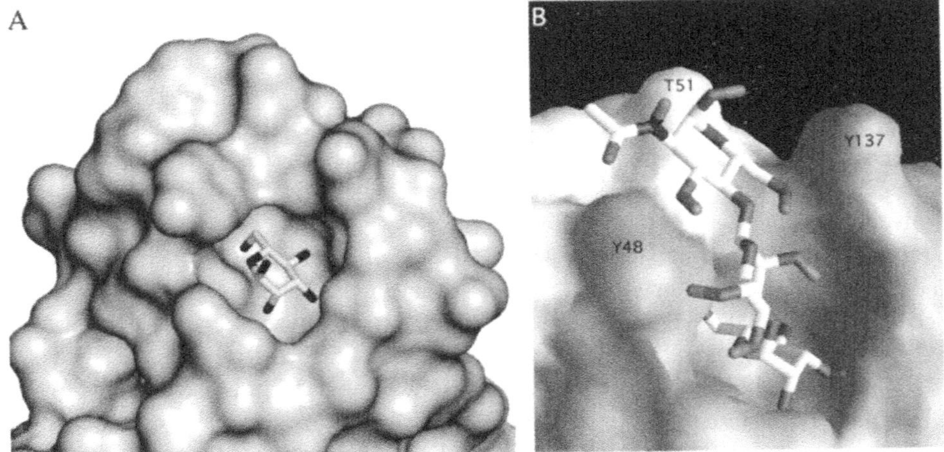

Figure 4. (A) Mannose bound in the monomannose-binding "C-HEGA" pocket at the tip of FimH. (B) $Man_2GlcNAc$ computationally docked in the C-HEGA pocket.

The most potent inhibitor of those studied consisted of two mannosyl residues linked by a chain of six methyl groups (Lindhorst, 2002; Lindhorst *et al.*, 1998). The accumulated data from these studies are not explained by the presence of a single monosaccharide binding-site and suggest the presence of two or more distinct sugar binding sites on FimH (Lindhorst, 2002). Using a novel cavity-detection algorithm (Choudhury, 2001) in combination with computational docking we identified a number of potential secondary carbohydrate binding sites on the surface of FimH. Site directed mutagenesis of amino acids in one of the predicted secondary binding sites resulted in significantly reduced receptor binding as measured by yeast agglutination and binding to immobilized mannan (D. Choudhury, M. Schembri, J. Berglund, P. Klemm, & S. D. Knight, unpublished data). The results suggest that the tip region of FimH might contain two adjacent extended carbohydrate binding-sites and that FimH specificity might depend on the combined effect of binding to both of these sites.

Although FimH is a highly conserved protein, variants do exist. Analysis of fecal and UTI isolates revealed that minor sequence variations correlated with variations in receptor binding properties (Sokurenko *et al.*, 1994; Sokurenko *et al.*, 1995). In an elegant study, Thomas *et al.* (2002) recently showed that whereas binding of type-1-piliated fecal *E. coli* isolates to red blood cells was enhanced by shear, UPEC binding was shear-independent. The observed effects were correlated with amino acid substitutions near the linker region between the pilin and the lectin domain of FimH. Computer simulations of the effect of shear on FimH suggested that shear forces would result in FimH linker extension and that UPEC FimH variants would have a more flexible and extensible linker. Linker extension was proposed to enhance receptor binding either by affecting the receptor-binding site at the opposite end of the lectin domain or by exposing cryptic carbohydrate binding sites. A more rigid linker in non-UPEC FimH would explain why fecal strains bound tightly only in the presence of significant shear forces. Thomas *et al.* (2002) suggested that shear-independent binding might offer advantages for rapid colonization of uroepithelial surfaces irrespective of the shear forces present at any given time. On the other hand, since FimH variants providing shear-dependent binding also are more resistant to inhibition by soluble inhibitors they might be beneficial for colonization of the oropharyngeal mucosa where bacteria are subjected to high shear and a high concentration of soluble mannosylated compounds in the saliva.

We recently crystallized and solved the structure for the isolated FimH lectin domain bound to *D*-mannose (J. Berglund & S. D. Knight, manuscript in preparation). The conformation of residues in the C-HEGA pocket is virtually identical to that observed in the FimC : FimH structure in spite of different ligands in the pocket and different contexts of the lectin domain in the two structures. This shows that the sugar-binding cavity is quite rigid and does not change significantly as a result of ligand binding. The largest difference between the two structures occurs in the β9-β10 loop that packs against the domain linker. The difference in structure is probably a result of differences in crystal packing in the two structures. The fact that there are no significant changes in the C-HEGA pocket in spite of relatively large conformational changes in the β9-β10 loop argues that the two regions of the structure (which are separated by more than 40 Å) are quite independent. In the light of this it appears unlikely that the shear-dependent binding observed by Thomas *et al.* (2002) is due to propagation of conformational changes from the FimH linker region to the tip-located sugar binding site. An alternative explanation for the observed effect is that increased linker flexibility—caused either by mutations in the linker region or by shear-induced linker extension—enhances binding by increasing the rate of productive collisions between lectin and receptor. Pilus plasticity has previously been suggested to influence binding in the presence of shear by allowing more adhesin molecules to approach receptor in a binding-conducive orientation, thereby promoting multivalent binding (Bullitt and Makowski, 1995).

3.2. P Pili

The P pilus adhesin PapG recognizes glycosphingolipids of the globo-series present on the kidney epithelium (Leffler and Svanborg-Edén, 1980; Leffler and Svanborg-Edén, 1981; Lund *et al.*, 1987). These receptors posses a Galα1-4 Gal (galabiose) disaccharide linked *via* β-glucose to a ceramide group that anchors the glycosphingolipid in the membrane. Additions to the galabiose unit of this core structure (GbO3, globotriaosylceramide) generate GbO4 (GalNAcβ(1-3)Galα(1-4)Galβ(1-4)GlcCer; globoside) and GbO5 (GalNAcα(1-3)GalNAcβ(1-3)Galα(1-4)Galβ(1-4)GlcCer; the Forssman antigen). Three different variants of PapG with different but overlapping receptor specificities have been identified (Striker *et al.*, 1995; Stromberg *et al.*, 1990; Stromberg *et al.*, 1991). Of these, PapG-II and PapG-III are important for colonization of the human urinary tract. PapG-II preferentially binds GbO4 and is the predominant PapG variant found in pyelonephritis isolates. PapG-III preferentially recognizes GbO5. Sialosyl galactosyl globoside and disialosyl galactosyl globoside receptors expressed on the kidney epithelium of non-secretors are also efficiently bound by each of the three PapG variants which might account for the increased susceptibility of non-secretors to UTI (Stapleton *et al.*, 1998).

Structures for the lectin domain of PapG-II bound to a GbO4 analogue containing a βO-linked trimethylsilylethyl group instead of ceramide (Dodson *et al.*, 2001), and for the unbound apo form of the domain (Dodson *et al.*, 2001; Sung *et al.*, 2001) have been reported. In spite of no sequence similarity, the sugar-binding lectin domains of PapG and FimH are similarly folded (r.m.s.d. 2.2 Å for 74 equivalent Cα atoms) (Figure 6). Both

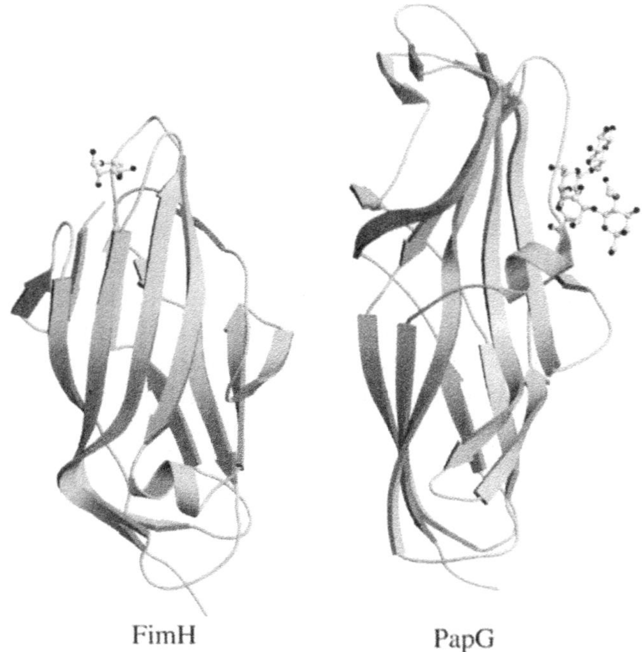

Figure 5. FimH (left) and PapG (right) lectin domains. Carbohydrates bound to the lectins in ball-and-stick.

```
FimH   FACKTANGTAIPIGGGSANVYVNLAPVVNVGQNL-VVDLSTQIFCHN-DYPETITDYVTLQR
FocH   LLCRNNQ-TGQTFQSGDSRFNITLSPTVQYDKAITVLDLNQLVLCQNEDASGQNYDYLRVRQ
SfaH   LLCRNNQ-TGQEFNSGDTSFRVNVSPVVEYDKSISVLDLSQLVSCQNEDSTGQNYDYLKILK
       *           *                                * **   * *

FimH   GSAYGGVLSN-------FSGTVKYSGSSYPFPTTSETPRVVYNSRTDKPWPVALYLTPVSSA
FocH   GTGFSPSLDAKTYGRLDFTNRLSGYSQTLPLQQDTKPTEAYWQYGVWKPFPAKMYLYPEPGV
SfaH   GSGFSPALDTKTYGRLDFTSRPTGYARQLPLQFDLQVTEAFYQYGVWKPFPAKLYLYPEPGV

FimH   GGVAIKAGSLIAVLILRQTNNYNSDD--FQFVWNIYANNDVVVP
FocH   FGKLIHAGELVATVYVNKFSTMGQEAGERNFTWRFYATNDVYIQ
SfaH   FGKVINNGDLLATLYVNKFSTKGQEAGERNFTWRFYATNDVHIQ
                       * * ** **   *
```

Figure 6. Sequence alignment of FimH, FocH, and SfaH, FimH. Identical residues shaded dark gray, homologous residues light gray. Residues in the C-HEGA pocket are indicated by "*" below the sequences.

domains are elongated jelly-roll-like ellipsoid-shaped structures with similar diameters (~20 Å), but the PapG lectin domain is significantly longer (~70 Å as compared to ~50 Å for FimH). The similar diameters might implicate shape constraints due to the size of the usher pore. In contrast to the tip-located mono-saccharide binding site of FimH, the PapG sugar binding site is located on the side of the lectin domain where the GbO4 tetrasaccharide binds side-ways in a shallow pocket (Figure 5). The sugar is bound in a bent V-shaped conformation with extensive interactions between the galabiose core residues and the protein. As is common for lectins, the structure of PapG is essentially unaffected by sugar binding and the structures of liganded and free PapG-II are virtually the same (Dodson *et al.*, 2001). The non-polar face of the Galβ(1-4)Glc reducing-end di-saccharide packs against the aromatic indole group of Trp 107. A number of charged residues (Glu 59, Glu 91, Arg 170, Lys 172) form hydrogen bonds to sugar hydroxyl groups. In addition to these direct contacts, several water molecules mediate interactions between the sugar and the protein. Site directed mutagenesis of binding-site residues completely abolished or significantly reduced P pilus-mediated haemagglutination (Dodson *et al.*, 2001) showing that the interactions formed by these residues are critical for binding in solution.

The differences in receptor-binding specificity exhibited by the different PapG variants is explained by structural variations in and near the GbO4 binding site (Dodson *et al.*, 2001). For example, PapG-I prefers a hydrophobic group attached to the reducing end of galabiose rather than glucose (Kihlberg *et al.*, 1989). Whereas the galabiose-binding residues are identical in PapG-I and PapG-II, several of the residues involved in binding the terminal Glc residue in the PapG-II crystal structure are different in the PapG-I variant. The preference of PapG-III for GbO5 appears to result from an extended carbohydrate-binding site with an additional GlcNAc-binding sub-pocket below the one used for GbO4 binding by PapG-II.

The side-ways binding mode of PapG may be important for establishing optimal lectin–receptor interactions since the tetra-saccharide portion of membrane-bound GbO4 is proposed to be oriented parallel to the membrane (Pascher *et al.*, 1992; Stromberg *et al.*, 1991). The presence of a shallow binding pocket on the side of PapG together with the inherent flexibility of the P pilus tip fibrillum might then be required in order for the adhesin to interact with its receptor (Dodson *et al.*, 2001).

3.3. F1C and S Pili

F1C pili target receptors in the kidney containing glycosphingolipids with a terminal glucose or galactose, with a preference for galactosyl ceramide and globotriaosyl ceramide (Bäckhed *et al.*, 2002; Khan *et al.*, 2000). The smallest carbohydrate unit that is recognised is β1-linked galactose or glucose. Binding apparently depends not only on the sugar component but also on the ceramide composition. Inhibition studies identified the neoglycoprotein GalNacβ1-4Gal-spacer-BSA as an efficient inhibitor of binding to asialo-GM_1 (GgO_4Cer) and asialo-GM_2 (GgO_3Cer) (Khan *et al.*, 2000). In contrast to P, type 1, and S pili, F1C pili do not mediate haemagglutination of red blood cells (Klemm *et al.*, 1982). Genetic and biochemical evidence have identified the receptor-binding subunit as one of the minor tip components FocF, FocG, or FocH (Klemm *et al.*, 1994). FocF and FocG are pilin subunits, FocH is an adhesin-like protein with an N-terminal domain similar to the lectin domain of FimH (Figure 6) fused to a C-terminal pilin domain. Complementation of an *E. coli* cell line expressing all of the type 1 subunits except FimH with a plasmid coding for FocH conferred typical F1C binding characteristics to the cells (Klemm *et al.*, 1994), arguing that FocH is the F1C adhesin. Homology modelling based on our FimH crystal structure suggests that the tip of FocH would also contain a tip-located cavity such as the C-HEGA pocket of FimH. The modelled FocH pocket is lined by charged and polar amino acids, suitable for carbohydrate binding. Hence, FocH, as FimH, might bind terminal sugars in a tip-located monosaccharide-binding pocket.

S pili are frequently found on *E. coli* strains associated with UTI and newborn meningitis (Korhonen *et al.*, 1985). They contribute to pyelonephritis by binding to sialylα(2-3)galactosides on human kidney epithelial cells (Korhonen *et al.*, 1986; Marre *et al.*, 1986) and also mediate haemagglutination of human and bovine erythrocytes by binding to O-linked sialylsaccharides on glycophorin A (Parkkinen *et al.*, 1986). Binding to kidney epithelium is inhibited by sialylα(2-3)lactose but not by free sialic acid or lactose. Sialylgalactoside binding depends on the minor pilin SfaS, located at the tip of S pili, but may be influenced by another minor tip component, SfaH (Hanisch *et al.*, 1993; Korhonen *et al.*, 1984; Morschhauser *et al.*, 1993; Parkkinen *et al.*, 1986; Schmoll *et al.*, 1989). S pili also mediate SfaS-independent binding to brain endothelial cells and to human milk glycoconjugates. The identified receptor contains a terminal galactose or sulphated galactose, with a preference for galactosyl ceramide (Prasadarao *et al.*, 1993; Schwertmann *et al.*, 1999). SfaS-independent binding has been attributed to SfaA or SfaH. SfaA is the major rod-forming S pilin; SfaH has the characteristics of a typical adhesin molecule with extensive homology to FimH and FocH (Figure 6). FocH is 74% identical to SfaH, and the residues in the predicted tip-located binding pockets are identical in the two proteins. Given the common galactosyl ceramide receptor of F1C and S pili, and the presence of highly homologous tip-located FimH-like subunits in both systems, it seems plausible that both types of pili bind to this receptor using a common binding paradigm.

3.4. Non-pilus Adhesins

Adhesins of the non-pilus Dr family (Nowicki *et al.*, 2001) mediate UPEC adherence to the urinary tract by binding to the Dr(a+) blood group antigen present on the complement regulatory molecule decay-accelerating factor (DAF). Children and pregnant women in particular appear to be at risk for infection by UPEC expressing Dr adhesins (Nowicki *et al.*, 2001). Residues important for DAF binding have been identified based on random mutagenesis of the Dr haemagglutinin subunit DraE (Van Loy *et al.*, 2002). No three-dimensional structural information is available for any of the Dr adhesins. However, we recently solved the structure of the *Y. pestis* F1 antigen subunit (Caf1) (Zavialov *et al.*, 2003) which serves as a prototype for the non-pilus class of organelles assembled *via* FGL chaperone/usher systems. Threading of the DraE sequence onto the structure of Caf1 allowed us to construct a homology model for DraE (Figure 7). Consistent with biochemical evidence (Carnoy and Moseley, 1997), the two cysteine residues common to Dr adhesins are juxtaposed in the resulting model and might form a disulfide bond linking strand A to strand B. All of the residues implicated in DAF binding except one cluster in two small areas that border a deep pocket on one side of the DraE model. One of the residues, Thr 10, is located in the predicted N-terminal donor strand of DraE. The predicted binding site is thus composed of residues from two consecutive subunits in a Dr haemagglutinin fiber, suggesting that the binding site is fully formed only after surface assembly of the organelle. In addition to the DAF-binding common to Dr adhesins, Dr haemagglutinin also mediates binding to type IV collagen (Westerlund *et al.*, 1989). The two binding specificities exhibited by DraE can be independently affected by mutagenesis and have been postulated to depend on two separate conformational binding sites (Carnoy and Moseley, 1997; Van Loy *et al.*, 2002). Consistent with this, residues implicated in type IV collagen binding (Carnoy and Moseley, 1997) are located on the opposite face of the DraE model, far away from the predicted DAF-binding site.

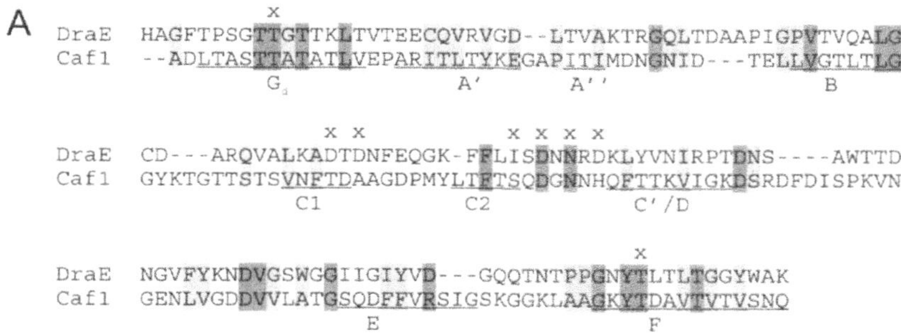

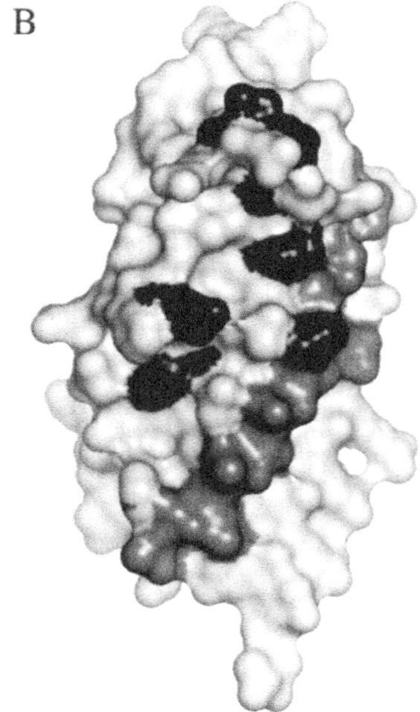

Figure 7. (A) Sequence alignment of DraE with Caf1 resulting from threading the DraE sequence onto the structure of Caf1. Underlined residues indicate secondary structure elements in Caf1. Residues implicated in DAF binding are indicated by "x" above the sequences. (B) Homology model of DraE. One "fiber module," consisting of the Ig-barrel of one DraE subunit complemented with the N-terminal donor strand (dark gray) from a neighbouring subunit, is shown. Residues implicated in DAF binding (black) cluster on the surface of the molecule.

4. OUTLOOK AND FUTURE PERSPECTIVES

Although in most cases not a deadly or seriously debilitating disease, UTI is a very unpleasant condition affecting a large number of people at a significant global cost, both in terms of direct health care expenditure and in terms of nonmedical costs associated with, for example travel, sick days, and morbidity. As for many other bacterial infections, the appearance of antibiotic resistant strains of uropathogenic bacteria underscores the importance of developing novel strategies for prevention and treatment. Since UPEC adhesive organelles are surface exposed and are critical for colonization of the urinary tract, adhesin-based vaccines offer one possible route for UTI prevention (Wizemann *et al.*, 1999). Although early vaccines based on purified intact pili were protective, they were of limited value since the major immunogenic component of pili is often antigenically highly variable, and protection was limited to a few bacterial strains. In contrast, since pilus adhesins are generally highly conserved, vaccines based on purified adhesin subunits generate good cross-reactivity. A vaccine based on the FimC:FimH chaperone:adhesin complex has been shown to protect against UPEC mucosal infection in both murine and primate models (Langermann *et al.*, 1997; Langermann *et al.*, 2000). *In vitro* binding data suggested that the ability of anti-FimH antibodies to block type-1 pilus adhesion contributed significantly to protection. The possibility that sugar binding sites of non-pilus adhesins might be formed only after linking of subunits *via* DSC suggests that self-complemented subunits (Barnhart *et al.*, 2000) might be a better alternative than chaperone:subunit complexes for generating blocking anti-bodies to this class of adhesins. An alternative to anti-adhesin vaccines that has attracted considerable interest is to design sugar derivatives and multi-antennary glycoclusters for use as adhesion inhibitors (Lindhorst, 2002; Nagahori *et al.*, 2002; Ohlsson *et al.*, 2002). It is commonly believed that bacterial resistance towards such inhibitors would evolve only slowly because infecting bacteria are dispelled rather than killed. A related approach that has recently been proposed is to prevent adhesion through receptor depletion (Svensson *et al.*, 2001b; Svensson *et al.*, 2003). In addition to the anti-adhesive strategies briefly summarized above that all interfere directly with receptor binding, strategies based on preventing the formation of adhesive organelles are also being developed. For example, based on structural information about binding of peptides and subunits to chaperones, pilicides that interfere with chaperone:subunit binding and prevent pilus formation have been designed (Svensson *et al.*, 2001a). The recent advances in understanding the structural basis for adhesion of UPEC in the urinary tract outlined here provide a solid ground for the continued development of novel ways of preventing and/or treating UTIs that hopefully will eventually lead to increased health and well-being, and to lowered health care costs.

5. ACKNOWLEDGMENTS

Work in this laboratory is supported by grants from the Swedish Foundation for Strategic Research, the Swedish Science Foundation, and the Wenner-Gren Foundations.

6. REFERENCES

Bäckhed, F., Alsen, B., Roche, N., Angstrom, J., von Euler, A., Breimer, M.E., Westerlund-Wikstrom, B., Teneberg, S., and Richter-Dahlfors, A., 2002, Identification of target tissue glycosphingolipid receptors for uropathogenic, F1C-fimbriated *Escherichia coli* and its role in mucosal inflammation, *J Biol Chem.* 277:18198–18205.

Bahrani-Mougeot, F.K., Buckles, E.L., Lockatell, C.V., Hebel, J.R., Johnson, D.E., Tang, C.M., and Donnenberg, M.S., 2002, Type 1 fimbriae and extracellular polysaccharides are preeminent uropathogenic *Escherichia coli* virulence determinants in the murine urinary tract, *Mol Microbiol.* 45:1079–1093.

Baorto, D.M., Gao, Z., Malaviya, R., Dustin, M., van der Merwe, A., Lublin, D.M., and Abraham, S.N., 1997, Survival of FimH-expressing enterobacteria in macrophages relies on glycolipid traffic, *Nature.* 389:636–639.

Barnhart, M.M., Pinkner, J.S., Soto, G.E., Sauer, F.G., Langermann, S., Waksman, G., Frieden, C., and Hultgren, S.J., 2000, PapD-like chaperones provide the missing information for folding of pilin proteins, *Proc Natl Acad Sci USA*. 97:7709–7714.

Brinton, C.C. Jr., 1965, The structure, function, synthesis and genetic control of bacterial pili and a molecular model for DNA and RNA transport in gram negative bacteria, *Trans N Y Acad Sci*. 27:1003–1054.

Bullitt, E., Jones, C.H., Striker, R., Soto, G., Jacob-Dubuisson, F., Pinkner, J., Wick, M.J., Makowski, L., and Hultgren, S.J., 1996, Development of pilus organelle subassemblies *in vitro* depends on chaperone uncapping of a beta zipper, *Proc Natl Acad Sci USA*. 93:12890–12895.

Bullitt, E. and Makowski, L., 1995, Structural polymorphism of bacterial adhesion pili, *Nature*. 373:164–167.

Carnoy, C. and Moseley, S.L., 1997, Mutational analysis of receptor binding mediated by the Dr family of *Escherichia coli* adhesins, *Mol Microbiol*. 23:365–379.

Chapman, D.A., Zavialov, A.V., Chernovskaya, T.V., Karlyshev, A.V., Zav'yalova, G.A., Vasiliev, A.M., Dudich, I.V., Abramov, V.M., Zav'yalov, V.P., and MacIntyre, S., 1999, Structural and functional significance of the FGL sequence of the periplasmic chaperone Caf1M of *Yersinia pestis*, *J Bacteriol*. 181:2422–2429.

Choudhury, D., 2001, Functional Implications of Macromolecular Recognition: Assembly of Adhesive Pili and Enzyme Substrate Interactions. Thesis. Swedish University of Agricultural Sciences, Uppsala.

Choudhury, D., Thompson, A., Stojanoff, V., Langermann, S., Pinkner, J., Hultgren, S.J., and Knight, S.D., 1999, X-ray structure of the FimC-FimH chaperone-adhesin complex from uropathogenic *Escherichia coli*, *Science*. 285:1061–1066.

Connell, H., Agace, W., Klemm, P., Schembri, M., Marild, S., and Svanborg, C., 1996, Type 1 fimbrial expression enhances *Escherichia coli* virulence for the urinary tract, *Proc Natl Acad Sci U S A*. 93:9827–9832.

Dodson, K.W., Jacob-Dubuisson, F., Striker, R.T., and Hultgren, S.J., 1993, Outer-membrane PapC molecular usher discriminately recognizes periplasmic chaperone-pilus subunit complexes, *Proc Natl Acad Sci U S A*. 90:3670–3674.

Dodson, K.W., Pinkner, J.S., Rose, T., Magnusson, G., Hultgren, S.J., and Waksman, G., 2001, Structural basis of the interaction of the pyelonephritic *E. coli* adhesin to its human kidney receptor, *Cell*. 105:733–743.

Firon, N., Ashkenazi, S., Mirelman, D., Ofek, I., and Sharon, N., 1987, Aromatic alpha-glycosides of mannose are powerful inhibitors of the adherence of type 1 fimbriated *Escherichia coli* to yeast and intestinal epithelial cells, *Infect Immun*. 55:472–476.

Firon, N., Ofek, I., and Sharon, N., 1982, Interaction of mannose-containing oligosaccharides with the fimbrial lectins of *Escherichia coli*, *Biochem Biophys Res Commun*. 105:1426–1432.

Firon, N., Ofek, I., and Sharon, N., 1983, Carbohydrate specificity of the surface lectins of *Escherichia coli*, *Klebsiella pneumoniae*, and *Salmonella typhimurium*. *Carbohydr Res*. 120:235–249.

Firon, N., Ofek, I., and Sharon, N., 1984, Carbohydrate-binding sites of the mannose-specific fimbrial lectins of enterobacteria, *Infect Immun*. 43:1088–1090.

Foxman, B., 2002, Epidemiology of urinary tract infections: incidence, morbidity, and economic costs, *Am J Med*. 113 Suppl 1A:5S–13S.

Foxman, B., Zhang, L., Palin, K., Tallman, P., and Marrs, C.F., 1995, Bacterial virulence characteristics of *Escherichia coli* isolates from first-time urinary tract infection. *J Infect Dis*. 171:1514–1521.

Gong, M. and Makowski, L., 1992, Helical structure of P pili from *Escherichia coli*. Evidence from X-ray fiber diffraction and scanning transmission electron microscopy, *J Mol Biol*. 228:735–742.

Goodsell, D.S. and Olson, A.J., 1990, Automated docking of substrates to proteins by simulated annealing, *Prot Struct Func Genet*. 8:1040–1045.

Gupta, K., Sahm, D.F., Mayfield, D., and Stamm, W.E., 2001, Antimicrobial resistance among uropathogens that cause community-acquired urinary tract infections in women: A nationwide analysis, *Clin Infect Dis*. 33:89–94.

Hahn, E., Wild, P., Hermanns, U., Sebbel, P., Glockshuber, R., Haner, M., Taschner, N., Burkhard, P., Aebi, U., and Muller, S.A., 2002, Exploring the 3D Molecular Architecture of *Escherichia coli* Type 1 Pili. *J Mol Biol*. 323:845–857.

Hanisch, F.G., Hacker, J., and Schroten, H., 1993, Specificity of S fimbriae on recombinant *Escherichia coli*: preferential binding to gangliosides expressing NeuGc alpha (2-3)Gal and NeuAc alpha (2-8)NeuAc, *Infect Immun*. 61:2108–2115.

Holmgren, A. and Branden, C.I., 1989, Crystal structure of chaperone protein PapD reveals an immunoglobulin fold, *Nature*. 342:248–251.

Hung, C.S., Bouckaert, J., Hung, D., Pinkner, J., Widberg, C., DeFusco, A., Auguste, C.G., Strouse, R., Langermann, S., Waksman, G., and Hultgren, S.J., 2002, Structural basis of tropism of *Escherichia coli* to the bladder during urinary tract infection, *Mol Microbiol*. 44:903–915.

Hung, D.L., Knight, S.D, and Hultgren, S.J., 1999, Probing conserved surfaces on PapD, *Mol Microbiol.* 31:773–783.

Hung, D.L., Knight, S.D., Woods, R.M., Pinkner, J.S., and Hultgren, S.J., 1996, Molecular basis of two subfamilies of immunoglobulin-like chaperones, *EMBO J.* 15:3792–3805.

Johnson, J.R., 1991, Virulence factors in *Escherichia coli* urinary tract infection, *Clin Microbiol Rev.* 4:80–128.

Johnson, J.R., Manges, A.R., O'Bryan, T.T., and Riley, L.W., 2002, A disseminated multidrug-resistant clonal group of uropathogenic *Escherichia coli* in pyelonephritis, *Lancet.* 359:2249–2251.

Jones, C.H., Danese, P.N., Pinkner, J.S., Silhavy, T.J., and Hultgren, S.J., 1997, The chaperone-assisted membrane release and folding pathway is sensed by two signal transduction systems, *EMBO J.* 16:6394–6406.

Jones, C.H., Dodson, K., and Hultgren, S.J., 1996, Structure, function and assembly of adhesive pili, in: *Urinary tract infection: molecular pathogenesis to clinical management*, H.L.T. Mobley and J.W. Warren, eds. (ASM), pp. 175–219.

Jones, C.H., Pinkner, J.S., Roth, R., Heuser, J., Nicholes, A.V., Abraham, S.N., and Hultgren, S.J., 1995, FimH adhesin of type 1 pili is assembled into a fibrillar tip structure in the *Enterobacteriaceae*, *Proc Natl Acad Sci U S A.* 92:2081–2085.

Khan, A.S., Kniep, B., Oelschlaeger, T.A., Van Die, I., Korhonen, T., and Hacker, J., 2000, Receptor structure for F1C fimbriae of uropathogenic *Escherichia coli*, *Infect Immun.* 68:3541–3547.

Kihlberg, J., Hultgren, S.J., Normark, S., and Magnusson, G., 1989, Probing the combining site of the PapG adhesin of uropathogenic *Escherichia coli* bacteria by synthetic analogs of galabiose, *J Am Chem Soc.* 111:6364–6368.

Klemm, P. and Christiansen, G., 1987, Three fim genes required for the regulation of length and mediation of adhesion of *Escherichia coli* type 1 fimbriae, *Mol Gen Genet.* 208:439–445.

Klemm, P. and Christiansen, G., 1990, The fimD gene required for cell surface localization of *Escherichia coli* type 1 fimbriae, *Mol Gen Genet.* 220:334–338.

Klemm, P., Christiansen, G., Kreft, B., Marre, R., and Bergmans, H., 1994, Reciprocal exchange of minor components of type 1 and F1C fimbriae results in hybrid organelles with changed receptor specificities, *J Bacteriol.* 176:2227–2234.

Klemm, P., Orskov, I., and Orskov, F., 1982, F7 and type 1-like fimbriae from three *Escherichia coli* strains isolated from urinary tract infections: protein chemical and immunological aspects, *Infect Immun.* 36:462–468.

Knight, S.D., Berglund, J., and Choudhury, D., 2000, Bacterial adhesins: structural studies reveal chaperone function and pilus biogenesis, *Curr Opin Chem Biol.* 4:653–660.

Knight, S.D., Choudhury, D., Hultgren, S., Pinkner, J., Stojanoff, V., and Thompson, A., 2002, Structure of the S pilus periplasmic chaperone SfaE at 2.2 A resolution. *Acta Crystallogr D Biol Crystallogr.* 58:1016–1022.

Korhonen, T.K., Parkkinen, J., Hacker, J., Finne, J., Pere, A., Rhen, M., and Holthofer, H., 1986, Binding of *Escherichia coli* S fimbriae to human kidney epithelium, *Infect Immun.* 54:322–327.

Korhonen, T.K., Vaisanen-Rhen, V., Rhen, M., Pere, A., Parkkinen, J., and Finne, J., 1984, *Escherichia coli* fimbriae recognizing sialyl galactosides, *J Bacteriol.* 159:762–766.

Korhonen, T.K., Valtonen, M.V., Parkkinen, J., Vaisanen-Rhen, V., Finne, J., Orskov, F., Orskov, I., Svenson, S.B., and Makela, P.H., 1985, Serotypes, hemolysin production, and receptor recognition of *Escherichia coli* strains associated with neonatal sepsis and meningitis, *Infect Immun.* 48:486–491.

Kuehn, M.J., Heuser, J., Normark, S., and Hultgren, S.J., 1992, P pili in uropathogenic *E. coli* are composite fibres with distinct fibrillar adhesive tips, *Nature.* 356:252–255.

Kuehn, M.J., Normark, S., and Hultgren, S.J., 1991, Immunoglobulin-like PapD chaperone caps and uncaps interactive surfaces of nascently translocated pilus subunits, *Proc Natl Acad Sci USA.* 88:10586–10590.

Kuehn, M.J., Ogg, D.J., Kihlberg, J., Slonim, L.N., Flemmer, K., Bergfors, T., and Hultgren, S.J., 1993, Structural basis of pilus subunit recognition by the PapD chaperone, *Science.* 262:1234–1241.

Kukkonen, M., Raunio, T., Virkola, R., Lähteenmäki, K., Mäkelä, P.H., Klemm, P., Clegg, S., and Korhonen, T.K., 1993, Basement membrane carbohydrate as a target for bacterial adhesion: binding of type I fimbriae of *Salmonella enterica* and *Escherichia coli* to laminin, *Mol Microbiol.* 7:229–237.

Langermann, S., Mollby, R., Burlein, J.E., Palaszynski, S.R., Auguste, C.G., DeFusco, A., Strouse, R., Schenerman, M.A., Hultgren, S.J., Pinkner, J.S. *et al.*, 2000, Vaccination with FimH adhesin protects cynomolgus monkeys from colonization and infection by uropathogenic *Escherichia coli*, *J Infect Dis.* 181:774–778.

Langermann, S., Palaszynski, S., Barnhart, M., Auguste, G., Pinkner, J.S., Burlein, J., Barren, P., Koenig, S., Leath, S., Jones, C.H., and Hultgren, S.J., 1997, Prevention of mucosal *Escherichia coli* infection by FimH-adhesin-based systemic vaccination, *Science.* 276:607–611.

Leffler, H. and Svanborg-Eden, C., 1981, Glycolipid receptors for uropathogenic *Escherichia coli* on human erythrocytes and uroepithelial cells, *Infect Immun.* 34:920–929.

Leffler, H. and Svanborg-Edén, C., 1980, Chemical identification of a glycosphingolipid receptor for *Escherichia coli* attaching to human urinary tract epithelial cells and agglutinating human erythrocytes, *FEMS Microbiol Lett.* 8:127–134.

Lindberg, F., Lund, B., Johansson, L., and Normark, S., 1987, Localization of the receptor-binding protein adhesin at the tip of the bacterial pilus, *Nature.* 328:84–87.

Lindhorst, T.K., 2002, Artificial multivalent sugar ligands to understand and manipulate carbohydrate-protein interaction, *Topics Curr Chem.* 218:201–235.

Lindhorst, T.K., Kieburg, C., and Krallmann-Wenzel, U., 1998, Inhibition of the type 1 fimbriae-mediated adhesion of *Escherichia coli* to erythrocytes by multiantennary alpha-mannosyl clusters: the effect of multivalency, *Glycoconj J.* 15:605–613.

Lund, B., Lindberg, F., Marklund, B.I., and Normark, S., 1987, The PapG protein is the alpha-D-galactopyranosyl-(1——4)-beta-D-galactopyranose-binding adhesin of uropathogenic *Escherichia coli*, *Proc Natl Acad Sci USA.* 84:5898–5902.

MacIntyre, S., Zyrianova, I.M., Chernovskaya, T.V., Leonard, M., Rudenko, E.G., Zav'Yalov, V.P., and Chapman, D.A., 2001, An extended hydrophobic interactive surface of *Yersinia pestis* Caf1M chaperone is essential for subunit binding and F1 capsule assembly, *Mol Microbiol.* 39:12–25.

Malagolini, N., Cavallone, D., Wu, X.-R., and Serafini-Cessi, F., 2000, Terminal glycosylation of bovine uroplakin III, one of the major integral-membrane glycoproteins of mammalian bladder, *Biochim Biophys Acta.* 1475:231–237.

Malaviya, R., Gao, Z., Thankavel, K., van der Merwe, P., and Abraham, S.N., 1999, The mast cell tumor necrosis factor a response to FimH-expressing *Escherichia coli* is mediated by the glycosylphosphatidylinositylanchored molecule CD48, *Proc Natl Acad Sci USA.* 96:8110–8115.

Marre, R., Hacker, J., Henkel, W., and Goebel, W., 1986, Contribution of cloned virulence factors from uropathogenic *Escherichia coli* strains to nephropathogenicity in an experimental rat pyelonephritis model, *Infect Immun.* 54:761–767.

Martinez, J.J., Mulvey, M.A., Schilling, J.D., Pinkner, J.S., and Hultgren, S.J., 2000, Type 1 pilus-mediated bacterial invasion of bladder epithelial cells, *EMBO J.* 19:2803–2812.

Morris, G.M., Goodsell, D.S., Halliday, R.S., Huey, R., Hart, W.E., Belew, R.K., and Olson, A.J., 1998, Automated docking using a Lamarckian genetic algorithm and an empirical binding free energy function, *J Comput Chem.* 19:1639–1662.

Morschhauser, J., Vetter, V., Korhonen, T., Uhlin, B.E., and Hacker, J., 1993, Regulation and binding properties of S fimbriae cloned from *E. coli* strains causing urinary tract infection and meningitis, *Zentralbl Bakteriol.* 278:165–176.

Mulvey, M.A., 2002, Adhesion and entry of uropathogenic *Escherichia coli*, *Cell Microbiol.* 4:257–271.

Mulvey, M.A., Lopez-Boado, Y.S., Wilson, C.L., Roth, R., Parks, W.C., Heuser, J., and Hultgren, S.J., 1998, Induction and evasion of host defenses by type 1-piliated uropathogenic *Escherichia coli*, *Science* 282:1494–1497.

Mulvey, M.A., Schilling, J.D., and Hultgren, S.J., 2001, Establishment of a persistent *Escherichia coli* reservoir during the acute phase of a bladder infection, *Infect Immun.* 69:4572–4579.

Nagahori, N., Lee, R.T., Nishimura, S., Page, D., Roy, R., and Lee, Y.C., 2002, Inhibition of adhesion of Type 1 fimbriated *Escherichia coli* to highly mannosylated ligands, *Chembiochem.* 3:836–844.

Neeser, J.-R., Koellreutter, B., and Wuersch, P., 1986, Oligomannoside-type glycopeptides inhibiting adhesion of *Escherichia coli* strains mediated by type 1 pili: Preparation of potent inhibitors from plant glycoproteins, *Infect Immun.* 52:428–436.

Nicolle, L.E., 2002, Resistant pathogens in urinary tract infections, *J Am Geriatr Soc.* 50:S230–235.

Nowicki, B., Selvarangan, R., and Nowicki, S., 2001, Family of *Escherichia coli* Dr adhesins: decay-accelerating factor receptor recognition and invasiveness, *J Infect Dis.* 183 Suppl 1:S24–27.

Nowicki, B., Svanborg-Eden, C., Hull, R., and Hull, S., 1989, Molecular analysis and epidemiology of the Dr hemagglutinin of uropathogenic *Escherichia coli*, *Infect Immun.* 57:446–451.

Ohlsson, J., Jass, J., Uhlin, B.E., Kihlberg, J., and Nilsson, U.J., 2002, Discovery of potent inhibitors of PapG adhesins from uropathogenic *Escherichia coli* through synthesis and evaluation of galabiose derivatives, *Chembiochem.* 3:772–779.

Pak, J., Yongbing, P., Zhang, Z.-T., Hasty, D.L., and Wu, X.-R., 2001, Tamm-Horsfall protein binds to type 1 fimbriated *Escherichia coli* and prevents *E. coli* from binding to uroplakin Ia and Ib receptors, *J Biol Chem.* 276:9924–9930.

Parkkinen, J., Rogers, G.N., Korhonen, T., Dahr, W., and Finne, J., 1986, Identification of the O-linked sialyloligosaccharides of glycophorin A as the erythrocyte receptors for S-fimbriated *Escherichia coli*, *Infect Immun.* 54:37–42.

Pascher, I., Lundmark, M., Nyholm, P.G., and Sundell, S., 1992, Crystal structures of membrane lipids, *Biochim Biophys Acta.* 1113:339–373.

Pellecchia, M., Guntert, P., Glockshuber, R., and Wuthrich, K., 1998, NMR solution structure of the periplasmic chaperone FimC, *Nat Struct Biol.* 5:885–890.

Pellecchia, M., Sebbel, P., Hermanns, U., Wuthrich, K., and Glockshuber, R., 1999, Pilus chaperone FimC-adhesin FimH interactions mapped by TROSY-NMR, *Nat Struct Biol.* 6:336–339.

Pouttu, R., Puustinen, T., Virkola, R., Hacker, R., Klemm, P., and Korhonen, T.K., 1999, Amino acid residue Ala-62 in the FimH fimbrial adhesin is critical for the adhesiveness of meningitis-associated *Escherichia coli* to collagens, *Mol Microbiol.* 31:1747–1757.

Prasadarao, N.V., Wass, C.A., Hacker, J., Jann, K., and Kim, K.S., 1993, Adhesion of S-fimbriated *Escherichia coli* to brain glycolipids mediated by sfaA gene-encoded protein of S-fimbriae, *J Biol Chem.* 268:10356–10363.

Pratt, L.A. and Kolter, R., 1998, Genetic analysis of *Escherichia coli* biofilm formation: roles of flagella, motility, chemotaxis and type I pili, *Mol Microbiol.* 30:285–293.

Riegman, N., Kusters, R., Van Veggel, H., Bergmans, H., Van Bergen en Henegouwen, P., Hacker, J., and Van Die, I., 1990, F1C fimbriae of a uropathogenic *Escherichia coli* strain: genetic and functional organization of the foc gene cluster and identification of minor subunits, *J Bacteriol.* 172:1114–1120.

Roberts, J.A., Marklund, B.I., Ilver, D., Haslam, D., Kaack, M.B., Baskin, G., Louis, M., Mollby, R., Winberg, J., and Normark, S., 1994, The Gal(alpha 1-4)Gal-specific tip adhesin of *Escherichia coli* P-fimbriae is needed for pyelonephritis to occur in the normal urinary tract, *Proc Natl Acad Sci USA.* 91:11889–11893.

Ronald, A., 2002, The etiology of urinary tract infection: traditional and emerging pathogens, *Am J Med.* 113:14S–19S.

Rosenstein, I.J., Mizuochi, T., Hounsell, E.F., Stoll, M.S., Childs, R.A., and Feizi, T., 1988, New type of adhesive specificity revealed by oligosaccharide probes in *Escherichia coli* from patients with urinary tract infections, *Lancet.* 2:1327–1330.

Sauer, F.G., Barnhart, M., Choudhury, D., Knight, S.D., Waksman, G., and Hultgren, S.J., 2000a, Chaperone-assisted pilus assembly and bacterial attachment, *Curr Opin Struct Biol.* 10:548–556.

Sauer, F.G., Futterer, K., Pinkner, J.S., Dodson, K.W., Hultgren, S.J., and Waksman, G., 1999, Structural basis of chaperone function and pilus biogenesis. *Science.* 285:1058–1061.

Sauer, F.G., Knight, S.D., Waksman, Gj and Hultgren, S.J., 2000b, PapD-like chaperones and pilus biogenesis, *Semin Cell Dev Biol.* 11:27–34.

Sauer, F.G., Pinkner, J.S., Waksman, G., and Hultgren, S.J., 2002, Chaperone priming of pilus subunits facilitates a topological transition that drives fiber formation, *Cell.* 111:543–551.

Saulino, E.T., Bullitt, E., and Hultgren, S.J., 2000, Snapshots of usher-mediated protein secretion and ordered pilus assembly, *Proc Natl Acad Sci USA.* 97:9240–9245.

Saulino, E.T., Thanassi, D.G., Pinkner, J.S., and Hultgren, S.J., 1998, Ramifications of kinetic partitioning on usher-mediated pilus biogenesis, *EMBO J.* 17:2177–2185.

Schembri, M.A. and Klemm, P., 2001, Biofilm formation in a hydrodynamic environment by novel FimH variants and ramifications for virulence, *Infect Immun.* 69:1322–1328.

Schembri, M.A., Sokurenko, E.V., and Klemm, P., 2000, Functional flexibility of the FimH adhesin: insights from a random mutant library, *Infect Immun.* 68:2638–2646.

Schmoll, T., Hoschutzky, H., Morschhauser, J., Lottspeich, F., Jann, K., and Hacker, J., 1989, Analysis of genes coding for the sialic acid-binding adhesin and two other minor fimbrial subunits of the S-fimbrial adhesin determinant of *Escherichia coli*, *Mol Microbiol.* 3:1735–1744.

Schmoll, T., Morschhauser, J., Ott, M., Ludwig, B., van Die, I., and Hacker, J., 1990, Complete genetic organization and functional aspects of the *Escherichia coli* S fimbrial adhesion determinant: nucleotide sequence of the genes sfa B, C, D, E, F, *Microb Pathog.* 9:331–343.

Schwertmann, A., Schroten, H., Hacker, J., and Kunz, C., 1999, S-fimbriae from *Escherichia coli* bind to soluble glycoproteins from human milk, *J Pediatr Gastroenterol Nutr.* 28:257–263.

Sharon, N., 1987, Bacterial lectins, cell-cell recognition and infectious disease, *FEBS Lett.* 217:145–157.

Slonim, L.N., Pinkner, J.S., Branden, C.I., and Hultgren, S.J., 1992, Interactive surface in the PapD chaperone cleft is conserved in pilus chaperone superfamily and essential in subunit recognition and assembly, *EMBO J.* 11:4747–4756.

Sokurenko, E.V., Courtney, H.S., Abraham, S.N., Klemm, P., and Hasty, D.L., 1992, Functional heterogeneity of type 1 fimbriae of *Escherichia coli, Infect Immun.* 60:4709–4719.

Sokurenko, E.V., Courtney, H.S., Maslow, J., Siitonen, A., and Hasty, D.L., 1995, Quantitative differences in adhesiveness of type 1 fimbriated *Escherichia coli* due to structural differences in fimH genes, *J Bacteriol.* 177:3680–3686.

Sokurenko, E.V., Courtney, H.S., Ohman, D.E., Klemm, P., and Hasty, D.L., 1994, FimH family of type 1 fimbrial adhesins: functional heterogeneity due to minor sequence variations among fimH genes, *J Bacteriol.* 176:748–755.

Soto, G.E., Dodson, K.W., Ogg, D., Liu, C., Heuser, J., Knight, S., Kihlberg, J., Jones, C.H., and Hultgren, S.J., 1998, Periplasmic chaperone recognition motif of subunits mediates quaternary interactions in the pilus, *EMBO J.* 17:6155–6167.

Stapleton, A.E., Stroud, M.R., Hakomori, S.I., and Stamm, W.E., 1998, The globoseries glycosphingolipid sialosyl galactosyl globoside is found in urinary tract tissues and is a preferred binding receptor *in vitro* for uropathogenic *Escherichia coli* expressing pap-encoded adhesins, *Infect Immun.* 66:3856–3861.

Striker, R., Nilsson, U., Stonecipher, A., Magnusson, G., and Hultgren, S.J., 1995, Structural requirements for the glycolipid receptor of human uropathogenic *Escherichia coli, Mol Microbiol.* 16:1021–1029.

Stromberg, N., Marklund, B.I., Lund, B., Ilver, D., Hamers, A., Gaastra, W., Karlsson, K.A., and Normark, S., 1990, Host-specificity of uropathogenic *Escherichia coli* depends on differences in binding specificity to Gal alpha 1-4Gal-containing isoreceptors, *EMBO J.* 9:2001–2010.

Stromberg, N., Nyholm, P.G., Pascher, I., and Normark, S., 1991, Saccharide orientation at the cell surface affects glycolipid receptor function, *Proc Natl Acad Sci USA.* 88:9340–9344.

Sung, M.A., Fleming, K., Chen, H.A., and Matthews, S., 2001, The solution structure of PapGII from uropathogenic *Escherichia coli* and its recognition of glycolipid receptors, *EMBO Rep.* 2:621–627.

Svensson, A., Larsson, A., Emtenas, H., Hedenstrom, M., Fex, T., Hultgren, S.J., Pinkner, J.S., Almqvist, F., and Kihlberg, J., 2001a, Design and evaluation of pilicides: potential novel antibacterial agents directed against uropathogenic *Escherichia coli, Chem bio chem.* 2:915–918.

Svensson, M., Frendeus, B., Butters, T., Platt, F., Dwek, R., and Svanborg, C., 2003, Glycolipid depletion in antimicrobial therapy, *Mol Microbiol.* 47:453–461.

Svensson, M., Platt, F., Frendeus, B., Butters, T., Dwek, R., and Svanborg, C., 2001b, Carbohydrate receptor depletion as an antimicrobial strategy for prevention of urinary tract infection, *J Infect Dis.* 183 Suppl 1:S70–73.

Swanson, T.N., Bilge, S.S., Nowicki, B., and Moseley, S.L., 1991, Molecular structure of the Dr adhesin: nucleotide sequence and mapping of receptor-binding domain by use of fusion constructs, *Infect Immun.* 59:261–268.

Thanassi, D.G. and Hultgren, S.J., 2000, Assembly of complex organelles: pilus biogenesis in gram-negative bacteria as a model system, *Methods.* 20:111–126.

Thanassi, D.G., Saulino, E.T., and Hultgren, S.J., 1998a, The chaperone/usher pathway: a major terminal branch of the general secretory pathway, *Curr Opin Microbiol.* 1:223–231.

Thanassi, D.G., Saulino, E.T., Lombardo, M.J., Roth, R., Heuser, J., and Hultgren, S.J., 1998b, The PapC usher forms an oligomeric channel: implications for pilus biogenesis across the outer membrane, *Proc Natl Acad Sci USA.* 95:3146–3151.

Thanassi, D.G., Stathopoulos, C., Dodson, K., Geiger, D., and Hultgren, S.J., 2002, Bacterial outer membrane ushers contain distinct targeting and assembly domains for pilus biogenesis, *J Bacteriol.* 184:6260–6269.

Thankavel, K., Madison, B., Ikeda, T., Malaviya, R., Shah, A.H., Arumugam, P.M., and Abraham, S.N., 1997, Localization of a domain in the FimH adhesin of *Escherichia coli* type 1 fimbriae capable of receptor recognition and use of a domain-specific antibody to confer protection against experimental urinary tract infection, *J Clin Invest.* 100:1123–1136.

Thomas, W.E., Trintchina, E., Forero, M., Vogel, V., and Sokurenko, E.V., 2002, Bacterial adhesion to target cells enhanced by shear force, *Cell.* 109:913–923.

Van Loy, C.P., Sokurenko, E.V., Samudrala, R., and Moseley, S.L., 2002, Identification of amino acids in the Dr adhesin required for binding to decay-accelerating factor, *Mol Microbiol.* 45:439–452.

Wang, M.C., Tseng, C.C., Chen, C.Y., Wu, J.J., and Huang, J.J., 2002, The role of bacterial virulence and host factors in patients with *Escherichia coli* bacteremia who have acute cholangitis or upper urinary tract infection, *Clin Infect Dis.* 35:1161–1166.

Westerlund, B., Kuusela, P., Risteli, J., Risteli, L., Vartio, T., Rauvala, H., Virkola, R., and Korhonen, T.K., 1989, The O75X adhesin of uropathogenic *Escherichia coli* is a type IV collagen-binding protein, *Mol Microbiol.* 3:329–337.

Wizemann, T.M., Adamou, J.E., and Langermann, S., 1999, Adhesins as targets for vaccine development, *Emerg Infect Dis.* 5:395–403.

Wu, X.-R. and Sun, T.-T., 1993, Molecular cloning of a 47 kDa tissue-specific and differentiation-dependent urothelial cell surface glycoprotein, *J Cell Sci.* 106:31–43.

Zavialov, A.V., Kersley, J., Korpela, T., Zav'yalov, V.P., MacIntyre, S., and Knight, S.D., 2002, Donor strand complementation mechanism in the biogenesis of non-pilus systems, *Mol Microbiol.* 45:983–995.

Zavialov, A. V., Berglund, J., Pudney, A. F., Fooks, L. J., Ibrahim, T. M., MacIntyre, S., and Knight, S. D., 2003, Structure and biogenesis of the capsular F1 antigen from Yersinia pestis: preserved folding energy drives fiber formation, *Cell* 113:587–596.

Zhou, G., Mo, W.J., Sebbel, P., Min, G., Neubert, T.A., Glockshuber, R., Wu, X.R., Sun, T.T., and Kong, X.P., 2001, Uroplakin Ia is the urothelial receptor for uropathogenic *Escherichia coli*: evidence from *in vitro* FimH binding, *J Cell Sci.* 114:4095–4103.

GLYCOSYLATION DEPENDENT VIRAL INFECTIONS

4

CRYSTAL STRUCTURE OF AN INTACT HUMAN IgG: ANTIBODY ASYMMETRY, FLEXIBILITY, AND A GUIDE FOR HIV-1 VACCINE DESIGN

Erica Ollmann Saphire[1,2], Robyn L. Stanfield[1], M. D. Max Crispin[1,4], Garrett Morris[1], Michael B. Zwick[2], Ralph A. Pantophlet[2], Paul W. H. I. Parren[5], Pauline M. Rudd[4], Raymond A. Dwek[4], Dennis R. Burton[1,2], and Ian A. Wilson[1,3]

Depts. of [1]Molecular Biology, [2]Immunology, and
[3]The Skaggs Institute for Chemical Biology
The Scripps Research Institute
10550 North Torrey Pines
La Jolla, California 92037 USA
[4]Department of Biochemistry
University of Oxford
South Parks Road, Oxford OX1 3QU, United Kingdom
[5]Genmab, Jenalaan 18a
3584 CK Utrecht
The Netherlands

1. INTRODUCTION

Antibodies link antigen and immunological effector systems through the use of highly flexible linkers that connect the hypervariable antigen binding sites (Fabs) to the effector domain (Fc). The extensive flexibility of the antibody molecule permits antibodies to adapt to a vast array of antigen shapes and sizes while retaining a covalent link between the Fab domains and the conserved Fc region that interacts with a limited number of effector systems, such as Fc receptor and complement (Burton, 1985; Burton, 1990). However, this inherent molecular flexibility of intact antibodies has hindered their crystallization. Although over 200 structures of antibody fragments, mainly Fab and Fab′ fragments, have been determined, entire structures of IgGs with full length hinges have only been reported three times: two structures of murine mAbs (Harris *et al.*, 1995; Harris *et al.*, 1997) and now the structure of human IgG1 b12, directed against

Glycobiology and Medicine, edited by John S. Axford
Kluwer Academic / Plenum Publishers, New York, 2003

HIV-1 gp120 (Saphire *et al.*, 2001a; Saphire *et al.*, 2001b). The structure of IgG1 b12 represents the first structure of an intact human antibody with a full-length hinge for which all domains are visible in the electron density map.

IgG1 b12 is of interest as it is one of only a very few antibodies able to potently neutralize a broad array of primary isolates of HIV-1. Hence, the crystal structure of IgG1 b12, and its docking onto gp120 also provides a valuable guide for design of HIV-1 vaccines.

2. DIFFICULTIES IN GENERATING NEUTRALIZING ANTIBODIES AGAINST HIV-1

HIV-1 is able to evade the humoral immune response by rapid and extensive variation of its surface glycoproteins gp120 and gp41 in response to selective pressure, cloaking of much of the protein surface in carbohydrate, and poor immunogenicity of critical conserved epitopes (Parren *et al.*, 1999; Wyatt and Sodroski, 1998). To date, sera from vaccination and natural infection have generally not exhibited antibody titers capable of neutralizing primary viral isolates (Parren *et al.*, 1999). In addition, polyclonal antibody preparations and pooled hyperimmune human immunoglobulin (HIVIG) are generally nonprotective against viral challenge in animals (Gauduin *et al.*, 1997; Mascola *et al.*, 1999).

However, a few individual monoclonal antibodies have been shown to be highly effective against HIV-1 but difficult to elicit (D'Souza *et al.*, 1991; Moulard *et al.*, 2002; Zwick *et al.*, 2001b). These antibodies are IgG b12 which is directed against the CD4 binding site of gp120 (Burton *et al.*, 1994), 2G12 which is directed against an oligomannose epitope on gp120 (Sanders *et al.*, 2002; Scanlan *et al.*, 2002; Trkola *et al.*, 1996), Fab X5 which is directed against a gp120 epitope enhanced by CD4 binding (Moulard *et al.*, 2002), and 2F5, 4E10, and Z13 which are directed against two neighboring epitopes at the base of the ectodomain of gp41 (Muster *et al.*, 1993; Trkola *et al.*, 1995; Zwick *et al.*, 2001b). Crystal structures of these antibodies should provide valuable insights into their interactions with the HIV envelope proteins and suggest how to optimally design a vaccine capable of eliciting an effective immune response. Of this set of antibodies, IgG1 b12 is the first for which a structure is publicly available.

3. THE POTENT, BROADLY NEUTRALIZING ANTIBODY IgG1 b12

IgG1 b12 was identified from a combinatorial phage display library developed from bone marrow donated by a 31-year-old homosexual male who had been seropositive, but asymptomatic, for six years (Burton *et al.*, 1994). This antibody recognizes a highly conserved epitope overlapping the CD4 binding region of gp120. IgG1 b12 neutralizes roughly 80% of clade B primary viruses (Burton *et al.*, 1994; Kessler *et al.*, 1997) and a similar, or somewhat lesser, proportion of other clades (Burton *et al.*, 1994; Trkola *et al.*, 1995). In addition, antibody b12 can protect hu-PBL-SCID mice (Gauduin *et al.*, 1997; Parren *et al.*, 1995) and macaques (Parren *et al.*, 2001) from viral challenge. This combination of potency and broad specificity suggests that the b12 epitope on gp120 may be an effective target for vaccine design.

4. CRYSTALLIZATION OF IGG1 B12

Structural studies of antibodies have traditionally focused on the Fab and Fc fragments of antibodies due to the large size and extreme mobility of the intact IgG. Hence, we originally set out to determine the structure of the Fab fragment of b12. However, despite exhaustive effort, crystals of the Fab fragment were very difficult to generate and diffracted very poorly. Surprisingly, the intact IgG1 b12 crystallized rapidly with diffraction superior to that of the Fab alone (Saphire *et al.*, 2001a).

Problems in the crystallization of IgG have previously been ascribed to the highly flexible hinge region that connects the Fab domains to the Fc region. Consequently, initial views of intact immunoglobulins were based upon crystal structures of the hinge-deleted, conformationally-rigid IgGs Dob and Mcg (Rajan *et al.*, 1983; Silverton *et al.*, 1977; Terry *et al.*, 1968). The hinge-containing human IgGs Kol and Zie were crystallized later, but were characterized by a lack of structured electron density for the Fc domain and lower hinge regions indicating extreme disorder in the crystal (Colman *et al.*, 1976; Ely *et al.*, 1978; Marquart *et al.*, 1980). Nevertheless, crystallization of a molecule as flexible as IgG can be achieved if the crystal lattice is able to stabilize a single conformer. Two intact murine antibodies were recently crystallized and their complete structures determined. These antibodies are mAb 231, a murine IgG2a directed against an unknown antigen on the surface of canine lymphoma cells (Harris *et al.*, 1995; Larson *et al.*, 1991) and mAb 61.1.3, a murine IgG1 directed against the small molecule phenobarbital (Harris *et al.*, 1998). The crystal structure of IgG1 b12 now provides the first example structure of an intact human antibody for which all domains are ordered and visible. The antibody combining site provides a template for HIV-1 vaccine design and reveals key structural elements that are capable of productive interaction with the CD4 binding site of gp120.

5. THE CRYSTAL STRUCTURE OF IGG1 B12 IS ASYMMETRIC

The IgG is highly asymmetric overall, and the structure can be considered a "snapshot" of the broad range of conformations available to the IgG in solution (Figure 1). The Fc is shifted some 32 Å from the central dyad relating the two Fabs and packs underneath one of the Fabs (for details of calculations, please see Saphire *et al.*, 2002). In addition, one Fab is shifted 16 Å vertically upwards from the other Fab. The overall shape is between a Y and a T, with a 143° angle between the major axes of the two Fabs. The IgG spans 171 Å from the apex of one antigen binding site to the other. The Fc region is twisted nearly perpendicularly to the planes of the Fabs. The Fabs are rotated 158° relative to each other so that in Figure 1, the light chain (light gray) is in front of the heavy chain (black) in one of the Fabs while the light chain (light gray) is in the back of the heavy chain (black) of the other Fab. The elbow angles also differ slightly between the two Fabs (170° and 174°). A similarly defined elbow angle between the Fc C_H2 domains and C_H3 domains is 175°, revealing some asymmetry in the two halves of the Fc. Similar Fc asymmetry is also found in an IgG Fc-peptide (DeLano *et al.*, 2000) and an IgE Fc-Fc receptor complex (Sondermann *et al.*, 2000).

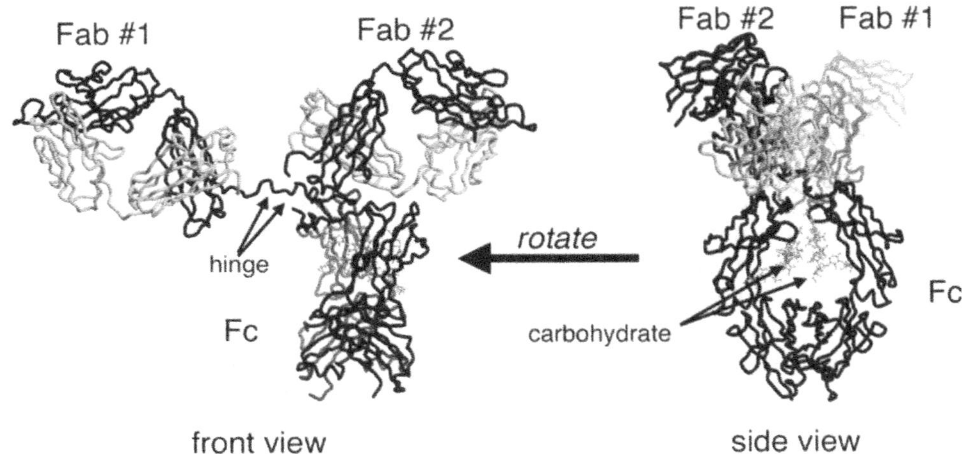

Figure 1. Two views of IgG1 b12. **Left, front view.** In this Cα ribbon representation of the intact antibody, the light chains are colored gray and heavy chains colored black so that each Fab is half black and half gray. The Fc is twisted nearly perpendicular to the Fabs and is shifted from center to pack underneath Fab #2. The Fc is comprised of two heavy chains (black) but perspective causes the heavy chain further in the background to appear gray. The two hinge regions are each 17 amino acids in length. One hinge is fully visible in electron density maps and extends from the Fc to Fab #1, while three central residues of the other hinge region are disordered. **Right, side view.** The Fc (two black heavy chains) is fully visible as are the two glycan moieties attached to the Fc heavy chains (gray ball-and-stick representation). Fab #2 projects towards the viewer while Fab #1 extends away from the viewer into the background.

6. ASYMMETRIC CARBOHYDRATE STRUCTURES

Some differences also exist in the corresponding electron density maps of the two carbohydrate chains attached to the Fc (Figure 2). These differences were further explored by release and analysis of the attached N-linked glycans (Saphire *et al.*, 2002). The IgG1 b12 sample used in these crystallographic studies was produced in CHO K1 cells, resulting in some differences between this IgG and normal human serum IgG. For example, there are no terminal sialic acids or bisecting *N*-acetyl glucosamine residues on IgG1 b12. Furthermore, in contrast to normal human serum polyclonal IgG, which contains approximately 19% of a-, 34% of mono-, and 47% of di-galactosylated biantennary glycans, glycan analysis indicated that biantennary glycans attached to IgG1 b12 are 51% a-, 41% mono-, and 8% di-galactosylated. Of the mono-galactosylated glycans present, 33% are on the 1,3 arm and 67% are on the 1,6 arm (Saphire *et al.*, 2002).

In general, at the protein level, glycan processing is controlled by the local three-dimensional structure of the protein around the glycosylation site and in the case of IgG, the accessibility of the glycan chain within the C_H2 domains (Rudd and Dwek, 1997). The two heavy chains of the IgG1 b12 are identical in sequence, but distinct in structure and have been named "H" and "K" to distinguish them in the deposited Protein Data Bank coordinates (accession code 1HZH). The carbohydrate moiety of the "K" heavy chain of b12 contains clear electron density for only the 1,6 arm galactose residue, while the glycan attached to the "H" chain has clear electron density for both galactose residues (Figure 2). Preferential asymmetric glycan pairing has been previously noted. For example,

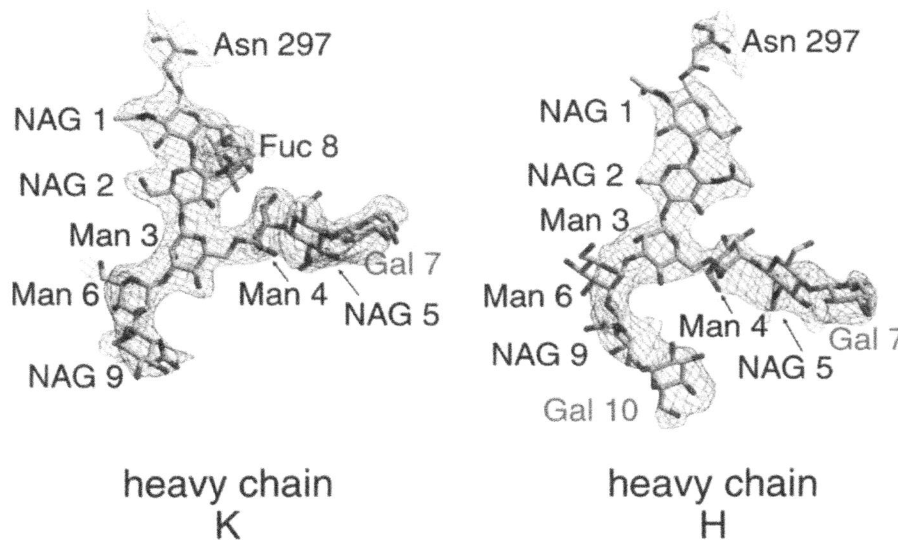

Figure 2. Carbohydrate moieties attached to IgG1 b12. The two attached glycans have been roughly aligned for comparison and are illustrated in ball-and-stick representation. 2Fo-Fc electron density corresponding to the attached glycans is illustrated in a "cage" representation around the glycan models. Note that only one terminal galactose residue is visible in electron density maps of the glycan attached to heavy chain K, while two terminal galactose residues are visible for the glycan attached to heavy chain H. Also note that the fucose residue is visible on the glycan attached to heavy chain K, but not on the glycan attached to heavy chain H.

bi-galactosylated glycans predominantly pair with 1,6 arm mono-galactosylated glycans (Masuda *et al.*, 2000) as in the b12 structure.

Interestingly, electron density of the di-galactosylated species is visible on the "H" chain. Given the small proportion (8%) of di-galactosylated glycans, this may represent the entire population, again suggesting that there has been asymmetric processing. Glycan analysis indicates that about 80% of the oligosaccharides contain core fucose, however, in the crystal structure, the fucose residue is visible on only one of the two carbohydrate chains. This further implies that there is a difference in the environment of the sugars in the "H" and "K" chains, which affects the core region of the sugars, allowing increased mobility of the "H" chain of the flexible α1,6 linked fucose. Moreover, the difference in angle of the 1,6 arm of Man 3 between H and K is a likely consequence of the asymmetry of the Fc C_H2 domains which contact Gal 7 terminating the 1,6 arm. These data are therefore consistent with the finding that b12 is highly asymmetric and imply that the two sugar chains are presented to the galactosyl transferase in significantly different context. Other IgG antibodies may similarly assume a variety of asymmetric conformations in solution and attached glycans may be presented to processing machinery in separate environments.

7. STRUCTURE OF THE B12 ANTIGEN BINDING SITE

The b12 antigen binding site displays an unusually long complementarity determining region (CDR) H3 which contains a ten residue insertion and projects vertically above

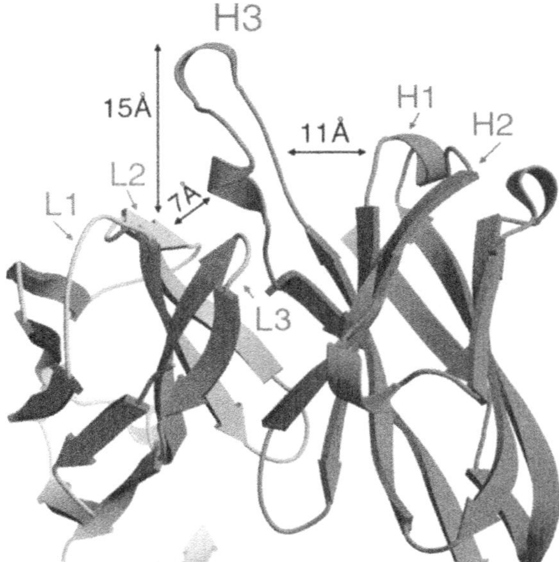

Figure 3. The antigen binding site of IgG1 b12 is oriented such that the heavy chain is on the right and the light chain on the left. The six complementarity determining regions (CDRs) (H1, H2, H3, L1, L2, and L3) are indicated. The CDR H3 of b12 projects vertically upwards from center, 15 Å above the light chain and 10 Å above the heavy chain. The CDR H3 is also flanked by a canyon on either side. The canyon between CDR H3 and the light chain is 7 Å in width while the canyon between CDR H3 and the rest of the heavy chain is 11 Å in width.

the rest of the antigen binding site (Figure 3). In the b12 structure, the CDR H3 is flanked by a canyon on either side. The canyon between CDR H3 and CDRs H1 and H2 is roughly 11 Å wide, and the canyon between CDR H3 and CDRs L1 and L2 is roughly 7 Å wide. The light chain CDRs L1, L2, and L3 together form a low platform roughly 15 Å below CDR H3, while the heavy chain CDRs H1 and H2 form a somewhat higher surface about 10 Å lower than CDR H3 (Figure 3).

Traditional views suggest that antigen binding sites for anti-protein antibodies are relatively flat or undulating. However, extended H3 loops are frequently seen in human antibodies directed against pathogens (Barbas *et al.*, 1992, 1993; Kunert *et al.*, 1998; Sanna *et al.*, 1995) allowing them to potentially access canyons and clefts on the viral surface (Smith *et al.*, 1996). Interestingly, mouse antibodies do not normally exhibit such long CDR H3 loops and, indeed, few murine anti-CD4 binding site antibodies have been described relative to the plethora of human monoclonal antibodies of this specificity.

This extended CDR H3 loop likely reflects a key structural feature for probing the recessed CD4 binding site of gp120; all of the members of a panel of 32 anti-CD4 binding site antibodies developed from phage display have an extended CDR H3 (Barbas *et al.*, 1993). The crystal structure of gp120 in complex with CD4 and an Fab fragment (Kwong *et al.*, 1998, 2000b) demonstrates that CD4 inserts a loop terminating in Phe 43 into a recessed pocket in gp120 in order to achieve complementarity. Our docking experiments guided by geometric fit and mutagenesis strongly suggest that the long CDR H3 finger of b12 also penetrates this cleft (Pantophlet *et al.*, 2003; Saphire *et al.*, 2001b; Zwick *et al.*, 2003). In addition, synthetic peptides corresponding to the CDR H3 sequence linked to BSA are able to neutralize laboratory-adapted strains of HIV-1 (Saphire *et al.*, 2001b).

8. DOCKING OF B12 ONTO HIV-1 GP120

Ideally, we would like to perform docking studies of b12 onto the structure of a complete envelope trimer. However, the only available structure is that of a truncated, monomeric gp120 core bound to CD4 and an Fab fragment (Kwong *et al.*, 1998, 2000a). One interpretation of thermodynamic studies on recombinant monomeric gp120 is that CD4 binding could be accompanied by structural rearrangement of gp120 (Myszka *et al.*, 2000). Nevertheless, since earlier studies indicated that b12 and CD4 are sensitive to the same mutations in gp120, it is not unreasonable to use the CD4-bound core gp120 structure as a starting model, and further test the docking model by mutagenesis.

An antibody is twice as wide as the CD4 receptor (two immunoglobulin domains in width versus one). Thus, the imprint of an Fab onto the neutralizing, non-glycosylated, CD4-binding face of gp120 is extremely limited by geometric fit. The opening of the CD4-binding face of gp120 between the V1/V2 loop stem and constant region 4 (C4) is 35 Å wide, while the width of the Fab combining site is 31 Å wide. For gp120 in the envelope trimer, space available for b12 binding in the lateral direction is bound by the trimer interface and the 2G12 epitope, as it is known that b12 is able to bind native trimeric spikes (Roben *et al.*, 1994), and is also able to bind concurrently with the 2G12 antibody (Moore and Sodroski, 1996). Computational docking studies performed using AutoDock arrive at a clear solution (Saphire *et al.*, 2001b): that b12 may fit snugly onto gp120 by binding an epitope extending from the V1/V2 loop stem across the neutralizing face, with Trp100 at the tip of the H3 loop penetrating a pocket in the recessed CD4 binding site (Figure 4).

Physical models generated from the b12 and gp120 coordinates demonstrate contact surfaces that are strikingly complementary, like fingers fitting into a glove. The protruding ridge formed by Ser 364 through Asp 368 of gp120 can nestle into a cleft between CDRs H3 and H2 of b12. The protruding D loop of gp120 can fit into a depression formed between CDRs H3, L1, and L3 of b12. Indeed, a phage-display-derived peptide that binds b12

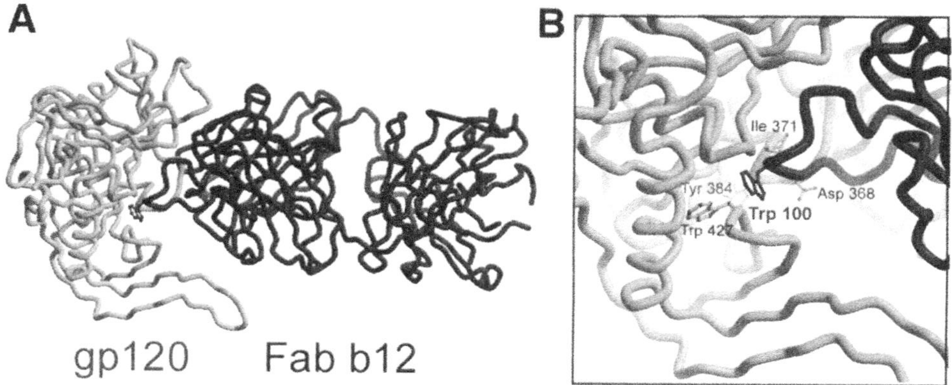

Figure 4. b12-gp120 docking. **A.** Fab b12 (black Cα ribbon) docked onto the monomeric gp120 core (gray Cα ribbon). Note the extended CDR H3 with a Trp residue (black ball-and-stick) at the apex of the loop that binds into the recessed cavity in the CD4 binding site of gp120. **B.** Close up view of the recessed CD4 binding site of gp120 with certain residues key for interaction of gp120 with CD4 and b12 illustrated in gray ball and stick. The Trp 100 residue of the b12 CDR H3 (black) binds into this pocket.

has several sequence elements in common with the D loop of gp120 (Zwick *et al.*, 2001a). Electron density maps of this peptide in complex with b12 Fab illustrate clear density for the peptide in a depression formed between CDRs H3, L1, and L3 (Saphire *et al.*, 2003).

The docking model suggests that approximately 1960 $Å^2$ (970 $Å^2$ on gp120 and 990 $Å^2$ on b12) of solvent accessible surface is buried. The b12 epitope, like the footprint of CD4, is centered on the outer domain of gp120 with additional contact extending to the V1/V2 loop stem, with minimal contact to the inner domain. This region is conserved in sequence, and b12 appears to form many contacts to main-chain atoms of gp120, which may explain the broad cross-reactivity of the b12 antibody (Saphire *et al.*, 2001b).

Why does b12 neutralize primary viruses whereas other CD4 binding site antibodies do not? The difference is that b12 is able to bind to trimeric as well as to monomeric gp120, whereas other antibodies of similar specificity bind only to monomeric gp120. The antibody b12 is also unique in its sensitivity to mutations associated with the V1/V2 loop (Roben *et al.*, 1994) and, in particular, to changes in the V2 loop stem structure (Binley *et al.*, 1998). In our model, the b12 Fab fits onto gp120 by contacting the inside face of the V1/V2 loop stem. This mode of interaction angles the rest of the antibody bulk away from the trimer interface so that IgG1 b12 could recognize both monomeric gp120 as well as the oligomeric gp120 on the viral surface. Extensive alanine-scanning mutagenesis of gp120 supports this model (Pantophlet *et al.*, 2003). Thus, the antibody b12 is probably capable of potent neutralization of a broad array of HIV-1 isolates as its epitope is conserved and angled in such a way that the antibody can access its epitope on the native viral surface.

9. A MODEL FOR NEUTRALIZATION OF HIV-1 BY IGG1 B12

In order to create a visual picture of how IgG1 b12 may neutralize HIV-1, we docked b12 onto a model of the native envelope spike created through addition of carbohydrate and gp41 (Chan *et al.*, 1997; Weissenhorn *et al.*, 1997) onto a predicted structure of a gp120 trimer (Kwong *et al.*, 2000b) (Figure 5). This model further confirms that the majority of the surface of the HIV-1 envelope spike is masked by carbohydrate with the exception of the CD4 binding site, which is the site of interaction with the b12 antibody (Figure 5). Trimeric envelope spikes have three equivalent CD4 binding sites. Although the 170 Å reach of the IgG indicates that it could bivalently span two different envelope spikes, b12 would probably not be able to bind simultaneously to two binding sites on the same trimeric spike.

HIV-1 is neutralized at an occupancy of approximately one IgG per viral envelope spike (Burton *et al.*, 2001; Schønning *et al.*, 1999). Thus, at the coating density of one IgG molecule per spike, the large mass (150 kDa) of the IgG compared to the envelope trimer will sterically block attachment of the virus to the target cell and/or fusion of viral and cell membranes. Multiple contacts between viral spikes and cellular receptors within a confined area may be necessary for attachment and fusion of the viral and host membranes (Kuhmann *et al.*, 2000). Hence, an IgG attached to one spike might also interfere with optimal functioning of a neighboring spike. This mode of neutralization, based on steric hindrance, is consistent with data showing that neutralization depends primarily upon antibody occupancy of envelope spikes irrespective of the particular epitope recognized (Parren *et al.*, 1998). It is possible that the flexibility of IgG molecules could allow them to occlude a somewhat larger region of viral surface through multiple domain arrangements and could

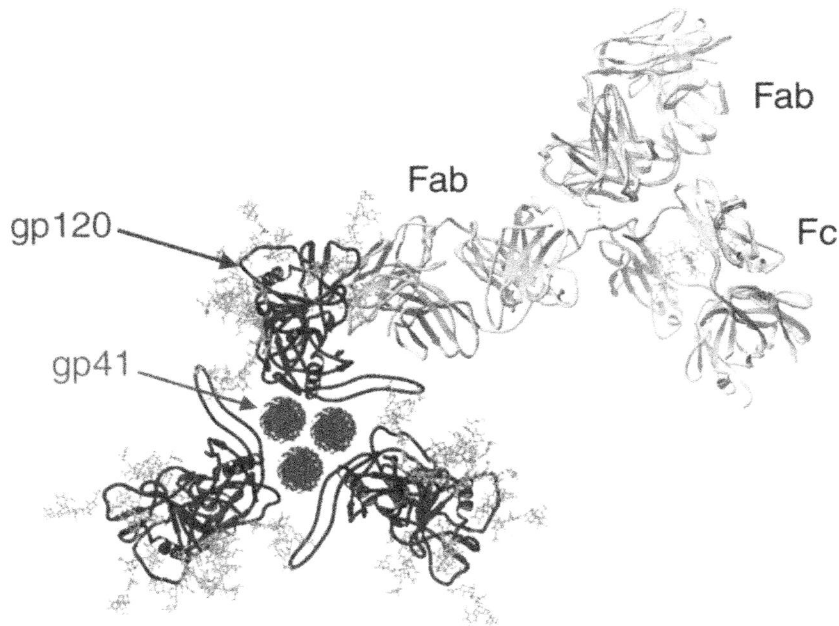

Figure 5. Model of IgG1 b12 neutralizing a trimeric HIV-1 envelope spike. Shown here is the crystal structure of IgG1 b12 (light gray) docked onto gp120 (identical to that seen in Figure 4). N-linked glycans have been modeled onto each of the three gp120 monomers arranged into a trimeric orientation according to a separate modeling study (Kwong *et al.*, 2000b). A representation of the trimeric gp41 molecule is shown at the center of the spike. Note that much of the surface of the envelope spike is cloaked in carbohydrate with the exception of the recessed CD4 binding site. The b12 antibody is probably able to potently neutralize HIV-1 as its epitope is accessible on both monomeric and trimeric gp120.

allow them to adopt conformations required for bivalent binding of two neighboring envelope spikes. Thus, the IgG1 b12 structure now provides a much clearer view of how a human antibody can effectively neutralize primary isolates of HIV-1.

10. CONCLUSIONS

In conclusion, we present here the first structure of an intact human antibody with a full-length hinge. This structure is highly asymmetric in both protein and attached carbohydrate moieties. The b12 crystal structures also offer an illustration of the antigen binding site surface topography of rare monoclonal antibodies capable of potently neutralizing a broad range of primary HIV-1 isolates, the identification of a highly extended CDR H3 that may provide leads for anti-viral compounds or peptides, and the docking of b12 onto gp120 supported by geometric fit and extensive mutagenesis. The difficulty in generating an effective humoral immune response to HIV-1 underscores the need to elucidate the structures of the epitopes of those rare, highly effective antibodies, in order to develop immunogens that can elicit more antibodies like them. Fine mapping of the b12 epitope can now facilitate the design of minimized gp120 cores or peptidomimetics, and such structural information should open new doors in the global effort to design an effective HIV-1 vaccine.

11. ACKNOWLEDGMENTS

We thank Mark Ultsch and Abraham de Vos at Genentech, Inc. for unpublished high-resolution human Fc coordinates, the staff of Stanford Synchrotron Radiation Laboratory Beamline 7-1. We gratefully acknowledge NIH grants GM46192 (IAW), AI33292 (DRB), AI40377 (PWHIP), the Universitywide AIDS Research Program FR1-SRI-121 (EOS), and The Skaggs Institute for Chemical Biology for support. This is publication No. 15501-MB from the Scripps Research Institute. The IgG1 b12 coordinates and structure factors have been deposited in the Protein Data Bank under accession number 1HZH.

12. REFERENCES

Barbas, C.F., III, Collet, T.A., Amberg, W., Roben, P., Binley, J.M., Hoekstra, D., Cababa, D., Jones, T.M., Williamson, R.A., Pilkington, G.R. *et al.*, 1993, Molecular profile of an antibody response to HIV-1 as probed by combinatorial libraries, *JMB.* 230:812–823.

Barbas, C.F., III, Crowe, J.E., Jr., Cababa, D., Jones, T.M., Zebedee, S.L., Murphy, B.R., Chanock, R.M., and Burton, D.R., 1992, Human monoclonal Fab fragments derived from a combinatorial library bind to respiratory syncytial virus F glycoprotein and neutralize infectivity, *PNAS.* 89:10164–10168.

Binley, J.M., Wyatt, R., Desjardins, E., Kwong, P.D., Hendrickson, W., Moore, J.P., and Sodroski, J., 1998, Analysis of the interaction of antibodies with a conserved enzymatically deglycosylated core of the HIV type 1 envelope glycoprotein 120, *AIDS Res Hum Retroviruses.* 14:191–198.

Burton, D.R., 1985, Immunoglobulin G: Functional Sites, *Ml Immunol.* 22:161–206.

Burton, D.R., 1990, Antibody: The flexible adaptor molecule, *Trends Bioch Sci.* 15:64–69.

Burton, D.R., Pyati, J., Koduri, R., Sharp, S.J., Thornton, G.B., Parren, P.W., Sawyer, L.S., Hendry, R.M., Dunlop, N., Nara, P.L. *et al.*, 1994, Efficient neutralization of primary isolates of HIV-1 by a recombinant human monoclonal antibody, *Science.* 266:1024–1027.

Burton, D.R., Saphire, E.O., and Parren, P.W.H.I., 2001, A model for neutralization of viruses based on antibody coating of the virion surface, *Curr Topics in Microbiology and Immunology.* 260:109–143.

Chan, D.C., Fass, D., Berger, J.M., and Kim, P.S., 1997, Core structure of gp41 from the HIV envelope glycoprotein, *Cell.* 89:263–273.

Colman, P.M., Deisenhofer, J., Huber, R., and Palm, W., 1976, Structure of the Human Antibody Molecule Kol (Immunoglobulin G1): An Electron Density Map at 5 Å Resolution, *JMB.* 100:257–278.

D'Souza, M.P., Durda, P., Hanson, C.V., and Milman, G., 1991, Evaluation of monoclonal antibodies to HIV-1 by neutralization and serological assays: an international collaboration. Collaborating Investigators, *Aids.* 5:1061–1070.

DeLano, W.L., Ultsch, M.H., de Vos, A.M., and Wells, J.A., 2000, Convergent solutions to binding at a protein-protein interface, *Science.* 287:1279–1283.

Ely, K.R., Colman, P.M., Abola, E.E., Hess, A.C., Peabody, D.S., Parr, D.M., Connell, G.E., Laschinger, C.A., and Edmundson, A.B., 1978, Mobile Fc Region in the Zie IgG Cryoglobulin: Comparison of Crystals of the $F(ab')^2$ Fragment and the Intact Immunoglobulin, *Biochemistry.* 17:820–823.

Gauduin, M.C., Parren, P.W., Weir, R., Barbas, C.F., Burton, D.R., and Koup, R.A., 1997, Passive immunization with a human monoclonal antibody protects hu-PBL- SCID mice against challenge by primary isolates of HIV-1, *Nat Med.* 3:1389–1393.

Harris, L.J., Larson, S.B., Hasel, K.W., Day, J., Greenwood, A., and McPherson, A., 1995, The three-dimensional structure of an intact monoclonal antibody for canine lymphoma, *Nature.* 360:369–372.

Harris, L.J., Larson, S.B., Hasel, K.W., and McPherson, A., 1997, Refined structure of an intact IgG2a monoclonal antibody, *Biochemistry.* 36:1581–1597.

Harris, L.J., Skaletsky, E., and McPherson, A., 1998, Crystallographic structure of an intact IgG1 monoclonal antibody, *JMB.* 275:861–872.

Kessler, J.A., 2nd, McKenna, P.M., Emini, E.A., Chan, C.P., Patel, M.D., Gupta, S.K., Mark, G.E., 3rd, Barbas, C.F., 3rd, Burton, D.R., and Conley, A.J., 1997, Recombinant human monoclonal antibody IgG1b12 neutralizes diverse human immunodeficiency virus type 1 primary isolates, *AIDS Res Hum Retroviruses*. 13:575–582.

Kuhmann, S.E., Platt, E.J., Kozak, S.L., and Kabat, D., 2000, Cooperation of multiple CCR5 coreceptors is required for infections by human immunodeficiency virus type 1, *J Virol.* 74:7005–7015.

Kunert, R., Ruker, F., and Katinger, H., 1998, Molecular characterization of five neutralizing anti–HIV type 1 antibodies: identification of nonconventional D segments in the human monoclonal antibodies 2G12 and 2F5, *AIDS Res Hum Retroviruses*. 14:1115–1128.

Kwong, P.D., Wyatt, R., Majeed, S., Robinson, J., Sweet, R.W., Sdroski, J., and Hendrickson, W.A., 2000a, Structures of HIV-1 gp120 envelope glycoproteins from laboratory-adapted and primary isolates, *Structure Fold Des.* 8:1329–1339.

Kwong, P.D., Wyatt, R., Robinson, J., Sweet, R.W., Sodroski, J., and Hendrickson, W.A., 1998, Structure of an HIV gp120 envelope glycoprotein in complex with the CD4 receptor and a neutralizing human antibody, *Nature*. 393:648–659.

Kwong, P.D., Wyatt, R., Sattentau, Q.J., Sodroski, J., and Hendrickson, W.A., 2000b, Oligomeric modeling and electrostatic analysis of the gp120 envelope glycoprotein of human immunodeficiency virus, *J Virol.* 74:1961–1972.

Larson, S., Day, J., Greenwood, A., Skaletsky, E., and McPherson, A., 1991, Characterization of crystals of an intact monoclonal antibody for canine lymphoma, *JMB*. 222:17–19.

Marquart, M., Deisenhofer, J., and Huber, R., 1980, Crystallographic refinement and atomic models of the intact immunoglobulin molecule Kol and its antigen-binding fragment at 3.0 Å and 1.9 Å resolution, *JMB*. 141:369–391.

Mascola, J.R., Lewis, M.G., Stiegler, G., Harris, D., VanCott, T.C., Hayes, D., Louder, M.K., Brown, C.R., Sapan, C.V., Frankel, S.S. *et al.*, 1999, Protection of Macaques against pathogenic simian/human immunodeficiency virus 89.6PD by passive transfer of neutralizing antibodies, *J Virol.* 73:4009–4018.

Masuda, K., Yamaguchi, Y., Kato, K., Takahashi, N., Shimada, I., and Arata, Y., 2000, Pairing of oligosaccharides in the Fc region of immunoglobulin G, *FEBS Lett.* 473:349–357.

Moore, J.P. and Sodroski, J., 1996, Antibody cross-competition analysis of the human immunodeficiency virus type 1 gp120 exterior envelope glycoprotein, *J Virol.* 70:1863–1872.

Moulard, M., Phogat, S.K., Shu, Y., Labrijn, A.F., Xiao, X., Binley, J.M., Zhang, M.Y., Sidorov, I.A., Broder, C.C., Robinson, J. *et al.*, 2002, Broadly cross-reactive HIV-1-neutralizing human monoclonal Fab selected for binding to gp120-CD4-CCR5 complexes, *PNAS*. 99:6913–6918.

Muster, T., Steindl, F., Purtscher, M., Trkola, A., Klima, A., Himmler, G., Ruker, F., and Katinger, H., 1993, A conserved neutralizing epitope on gp41 of human immunodeficiency virus type 1, *J Virol.* 67:6642–6647.

Myszka, D.G., Sweet, R.W., Hensley, P., Brigham-Burke, M., Kwong, P.D., Hendrickson, W.A., Wyatt, R., Sodroski, J., and Doyle, M.L., 2000, Energetics of the HIV gp120-CD4 binding reaction, *PNAS*. 97:9026–9031.

Pantophlet, R.A., Saphire, E.O., Poignard, P., Parren, P.W.H.I., Wilson, I.A., and Burton, D.R., 2003, Fine mapping of the interaction of neutralizing and non-neutralizing monoclonal antibodies with the CD4 binding site of human immunodeficiency virus type I gp120, *J Virol.* 77:642–658.

Parren, P.W.H.I., Ditzel, H.J., Gulizia, R.J., Binley, J.M., Barbas, C.F., III, Burton, D.R., and Mosier, D.E., 1995, Protection against HIV-1 infection in hu-PBL-SCID mice by passive immunization with a neutralizing human monoclonal antibody against the gp120 CD4-binding site, *Aids*. 9:F1–6.

Parren, P.W.H.I., Marz, P., Cheng-Mayer, C., Harouse, J., Moore, J.P., and Burton, D.R., 2001, Antibody protection of macaques against vaginal challenge with a pathogenic R5 HIV-1/SIV chimeric virus requires complete neutralization of HIV-1. *J Virol.* in press.

Parren, P.W.H.I., Mondor, I., Naniche, D., Ditzel, H.J., Klasse, P.J., Burton, D.R., and Sattentau, Q.J., 1998, Neutralization of human immunodeficiency virus type 1 by antibody to gp120 is determined primarily by occupancy of sites on the virion irrespective of epitope specificity, *J Virol.* 72:3512–3519.

Parren, P.W.H.I., Moore, J.P., Burton, D.R., and Sattentau, Q.J., 1999, The neutralizing antibody response to HIV-1: viral evasion and escape from humoral immunity, *Aids*. 13:S137–162.

Rajan, S.S., Ely, K.R., Abola, E.E., Wood, M.K., Colman, P.M., Athay, R.J., and Edmundson, A.B., 1983, Three-dimensional structure of the Mcg IgG1 immunoglobulin, *Ml Immunol.* 20:787–799.

Roben, P., Moore, J.P., Thali, M., Sodroski, J., Barbas, C.F., 3rd, and Burton, D.R., 1994, Recognition properties of a panel of human recombinant Fab fragments to the CD4 binding site of gp120 that show differing abilities to neutralize human immunodeficiency virus type 1, *J Virol.* 68:4821–4828.

Rudd, P.M. and Dwek, R.A., 1997, Glycosylation: heterogeneity and the 3D structure of proteins, *Crit Rev Biochem Mol Biol.* 32:1–100.

Sanders, R.W., Venturi, M., Schiffner, L., Kalyanaraman, R., Katinger, H., Lloyd, K.O., Kwong, P.D., and Moore, J.P., 2002, The mannose-dependent epitope for neutralizing antibody 2G12 on human immunodeficiency virus type 1 glycoprotein gp120, *J Virol.* 76:7293–7305.

Sanna, P.P., Williamson, R.A., De Logu, A., Bloom, F.E., and Burton, D.R., 1995, Directed selection of recombinant human monoclonal antibodies to herpes simplex virus glycoproteins from phage display libraries, *PNAS.* 92:6439–6443.

Saphire, E.O., Montero, M., Menendez, A., Irving, M.B., Zwick, M.B., Parren, P.W.H.I., Burton, D.R., Scott, J.K., and Wilson, I.A., 2003, Crystal structure of a broadly neutralizing anti-HIV-1 antibody in complex with a peptide mimotope, submitted.

Saphire, E.O., Parren, P.W.H.I., Barbas, C.F., III, Burton, D.R., and Wilson, I.A., 2001a, Crystallization and preliminary structure determination of an intact human immunoglobulin b12: an antibody that broadly neutralizes primary isolates of HIV-1, *Acta Cryst D57*:168–171.

Saphire, E.O., Parren, P.W.H.I., Pantophlet, R., Zwick, M.B., Morris, G.M., Stanfield, R.L., Rudd, P.M., Dwek, R.A., Burton, D.R., and Wilson, I.A., 2001b, Crystal structure of an intact human IgG with broad and potent activity against primary HIV-1 isolates: A template for HIV vaccine design, *Science.* 293:1155–1159.

Saphire, E.O., Stanfield, R.L., Crispin, M.D., Rudd, P.M., Dwek, R.A., Parren, P.W.H.I., Burton, D.R., and Wilson, I.A., 2002, Contrasting IgG structures reveal extreme asymmetry and flexibility, *JMB.* 309:9–18.

Scanlan, C.N., Pantophlet, R., Wormald, M.R., Saphire, E.O., Stanfield, R., Wilson, I.A., Katinger, H., Dwek, R.A., Rudd, P.M., and Burton, D.R., 2002, The broadly neutralizing anti-human immunodeficiency virus type 1 antibody 2G12 recognizes a cluster of alpha1→ 2 mannose residues on the outer face of gp120, *J Virol.* 76:7306–7321.

Schønning, K., Lund, O., Lund, O.S., and Hansen, J.E., 1999, Stoichiometry of monoclonal antibody neutralization of T-cell line-adapted human immunodeficiency virus type 1, *J Virol.* 73:8364–8370.

Silverton, E.W., Navia, M.A., and Davies, D.R., 1977, Three-dimensional structure of an intact human immunoglobulin, *PNAS.* 74:5140–5144.

Smith, T.J., Chase, E.S., Schmidt, T.J., Olson, N.H., and Baker, T.S., 1996, Neutralizing antibody to human rhinovirus 14 penetrates the receptor-binding canyon, *Nature.* 383:350–354.

Sondermann, P., Huber, R., Oosthuizen, V., and Jacob, U., 2000, The 3.2-Å crystal structure of the human IgG1 Fc fragment-FcγRIII complex, *Nature.* 406:267–273.

Terry, W.D., Matthews, B.W., and Davies, D.R., 1968, Crystallographic studies of a human immunoglobulin, *Nature.* 220:239–241.

Trkola, A., Pomales, A.B., Yuan, H., Korber, B., Maddon, P.J., Allaway, G.P., Katinger, H., Barbas, C.F., 3rd, Burton, D.R., Ho, D.D. *et al.*, 1995, Cross-clade neutralization of primary isolates of human immunodeficiency virus type 1 by human monoclonal antibodies and tetrameric CD4-IgG, *J Virol.* 69:6609–6617.

Trkola, A., Purtscher, M., Muster, T., Ballaun, C., Buchacher, A., Sullivan, N., Srinivasan, K., Sodroski, J., Moore, J.P., and Katinger, H., 1996, Human monoclonal antibody 2G12 defines a distinctive neutralization epitope on the gp120 glycoprotein of human immunodeficiency virus type 1, *J Virol.* 70:1100–1108.

Weissenhorn, W., Dessen, A., Harrison, S.C., Skehel, J.J., and Wiley, D.C., 1997, Atomic structure of the ectodomain from HIV-1 gp41, *Nature.* 387:426–430.

Wyatt, R. and Sodroski, J., 1998, The HIV-1 envelope glycoproteins: fusogens, antigens, and immunogens, *Science.* 280:1884–1888.

Zwick, M.B., Bonnycastle, L.L.C., Menendez, A., Irving, M.B., Barbas, C.F., III, Parren, P.W.H.I., Burton, D.R., and Scott, J.K., 2001a, Identification and characterization of a peptide that specifically binds the broadly HIV-1 neutralizing, human antibody, b12. *J Virol.* 75:6692–6699.

Zwick, M.B., Labrijn, A.F., Wang, M., Spenlehauer, C., Saphire, E.O., Binley, J.M., Moore, J.P., Steigler, G., Katinger, H., Burton, D.R., and Parren, P.W.H.I., 2001b, Broadly neutralizing antibodies targeted to the membrane-proximal region of Human Immunodeficiency Virus Type 1 Glycoprotein GP41, *J Virol.* 75:10892–10905.

Zwick, M.B., Parren, P.W.H.I., Saphire, E.O., Wang, M., Scott, J.K., Dawson, P.E., Wilson, I.A., and Burton, D.R., 2003, Molecular features of the broadly neutralizing antibody b12 required for recognition of HIV-1 Gp120, *J Virol.* 77:5863–5876.

INFLAMMATION

5

REMNANT EPITOPES GENERATE AUTOIMMUNITY: FROM RHEUMATOID ARTHRITIS AND MULTIPLE SCLEROSIS TO DIABETES

Francis J. Descamps, Philippe E. Van den Steen, Inge Nelissen, Jo Van Damme, and Ghislain Opdenakker

Rega Institute for Medical Research
University of Leuven
Minderbroedersstraat 10
B-3000 Leuven
Belgium

1. ABSTRACT

Autoimmune diseases are characterized by inflammation and by the development and maintenance of antibodies and T lymphocytes against "self" antigens. Although the etiology of these diseases is unknown, they have a number of cellular and molecular mechanisms in common. Pro-inflammatory cytokines, such as interleukin-1 (IL-1) and tumor necrosis factor (TNF), are upregulated and activate the inflammatory process. Chemokines recruit and activate leukocytes to release proteases, including matrix metalloproteinases (MMPs). These proteases degrade proteins into remnant fragments, which often constitute immunodominant epitopes. Either by direct loading into major histocompatibility complex (MHC) molecules or after classical antigen uptake, processing and MHC presentation, these remnant epitopes are presented to autoreactive T lymphocytes. Also, posttranslationally modified remnant peptides may stimulate B cells to produce autoantibodies. This forms the basis of the "Remnant Epitopes Generate Autoimmunity" (REGA) model. We have documented evidences for this model in multiple sclerosis (MS), rheumatoid arthritis (RA) and diabetes, which are summarized here. Furthermore, three topics will be addressed to illustrate the importance of glycobiology in the pathogenesis of autoimmune diseases. In MS, gelatinase B or MMP-9 is a pathogenic glycoprotein of which the sugars contribute to its interactions with the tissue inhibitor of metalloproteinases-1 (TIMP-1) and thus assist in the determination of the enzyme activity. In RA, gelatinase B cleaves

Glycobiology and Medicine, edited by John S. Axford
Kluwer Academic / Plenum Publishers, New York, 2003

denatured type II collagen into remnant epitopes, some of which constitute immunodominant glycopeptides. This implies that immunodominant epitope scanning experiments should preferably be done with natural posttranslationally modified glycopeptides, rather than with unmodified (synthetic) peptides. Sugars can also be used as molecular probes to induce autoimmune diseases. One of the best examples is the induction of acute pancreatitis, insulitis and diabetes by streptozotocin. In addition, gelatinase B is upregulated in pancreatitis and cleaves insulin. The most efficient cleavage by gelatinase B leads to a major insulin remnant epitope.

2. INTRODUCTION

Research on autoimmune diseases has mainly been focussed on lymphocyte functions. The knowledge of cellular and humoral immunity has progressed exponentially, but insufficiently to fully explain the pathogenesis of autoimmunity. In addition, *specific* immune mechanisms and various antigens, involved in different autoimmune disorders, have been studied in detail. Many components of the adaptive immune system [e.g., major histocompatibility complex (MHC) proteins, T cell receptors, antibodies] and the innate immune system (e.g., complement proteins, cytokines) are glycoproteins (Rudd *et al.*, 2001). Although the protein functions have been studied extensively, knowledge of the functions of the attached oligosaccharides is still fragmentary. From our growing understanding of the working mechanisms of the adaptive immune system, it becomes clear that specific therapeutic strategies for autoimmune diseases (e.g., T cell vaccination, tolerance induction) will be patient-specific and perhaps not generally applicable.

We previously studied *aspecific* host defense or innate immune mechanisms and their role in autoimmunity. As a result of these investigations, we have proposed the REGA model (Remnant Epitopes Generate Autoimmunity) for the generation of autoantigens and their interaction with the T cell receptor complex (Opdenakker and Van Damme, 1994; Van den Steen *et al.*, 2002). In this model, cytokine- and chemokine-regulated proteases play a central role in the generation of autoimmune antigens. As a consequence, disease-promoting cytokines and proteases are suggested to be therapeutic targets, whereas disease-limiting cytokines and protease inhibitors may be products in treatment of autoimmune diseases. Most cytokines, proteases and protease inhibitors are glycoproteins, and their glycosylation often plays a role in molecular targeting or in fine-tuning their specific activity (Opdenakker *et al.*, 1995; Van den Steen *et al.*, 1998a).

Here, we will exemplify the glycoimmunology of gelatinase B/MMP-9. First, the oligosaccharides that are attached to MMP-9 influence its interaction with the tissue inhibitor of matrix metalloproteinases-1 (TIMP-1). Second, gelatinase B as an enzyme cleaves collagen II into natural immunodominant glycopeptide antigens. Third, the sugar derivative streptozotocin is a molecular probe for the induction of acute pancreatic insulitis and diabetes. In pancreatitis, gelatinase B is shown to destroy insulin into remnant peptides, which may enhance the autoimmune process.

3. GLYCOSYLATION OF GELATINASE B

Gelatinase B is a complex multidomain glycoprotein (Opdenakker *et al.*, 2001 and Figure 1). Like all MMPs, gelatinase B contains a propeptide, an active domain and

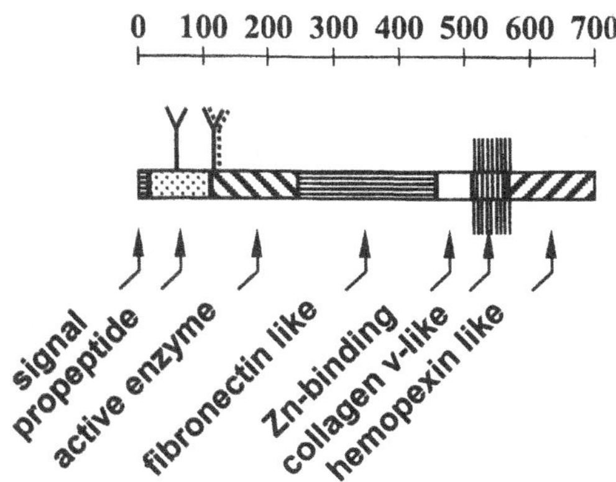

Figure 1. Domain structure and glycosylation of gelatinase B. Gelatinase B is constituted of several protein domains, some of which are common for most MMPs (signal peptide, propeptide, active and Zn^{2+}-binding domains, hemopexin domain), whereas the fibronectin domain is typical for gelatinases. The collagen type V domain is unique for gelatinase B. Three N-linked glycosylation consensus sequences are indicated (Y), one of which is conserved, but not occupied (dotted Y). Since the collagen type V domain is likely to contain clustered O-linked glycans, these are indicated with vertical lines. Isolated O-linked sugars may be attached to other domains as well (not indicated).

a Zn^{2+}-binding domain. These three domains together constitute the most essential functional unit of the enzyme. In addition, a hemopexin domain is present in most MMPs (except matrilysins) and is located at the carboxy-terminal part. A fibronectin domain, containing three type II fibronectin repeats, is inserted between the active and Zn^{2+}-binding domains of gelatinases, and confers highly efficient gelatinolytic activity. Unique for gelatinase B is the so-called collagen type V domain, which is inserted between the Zn^{2+}-binding and the hemopexin domains.

In addition to these protein domains, gelatinase B contains also voluminous glycans. Two N-linked glycans are attached to N-glycosylation consensus sequences (sequons Asn-Xaa-Ser/Thr, with Xaa denoting any amino acid except Pro) in the prodomain and in the active domain. A third N-glycosylation site, located at only 4 amino acids from the sequon in the active domain, is conserved in most MMPs throughout different species. Recently, this site has been reported not to be occupied in recombinant gelatinase B (Kotra *et al.*, 2002). A large number of O-linked glycans are also attached to gelatinase B, but the exact attachment sites of these sugars remain to be determined. However, the so-called collagen type V domain is composed of 11 repeats of the sequence T/SXXP, which are putative attachment sites for O-linked glycans (Van den Steen *et al.*, 1998a). The organization of this gelatinase B domain is different from that of collagen type V itself, which is composed of repeats of the sequence Gly-Xxx-Xxx (with Xxx often being Pro). In gelatinase B, this domain rather resembles the sequence of mucins, which are glycoproteins known to contain a large proportion of clustered O-linked sugars. Therefore, this domain can better be named mucin-like domain (Van den Steen *et al.*, 2001). The presence of clustered O-linked oligosaccharides is expected to result in the extension of this glycoprotein domain, leading to a rigid bottlebrush-like structure that acts as a spacer between the Zn^{2+}-binding and the hemopexin domain (Mattu *et al.*, 2000). In gelatinase A, the hemopexin domain makes extensive contacts with the catalytic site of the enzyme (Morgunova *et al.*, 1999) and has a strong influence on the substrate-specificity by its binding to substrates, for example, CC-chemokines (McQuibban *et al.*, 2000). To what extent the O-glycosylated mucin-type collagen type V domain modifies the substrate-specificity of gelatinase B has not yet been investigated, but it has been found that gelatinase B does not cleave CC-chemokines. In contrast, gelatinase B cleaves several CXC-chemokines, resulting in either potentiation or degradation of the chemokine (Van den Steen *et al.*, 2000).

The structures of the N-linked (Rudd *et al.*, 1999) and O-linked (Mattu *et al.*, 2000) sugars of natural gelatinase B from human neutrophils (after hydrazinolysis to separate the glycans from the protein) have been studied in great detail by NP-HPLC of the fluorescently labelled glycans and by on-line mass spectrometry analysis. Complex bi- and tri-antennary N-linked glycans were found with terminal sialylation and core fucosylation. The more abundant O-linked glycans have core 1 and core 2 structures with terminal sialylation, fucosylation and lactosamine extensions. The sugars of recombinant gelatinase B, expressed in the yeast *Pichia pastoris* (Van den Steen *et al.*, 1998b) or in human HeLa cells (Kotra *et al.*, 2002) have also been studied.

The functions of the sugars, attached to gelatinase B are still unclear, and several hypotheses have been formulated. For instance, the sugars may be protective against autoproteolysis of gelatinase B or proteolysis by other proteases. In general, oligosaccharides may also influence the specific activity of enzymes (Rudd *et al.*, 1994; Mori *et al.*, 1995). This might be particularly useful, since neutrophils secrete gelatinase B together with several other proteases. In addition, glycosylation is also important for the folding of and the recognition between glycoproteins (Rademacher *et al.*, 1988; Rudd and Dwek, 1997; Varki, 1993). Finally, a limited effect of the sialic acids on the inhibition of gelatinase B by TIMP-1 has been found (Van den Steen *et al.*, 2001).

4. GELATINASE B IN MULTIPLE SCLEROSIS

Gelatinase B has been detected in the cerebrospinal fluid of patients with multiple sclerosis and in experimental animal models of multiple sclerosis, for example, in experimental autoimmune encephalomyelitis. Because all neurological pathologies, in which gelatinase B was detected were characterized by demyelination, the effect of gelatinase B on myelin destruction was studied. It was found that human myelin basic protein (MBP) is a substrate for gelatinase B and that the cleavages in MBP resulted in peptide fragments which corresponded to the most important encephalitogenic autoantigens (Proost *et al.*, 1993). These data reinforced us to postulate the REGA model for autoimmune diseases (Opdenakker and Van Damme, 1994). Further proof of concept was obtained with the use of gelatinase B-deficient mice. In comparison with wild-type mice, young gelatinase B-deficient mice were found to be resistant against the development of experimental autoimmune encephalomyelitis and against an amputating form of tail necrosis (Dubois *et al.*, 1999).

5. GELATINASE B AND ARTHRITIS

Production of gelatinase B is induced at sites of inflammation by cytokines, for example TNF-α, interferon-γ (IFN-γ), IL-1 and chemokines. A typical example is the inflamed joint of patients with rheumatoid arthritis (Figure 2). In this autoimmune disease, inflammation is accompanied by the destruction of cartilage and an autoimmune reaction against the major cartilage constituent, type II collagen. In the synovial fluids of these patients, the levels of the chemokine IL-8 are increased, compared to controls (e.g., osteoarthritis patients) (Rampart *et al.*, 1992). IL-8 attracts neutrophils and stimulates these cells to release gelatinase B from their granules (Masure *et al.*, 1991). After activation of gelatinase B by removal of the propeptide, the protease participates in the degradation of the cartilage. A single cleavage of type II collagen by collagenases

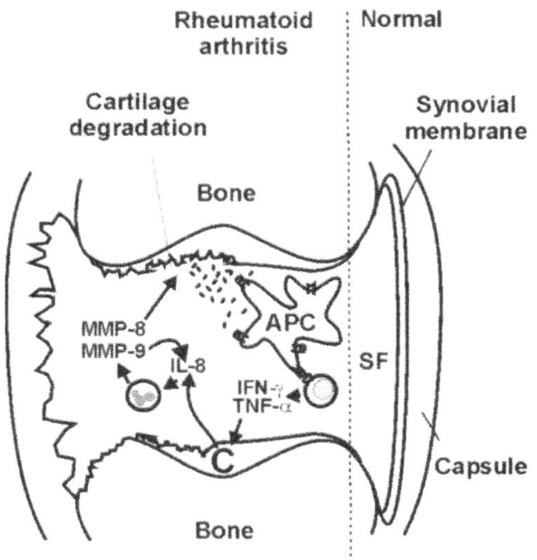

Figure 2. Role of gelatinase B in rheumatoid arthritis. Under influence of pro-inflammatory cytokines, IL-8 is produced and released in the synovial fluid (SF). As a consequence, neutrophils are chemoattracted and stimulated to release gelatinase B and neutrophil collagenase (MMP-8). A positive feedback loop occurs as gelatinase B potentiates IL-8 10-fold by aminoterminal processing. Gelatinase B and collagenases degrade the cartilage, in particular type II collagen, of which the immunodominant fragments activate autoreactive T lymphocytes after presentation by antigen-presenting cells (APC). APC may act locally or transfer the antigens to stimulate T cells in lymph nodes and follicles. This results in a second positive feedback loop, since the activated T cells produce pro-inflammatory cytokines.

(MMP-1, MMP-8, MMP-13) induces the partial unwinding of its triple helix structure (Gioia *et al.*, 2002), so that it becomes accessible to gelatinase B. Gelatinase B then cleaves the type II collagen into a large number of fragments. Determination of the cleavage sites by Edman degradation and mass spectrometry of the fragments yielded interesting data about the substrate specificity of the enzyme and the posttranslational modifications of the substrate (Van den Steen *et al.*, 2002). Thus it was observed that prolines in type II collagen are often hydroxylated, in particular when occurring at the third position of the $(\text{Gly-Xxx-Xxx})_n$ repeat. Lysine at the same position is also often hydroxylated, and hydroxylysine can be glycosylated with a single galactose (Galβ1-O-Lys), or with a galactose and glucose (Glcα1-2Galβ1-O-Lys). These posttranslational modifications enhance the solubility and stability of the collagen, and their exact localization has been found to influence the interactions with gelatinase B. In addition to known substrate sequence preferences (a hydrophobic residue at P1′, a small amino acid at P1 and proline at P3), it was found that gelatinase B has a preference for hydroxyprolines at the position P5′, while hydroxyprolines are not well tolerated at P2′. In addition, one of the two immunodominant epitopes of collagen II has been found to be modified by Lys-hydroxylation, and possibly also by glycosylation. Such modifications play a major role in the binding of the peptides to the MHC (Haurum *et al.*, 1995) and in the subsequent recognition of the MHC-peptide complex by T cells (Haurum *et al.*, 1994). Furthermore, it was shown that the severity of arthritis, induced in mice by immunization with type II collagen, correlates with the extent of modifications of type II collagen (Michaëlsson *et al.*, 1994). Recently, most of the autoreactive T cell clones from rheumatoid arthritis patients have been shown to react preferentially with the glycosylated form of the immunodominant type II collagen epitope (Bäcklund *et al.*, 2002). These findings imply that posttranslational modifications may play an essential role in autoimmune diseases.

From the positions of the cleavage sites in collagen II it can be inferred that gelatinase B does not destroy the immunodominant epitopes, but rather induces their release from the whole protein. This is also the case with other major autoantigens in other autoimmune diseases (Descamps *et al.*, 2003; Proost *et al.*, 1993). Gelatinase B thus participates in the generation of autoantigenic peptides, according to the REGA model (Opdenakker and Van Damme, 1994)

6. ACUTE PANCREATITIS, DIABETES AND ROLES OF GELATINASE B

Until recently, only correlative data existed about the role of gelatinase B in diabetes (Ebihara *et al.*, 1998; Maxwell *et al.*, 2001; Tomita and Iwata, 1997). We demonstrated by immunohistochemistry high expression levels of gelatinase B by neutrophils in acute pancreatitis and by ductular epithelial cells in chronic pancreatitis (Descamps *et al.*, 2003). With the use of double immunostainings for insulin and gelatinase B, it was shown that high expression levels of gelatinase B by the latter cells are maintained in the immediate proximity of insulin-secreting beta cells. Thus, after secretion of insulin by beta cells, both molecules may interact.

Ongoing research is directed to elucidate the role of gelatinase B in diabetes *in vivo*. For this purpose, we chose to use an animal model in which diabetes is chemically induced with multiple low doses of streptozotocin (Like and Rossini, 1976). As shown in Figure 3, streptozotocin is a glucose-derivative, which can be used as a molecular probe to induce diabetes. The immunology of the streptozotocin-induced diabetes model is well documented and is similar to human insulin-dependent diabetes mellitus (Lukic *et al.*, 1998). Another advantage is that the induction is suitable to establish diabetes in a variety of mouse strains, which allows the use of specific knockout mice, without the need of back-crossing to naturally sensitive strains such as the non-obese diabetic mice.

In the development of autoimmune diabetes, insulin is probably the most important autoantigen (Wegmann and Eisenbarth, 2000). The fragment of the insulin beta-chain that contains the residues 9 to 23, constitutes the immunodominant epitope (Wegmann *et al.*, 1994). Loss of tolerance to beta cell antigens can be mediated by genetic, endocrine and environmental factors. In some particular cases, diabetes may result from pancreatitis (Koizumi *et al.*, 1998; Wakasugi *et al.*, 1998). Autoimmunity then arises from the uptake of insulin or other beta cell antigens and their ferrying to the pancreatic lymph nodes, permitting presentation to naive circulating beta cell-reactive T lymphocytes.

We recently documented that gelatinase B degrades insulin into fragments and may assist in the generation of immunodominant insulin epitopes (Descamps *et al.*, 2003). Human insulin and gelatinase B were incubated at a molar substrate : enzyme ratio of 6.5 (or more). The reaction mixture of a complete digestion was separated by RP-HPLC on a

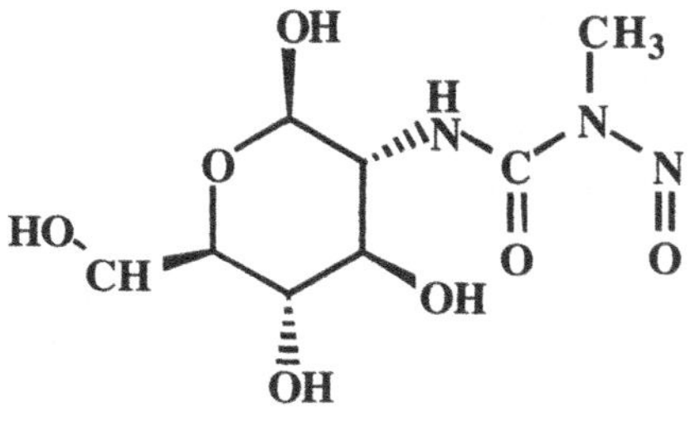

Figure 3. Structure of streptozotocin.

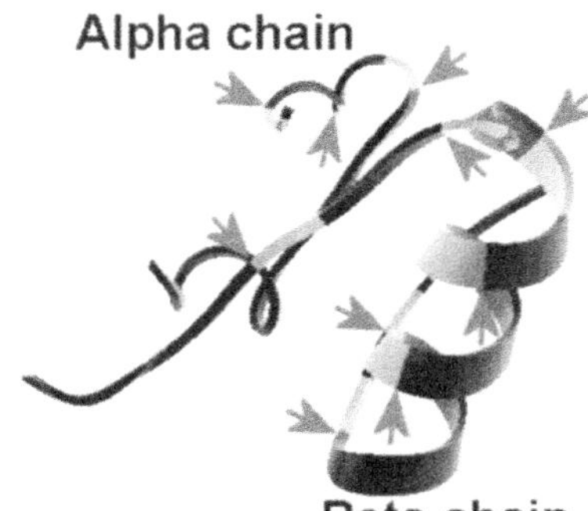

Figure 4. Three-dimensional view of the crystal structure of insulin. Cleavage sites by gelatinase B are indicated by means of arrows to demonstrate the accessibility of the insulin structure to the enzyme and the extent of insulin degradation.

C18 column and analyzed by on-line mass spectrometry. Exact molecular masses of peptide fragments were derived from the spectra and were fitted to those of virtual insulin fragments. Predicted amino-acid sequences were confirmed by tandem mass spectrometry analysis of the corresponding peptides. Identification of all degradation products of insulin denoted 10 cleavage sites, as illustrated on a three-dimensional model of insulin (Figure 4) (Adams *et al.*, 1969; Smith *et al.*, 2001).

We determined that the cleavage in front of residue 24 is the most efficient one. The beta chain fragment from residue 24 to 30 was the first degradation product that appeared as a function of time. This implies that the remnant fragment, containing residues 1 to 23 of the beta chain, may be presented in a complex with MHC class II, either after processing into the immunodominant epitope [residues 9 to 23] or after direct uptake by antigen-presenting cells (Pu *et al.*, 2002). However, if the insulin fragment 1 to 23 persists in the extracellular milieu for prolonged time intervals, it will be further degraded by the action of gelatinase B. In both cases, the REGA model is applicable to understand the development of the diabetogenic role of gelatinase B.

7. CONCLUSIONS AND PERSPECTIVES

The role of gelatinase B in the REGA model has been validated in three different autoimmune diseases: multiple sclerosis, rheumatoid arthritis and diabetes. Gelatinase B is a disease-promoting glycoprotein and forms a target for therapy. This implies that inhibitors of gelatinase B may become novel drugs for the treatment of autoimmune diseases.

8. ACKNOWLEDGMENTS

Supported by the Belgian Geconcerteerde OnderzoeksActies (GOA-11, 2002–2006), the National Fund for Scientific Research (FWO-Vlaanderen), Fortis AB, the Charcot Foundation Belgium and the British-Dutch Foundation for research on multiple sclerosis. We thank Ilse van Aelst, Pierre Fiten and Erik Martens for skillful technical assistance.

9. REFERENCES

Adams, M.J., Baker, E.N., Blundell, T.L., Harding, M.M., Dodson, E.J., Hodgkin, D.C., Dodson, G.G., Rimmer, B., Vijayan, M., and Sheats, S., 1969, Structure of rhombohedral 2 zinc insulin crystals, *Nature.* 224:491–495.

Bäcklund, J., Carlsen, S., Hoger, T., Holm, B., Fugger, L., Kihlberg, J., Burkhardt, H., and Holmdahl, R., 2002, Predominant selection of T cells specific for the glycosylated collagen type II epitope (263–270) in humanized transgenic mice and in rheumatoid arthritis, *Proc Natl Acad Sci USA.* 99:9960–9965.

Descamps, F.J., Van den Steen, P.E., Martens, E., Ballaux, F., Geboes, K., and Opdenakker G., 2003, Gelatinase B is diabetogenic in acute and chronic pancreatitis by cleaving insulin, *FASEB J.* 17:887–889.

Dubois, B., Masure, S., Hurtenbach, U., Paemen, L., Heremans, H., van den Oord, J., Sciot, R., Meinhardt, T., Hämmerling G., Opdenakker, G., and Arnold, B., 1999, Resistance of young gelatinase B-deficient mice to experimental autoimmune encephalomyelitis and necrotising tail lesions, *J Clin Invest.* 104:1507–1515.

Ebihara, I., Nakamura, T., Shimada, N., and Koide, H., 1998, Increased plasma metalloproteinase-9 concentrations precede development of microalbuminuria in non-insulin-dependent diabetes mellitus, *Am J Kidney Dis.* 32:544–550.

Gioia, M., Fasciglione, G.F., Marini, S., D'Alessio, S., De Sanctis, G., Diekmann, O., Pieper, M., Politi, V., Tschesche, H., and Coletta, M., 2002, Modulation of the catalytic activity of neutrophil collagenase MMP-8 on bovine collagen I. Role of the activation cleavage and of the hemopexin-like domain, *J Biol Chem.* 277:23123–23130.

Haurum, J.S., Arsequell, G., Lellouch, A.C., Wong, S.Y., Dwek, R.A., McMichael, A.J., and Elliott, T., 1994, Recognition of carbohydrate by major histocompatibility complex class I-restricted, glycopeptide-specific cytotoxic T lymphocytes, *J Exp Med.* 180:739–744.

Haurum, J.S., Tan, L., Arsequell, G., Frodsham, P., Lellouch, A.C., Moss, P.A., Dwek, R.A., McMichael, A.J., and Elliott, T., 1995, Peptide anchor residue glycosylation: effect on class I major histocompatibility complex binding and cytotoxic T lymphocyte recognition, *Eur J Immunol.* 25:3270–3276.

Koizumi, M., Yoshida, Y., Abe, N., Shimosegawa, T., and Toyota, T., 1998, Pancreatic diabetes in Japan, *Pancreas.* 16:385–391.

Kotra, L.P., Zhang, L., Fridman, R., Orlando, R., and Mobashery, S., 2002, N-Glycosylation pattern of the zymogenic form of human matrix metalloproteinase-9, *Bioorg Chem.* 30:356–370.

Like, A.A. and Rossini, A.A., 1976, Streptozotocin-induced pancreatic insulitis: new model of diabetes mellitus, *Science.* 30:415–417.

Lukic, M.L., Stosic-Grujicic, S., and Shahin, A., 1998, Effector mechanisms in low-dose streptozotocin-induced diabetes, *Dev Immunol.* 6:119–128.

Masure, S., Proost, P., Van Damme, J., and Opdenakker, G., 1991, Purification and identification of 91-kDa neutrophil gelatinase. Release by the activating peptide interleukin-8, *Eur J Biochem.* 198:391–398.

Mattu, T.S., Royle, L., Langridge, J., Wormald, M.R., Van den Steen, P.E., Van Damme, J., Opdenakker, G., Harvey, D.J., Dwek, R.A., and Rudd, P.M., 2000, O-Glycan analysis of natural human neutrophil gelatinase B using a combination of normal phase- HPLC and online tandem mass spectrometry: implications for the domain organization of the enzyme, *Biochemistry.* 39:15695–15704.

Maxwell, P.R., Timms, P.M., Chandran, S., and Gordon, D., 2001, Peripheral blood level alterations of TIMP-1, MMP-2 and MMP-9 in patients with type I diabetes, *Diabetic Med.* 18:777–780.

McQuibban, G.A., Gong, J.H., Tam, E.M., McCulloch, C.A., Clark-Lewis, I., and Overall, C.M., 2000, Inflammation dampened by gelatinase A cleavage of monocyte chemoattractant protein-3, *Science.* 289:1202–1206.

Michaëlsson, E., Malmstrom, V., Reis, S., Engstrom, A., Burkhardt, H., and Holmdahl, R., 1994, T cell recognition of carbohydrates on type II collagen, *J Exp Med.* 180:745–749.

Morgunova, E., Tuuttila, A., Bergmann, U., Isupov, M., Lindqvist, Y., Schneider, G., and Tryggvason, K., 1999, Structure of human pro-matrix metalloproteinase-2: activation mechanism revealed, *Science.* 284:1667–1670.

Mori, K., Dwek, R.A., Downing, A.K., Opdenakker, G., and Rudd, P.M., 1995, The activation of type 1 and type 2 plasminogen by type 1 and type 2 tissue plasminogen activator, *J Biol Chem.* 270:3261–3267.

Opdenakker, G., Rudd, P.M., Wormald, M., Dwek, R.A., and Van Damme, J., 1995, Cells regulate the activities of cytokines by glycosylation, *FASEB J.* 9:453–457.

Opdenakker, G., and Van Damme J., 1994, Cytokine-induced proteolysis in autoimmune diseases, *Immunol Today.* 15:104–107.

Opdenakker, G., Van den Steen P.E., and Van Damme, J., 2001, Gelatinase B: a tuner and amplifier of immune functions, *Trends Immunol.* 22:571–579.

Proost, P., Van Damme, J., and Opdenakker, G., 1993, Leukocyte gelatinase B cleavage releases encephalitogens from human myelin basic protein, *Biochem Biophys Res Commun.* 192:1175–1181.

Pu, Z., Carrero, J.A., and Unanue, E.R., 2002, Distinct recognition by two subsets of T cells of an MHC class II-peptide complex, *Proc Natl Acad Sci USA.* 99:8844–8849.

Rademacher, T.W., Parekh, R.B., and Dwek, R.A., 1988, Glycobiology, *Annu Rev Biochem.* 57:785–838.

Rampart, M., Herman, A.G., Grillet, B., Opdenakker, G., and Van Damme, J., 1992, Development and application of a radioimmunoassay for interleukin-8: detection of interleukin-8 in synovial fluids from patients with inflammatory joint disease. *Lab Invest.* 66:512–518.

Rudd, P.M. and Dwek, R.A., 1997, Glycosylation: heterogeneity and the 3D structure of proteins., *Crit Rev Biochem Mol Biol.* 32:1–100.

Rudd, P.M., Elliott, T., Cresswell, P., Wilson, I.A., and Dwek, R.A., 2001, Glycosylation and the immune system, *Science.* 291:2370–2376.

Rudd, P.M., Joao, H.C., Coghill, E., Fiten, P., Saunders, M.R., Opdenakker, G., and Dwek, R.A., 1994, Glycoforms modify the dynamic stability and functional activity af an enzyme, *Biochemistry.* 33:17–22.

Rudd, P.M., Mattu, T.S., Masure, S., Bratt, T., Van den Steen, P.E., Wormald, M.R., Küster, B., Harvey, D.J., Borregaard, N., Van Damme, J., Dwek, R.A., and Opdenakker, G., 1999, Glycosylation of natural human neutrophil gelatinase B and neutrophil gelatinase B-associated lipocalin, *Biochemistry.* 38:13937–13950.

Smith, G.D., Pangborn, W.A., and Blessing, R.H., 2001, Phase changes in T(3)R(3)(f) human insulin: temperature or pressure induced?, *Acta Crystallogr D Biol Crystallogr.* 57:1091–1100.

Tomita, T. and Iwata, K., 1997, Gelatinases and inhibitors of gelatinases in pancreatic islets and islet cell tumors, *Mod Pathol.* 10:47–54.

Van den Steen, P.E., Opdenakker, G., Wormald, M.R., Dwek, R.A., and Rudd, P.M., 2001, Matrix remodelling enzymes, the protease cascade and glycosylation, *Biochim Biophys Acta.* 1528:61–73.

Van den Steen, P.E., Proost, P., Grillet, B., Brand, D.D., Kang, A.H., Van Damme, J., and Opdenakker, G., 2002, Cleavage of denatured natural collagen type II by neutrophil gelatinase B reveals enzyme specificity, posttranslational modifications in the substrate, and the formation of remnant epitopes in rheumatoid arthritis, *FASEB J.* 16:379–389.

Van den Steen, P.E., Proost, P., Wuyts, A., Van Damme, J., and Opdenakker, G., 2000, Neutrophil gelatinase B potentiates interleukin-8 tenfold by aminoterminal processing, whereas it degrades CTAP-III, PF-4, and GRO-alpha and leaves RANTES and MCP-2 intact, *Blood.* 96:2673–2681.

Van den Steen, P., Rudd, P.M., Dwek, R.A., and Opdenakker, G., 1998a, Concepts and principles of O-linked glycosylation, *Crit Rev Biochem Mol Biol.* 33:151–208.

Van den Steen, P., Rudd, P.M., Proost, P., Martens, E., Paemen, L., Küster, B., Van Damme, J., Dwek, R.A., and Opdenakker, G., 1998b, Oligosaccharides of recombinant mouse gelatinase B variants, *Biochim Biophys Acta.* 1425:587–598.

Varki, A., 1993, Biological roles of oligosaccharides: all of the theories are correct, *Glycobiology.* 3:97–130.

Wakasugi, H., Funakoshi, A., and Igushi, H., 1998, Clinical assessment of pancreatic diabetes caused by chronic pancreatitis, *J Gastroenterol.* 33:254–259.

Wegmann, D.R. and Eisenbarth, G.S., 2000, It's insulin, *J Autoimmun.* 15:286–291.

Wegmann, D.R., Norbury-Glaser, M., and Daniel, D., 1994, Insulin-specific T cells are a predominant component of islet infiltrates in pre-diabetic NOD mice, *Eur J Immunol.* 24:1853–1857.

6

ENDOTHELIAL CELL GLYCOSYLATION: REGULATION AND MODULATION OF BIOLOGICAL PROCESSES

Claudine Kieda[1] and Danuta Dus[2]

[1]CNRS UPR 4301, Cell recognition group:
endogenous lectins. Centre de Biophysique Moléculaire
45071 Orléans CEDEX 2, France
[2]Department of Medical Immunology
Institute of Immunology and Experimental Therapy
Polish Academy of Sciences
Wroclaw, Poland

1. INTRODUCTION

The concept of homing tissue specialization and tissue-specific lymphocyte recirculation that is fundamental to acquired immunity raised from the demonstration, by Gowans and Knight in 1964, that circulating leukocytes enter secondary lymphoid organs at specialized endothelial sites (Picker and Butcher, 1992), which was the starting point for many studies to understand specificity of cell homing. T memory lymphocytes were first shown to be able to recirculate to tissues where they came from; further on, B immunoblasts were shown to migrate preferentially into mucosal tissues (for a review, see Kunkel and Butcher, 2002). The regio-selectivity is indeed favorable to perform an immune response (Butcher and Picker, 1996): it helps meetings between antigens and lymphocytes which carry the antigen specific receptor, reducing cross reactivity and allowing site specialization of the immune responses. Indeed, integration and control of systemic immune responses depend on the regulated trafficking of lymphocytes. Homing process disperses the immunological repertoire, directs lymphocyte subsets to the specialized microenvironments that control their differentiation, regulates their survival, and targets immune effector cells to sites of antigenic or microbial invasion.

Lymphocyte homing exquisite specificity is determined by combinatorial "decision processes" involving multistep sequential engagement of adhesion molecules and signalling receptors as described by Springer (Springer, 1990, 1994).

Glycobiology and Medicine, edited by John S. Axford
Kluwer Academic / Plenum Publishers, New York, 2003

Lymphocyte homing and recirculation processes represent a natural model for targeted trafficking, the specificity of which results from a refined regulation due to the controlled expression of adhesion molecules, cytokines, and chemokine receptors on the one hand (circulating cells) while the counterpart cells, that is the endothelial cells, express the corresponding ligands and present them in a proper way, at the proper time. It reflects the interaction of the lymphocytes with the microenvironment and is part of the overall response which results in the control of the lifespan of lymphocytes and the part played by the various related interacting cell populations.

Since very similar molecular mechanisms are used in pathological situations, the comparison of homing/metastases localization, recirculation/tumour cells escape (Sher *et al.*, 1988), tumour-mediated and inflammation-mediated angiogenesis (Achen *et al.*, 2002; Beasley *et al.*, 2002) can bring highly informative observations.

At the different successive steps of the recognitions which lead to invasion the participation of sugar protein recognitions is getting largely documented and bring explanation to molecular mechanisms either as a direct activity as in the case of selectins or as a co-recognition as in the case of integrins. Modulation factors such as cytokines are also frequently acting by a bi-functional process, one of which is involving a lectin activity which permits an efficient cytokine primary activity to be optimized, as documented by Zanetta and Vergotten in this volume. This bi-functionality is also a feature of chemokines, the activity of which is exerted through proper presentation by cell surface glycosaminoglycans.

Indeed, blood borne circulating cells recognize and, are recognized by vascular endothelial cells. The latter allow circulating cells to extravasate in a tissue-specific manner and contribute to defining the cell homing selectivity, which is fundamental in physiological normal, as well as pathological processes (inflammation, autoimmune diseases, metastasis). These properties led to express the endothelial organo-specificity concept. Consequently, knowing how endothelial cells are able to sort the circulating cells subpopulations, arrest them, and make them enter a tissue, is a key to immunotherapy (Li *et al.*, 2002; Wei *et al.*, 2000), targeted drug therapy (Staroselsky *et al.*, 1996) and cellular therapy designs. This review will focus on biological processes that are accomplished using the endothelial cell organo-specificity and the refinement brought by sugar specific recognitions.

2. THE ENDOTHELIAL CELL GLYCOSYLATION ORGANO-SPECIFICITY

The endothelium is structurally and functionally heterogeneous, as it has been long recognized (Pals *et al.*, 1989; Stoolman *et al.*, 1987). Microvascular endothelium indeed, controls the body compartmentation and homing of lymphocytes into lymphoid (Girard and Springer, 1995) as well as non-lymphoid sites, in a tissue-specific manner (see the review by Butcher and Picker, 1996). Microvascular endothelial cells recognize blood borne circulating cells and allow them to extravasate in a tissue-specific manner. As this property determines the selectivity of lymphocyte homing, it is fundamental in physiological as well as pathological processes: inflammation (Duijvestijn *et al.*, 1987), autoimmune diseases (Jalkanen *et al.*, 1986), metastasis (Sher *et al.*, 1988). The studies pointed out the existence of molecules that could distinguish one endothelium from another one and the mechanisms of their regulation. Such molecules are keys to understanding the selectivity mechanism. Homing was shown to be mediated by the spatio-temporarily regulated sequential expression of specific cell adhesion molecules, present both on the circulating leukocytes and on endothelial cells (Berg *et al.*, 1989). The requirement for multiple protein–protein interactions allows a genetically limited receptor–ligand repertoire to be used combinatorially to control the recirculation of different leukocytes and other circulating cells subsets. The observed final specificity results generally from several steps in the process, which is selective and distinct for a tissue site. That combination of several selective steps results in a high degree of specificity.

2.1. Endothelial Lectins

Because the recirculation begins with blood lymphocytes interacting transiently and reversibly with the vascular endothelium, an early step of the adhesion cascade is a decisive contact between circulating cell and microvascular endothelial cell, mediated by inducible endothelial cell adhesion molecules; E-selectin and P-selectin on the endothelial cells (Berg *et al.*, 1992; Bevilacqua *et al.*, 1994; Butcher *et al.*, 1999; Cummings and Smith, 1992; Fieger *et al.*, 2001; Gulubova, 2002; Li *et al.*, 2001; Picker *et al.*, 1991). Later steps involve other adhesion molecules, for example, intercellular adhesion molecule-1 (ICAM-1) (Ebnet *et al.*, 1996; Haraldsen *et al.*, 1996; Schurmann, 1997; Springer, 1990, 1994; Tanaka, 2000; Warnock *et al.*, 1998). E-selectin and P-selectin are quite generally distributed among endothelial cells and consequently can participate to a regio selective reaction through the modulation of their degree and time of expression. This prompted the search for distinctive signals which were evidenced by help of designing an *in vitro* model for endothelial cell comparisons (Bizouarne *et al.*, 1993; Kieda *et al.*, 2002). Figure 1 indicates that endothelial cells from peripheral lymph nodes selectively expressed

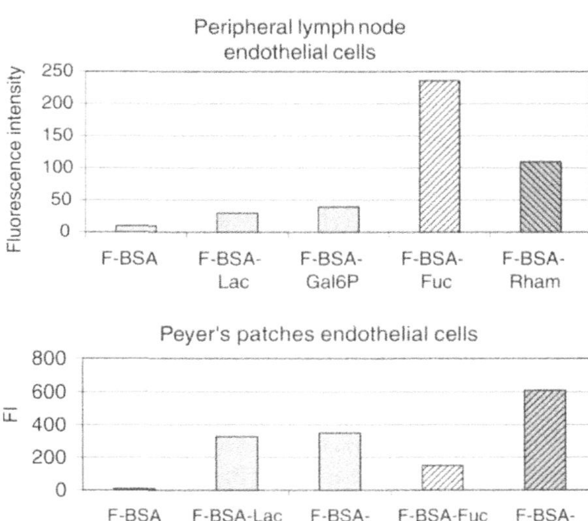

Figure 1. Endogenous lectins detection in endothelial cells lines.

fucose-specific receptors while rhamnose-specific receptors were detected on Peyer's patches endothelial cells.

The immortalization of human endothelial cells as well as the phenotype stabilization allowed to confirm such a gene selective expression as observed by differential display of gene expression.

2.2. Endothelial Addressins

Specific recognitions due to the leukocyte L-selectin (Bargatze *et al.*, 1994; Berg *et al.*, 1993a; Hasslen *et al.*, 1995) gets to be regio-selective despite its large distribution. This is attributed to vascular addressins-restricted expression among endothelial cells according to the tissues and vessels. Addressins are mucin-type or mixed mucin/immunoglobulin-type glycoproteins (Berg *et al.*, 1993b; Berg *et al.*, 1998; Hemmerich *et al.*, 1994; Michie *et al.*, 1995; von Andrian *et al.*, 1993; Yednock *et al.*, 1987).

Addressins provide the regio-selectivity to selectins recognition because they are differentially expressed on the endothelium, depending upon its tissue origin (Streeter *et al.*, 1988; Michie *et al.*, 1993; Michie *et al.*, 1995; Vestweber and Blanks, 1999). Furthermore, their structure is highly modulated by the microenvironment at the post-translational level (Denis *et al.*, 1996). To become high affinity L-selectin ligands, they must undergo proper glycosylation, especially with the most common sugar epitopes recognized by L-selectin: sialyl Lewis x (CD15s) and its sulfated form, presented by appropriate mucin-type proteins (GlyCAM-1, MAdCAM-1 or CD34). In the mouse, typical peripheral lymph node addressins (PNAds) are GlyCAM-1 (Streeter *et al.*, 1988; Imai *et al.*, 1992), CD34 and Sgp200 (sulphated glycoprotein of 200 kD) (Berg *et al.*, 1989; Berg *et al.*, 1991a; Streeter *et al.*, 1988). Glycosylated epitopes of PNAds (Berg *et al.*, 1998), are also presented on podocalyxin-like, CD34 and Sgp200 sialomucins (Puri *et al.*, 1995; Sassetti *et al.*, 1998; Sassetti *et al.*, 2000).

Specific mucosa-associated addressin, MAdCAM-1, has been found first in mouse (Berg *et al.*, 1989; Denis *et al.*, 1996; Nakache *et al.*, 1989; Picker *et al.*, 1989) and further in human endothelial cells (Briskin *et al.*, 1997). In addition to being recognized by L-selectin through its sialomucin sugar residues (Berg *et al.*, 1998; Briskin *et al.*, 1997; Morgan *et al.*, 1999), MAdCAM-1 molecule possesses immunoglobulin-like domain which interacts with the $\alpha_4\beta_7$ integrin homing receptor of lymphocytes (Berlin *et al.*, 1995).

An example of this Mad-CAM-1 tissue-restricted distribution is illustrated in Figure 2, which gives the expression profiles of Mad-CAM-1 detected on endothelial cell lines from various tissues by the binding of specific antibodies. Mad-CAM-1 is indeed expressed on mucosa associated endothelial cells and is hardly visible on peripheral tissue derived cells.

Selectin binding epitopes are also present on other mucin-type glycoproteins such as: P-selectin ligand-l (PSGL-l) (Dimitroff *et al.*, 2001; Ley *et al.*, 1995; Mehta *et al.*, 1998; Moore, 1998; Norman *et al.*, 1995; Pouyani and Seed, 1995; Tu *et al.*, 1999; Wardlaw, 2001), E-selectin ligand-1: ESL-l (Steegmaier *et al.*, 1995; Zollner and Vestweber, 1996), and cutaneous leukocyte antigen—CLA (Berg *et al.*, 1991b).

It must be emphasized that cellular glycoconjugates are decisive not only in selectin/addressin interactions (Berg *et al.*, 1998; Berg *et al.*, 1991a) but also in other lectin/glycoconjugate-mediated recognition phenomena in homing and invasion (Berg *et al.*, 1991b; Kieda and Monsigny, 1986; Kieda *et al.*, 1978).

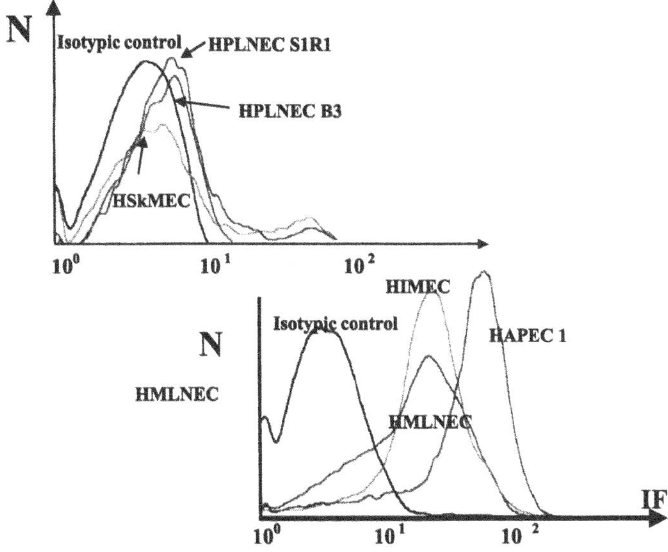

Figure 2. Mad-CAM1 addressin distribution among endothelial cells from various origins.
HAPEC.1, human appendix endothelial cells
HIMEC, human intestine endothelial cells
HMLNEC, human mesenteric lymph nodes endothelial cells
HSkMEC, human skin microvascular endothelial cells
HPLNEC, human peripheral lymph nodes endothelial cells.

3. ENDOTHELIAL CELLS AND METASTASES LOCALIZATION

The discriminative recognition and adhesion between leukocyte subpopulations and endothelial cells built up interest in view of understanding organ selective cancer metastasis. Adhesive interactions are involved in the control of proliferation, migration, differentiation, and other cell functions. As mentioned earlier, most information concerning adhesive molecules and their ligands involved in the adhesion process come from researches dealing with normal leukocytes: similar adhesive molecules and mechanisms could be involved in metastasis and other invasive pathologies.

Endothelial cells play fundamental roles in normal processes but also in pathological ones as wound healing and inflammation (Chin *et al.*, 1991; Arvilommi *et al.*, 1996; Augustin *et al.*, 1995; McEver, 1991; Seppo *et al.*, 1996; Steffen *et al.*, 1996; Yang *et al.*, 2002). Moreover, they are the soil for the seeding of secondary tumoral foci (metastases) (Pauli *et al.*, 1990) and their capacity of making angiogenesis in the proximity and under the influence of a tumor (Jain and Padera, 2002) makes them: (1) mediate tumor survival and (2) provide the tumor cells with the means to escape from the primary site into the circulation (Karkkainen *et al.*, 2002; Pauli *et al.*, 1990; Wang *et al.*, 2000; Zhu *et al.*, 1991).

In this context endothelial cells are the first target to aim for antiadhesion (Woynarowska *et al.*, 1996; Zeisig *et al.*, 2002) and/or gene targeting (Davidoff *et al.*, 2001;

Ojeifo *et al.*, 2001) therapies. In order to elaborate clinical protocols for such therapies, adequate cellular models with preserved regio-specificity are necessary (Aoudjit *et al.*, 1998; St-Pierre *et al.*, 1999). In this case again, established microvascular endothelial cell lines from primary cell cultures, were isolated, cultured, and further immortalized in a similar way. As such, the resulting cell lines were comparable endothelial cell lines preserved partly tissue-specific phenotypes (Haraldsen *et al.*, 1996) and they displayed tissue-specific behavior in lymphoid cells and myeloid cells (Kieda *et al.*, 2002). This makes them tools for advanced *in vitro* studies on molecular mechanisms of cell–cell interactions.

Along this line of evidence, addressins (and selectins) are actively studied because they are significant adhesion and organ-selective molecules expressed on ECs and highly dependent upon the microenvironment biological conditions. During inflammation selectin molecules appear rapidly on endothelial cell surface (P-selectin, E-selectin) and the expression of others increases (as ICAM-1), to mediate competent cells influx into tumor-inflamed tissue (Steeber *et al.*, 1999).

E-selectin seems to be involved in metastasis of colon cancer since *in vitro* adhesion of several colon cell lines expressing E-selectin ligands: sialyl Le^x and sialyl Le^a, was shown to depend upon the presence of E-selectin on activated endothelium (Nemoto *et al.*, 1998; Welply *et al.*, 1994). It was also shown, that colon cancer cells adhesion to E-selectin occurs under flow conditions and leads to their complete arrest (Mannori *et al.*, 1995; Matsushita *et al.*, 1998).

Other lectins like galectin-1, expressed on the extracellular surface of endothelial cells may also be involved in tumor cell adhesion (Lotan *et al.*, 1994) while galectin-3 (Deininger *et al.*, 2002) displays an inversely correlated distribution among endothelial cells and colon cancer cells as far as invasiveness is concerned.

Such regulated and correlated expressions can be highly significant because some types of cancers metastasize preferentially into selected organs: for example prostate cancer cells into bones (Cooper *et al.*, 2002; Scott *et al.*, 2001) or melanoma cells into lungs (Bastias *et al.*, 2002; Irimura and Nicolson, 1981; Saiki *et al.*, 1996; Zhang *et al.*, 1995).

We described (Paprocka *et al.*, submitted), a cytofluorimetric method to demonstrate specific and discriminative cell-to-cell adhesions occurring between human microvascular endothelial cell lines of distinct tissue origins and lymphocytes as well as colon carcinoma cells. This quantitative assay was proven to be useful to study the molecular mechanisms involved in tissue specific cell-to-cell interactions. Indeed, the labelling of one adhesion partner with a red fluorescent label, chosen because of its stability and its non-exchangeability between cells (Horan and Slezak, 1989), allowed a further characterization of the adherent cellular partners during this assay. Consequently, this assay is potentially useful to identify the molecules involved in the adhesion process (Opolski *et al.*, 1998); it demonstrated *in vitro* the exclusive capacity of endothelial cells in selecting the cell population(s) that they are capable to recognize and sort.

It can be assumed that such a sorting is likely to be more precisely achieved *in vivo* because of the microenvironmental controls and regulations, which external stimuli endothelial cells are extremely reactive to. This is part of their main characteristic properties and results in their extreme selectivity.

Adhesion to, and invasion of endothelial cells layer by tumor cells is one aspect on this cell-to-cell interactive process. It is further illustrated by the neovascularization that is induced by the tumor environment.

4. CYTOKINES/CHEMOKINES PRESENTATION BY ENDOTHELIAL CELL GLYCOCONJUGATES

IL-7 is documented as one of the lectin-active cytokines that achieve their cytokine role together with a lectin activity (Borghesi *et al.*, 1999) by being presented on the endothelial surface upon binding to glycosaminoglycans (Adachi *et al.*, 1998). This cytokine can function as a cofactor during myelopoiesis, generation of cytotoxic T cells (Widmer *et al.*, 1990; Zoll *et al.*, 1998) NK cells (Alderson *et al.*, 1990; Zoll *et al.*, 1998) and activated monocytes (Alderson *et al.*, 1991), it promotes formation of some organs that is Peyer's patch anlage and germinal centre organization (Yoshida *et al.*, 1999).

Bizouarne *et al* (Bizouarne *et al.*, 1993a; Bizouarne *et al.*, 1993b; Denis *et al.*, 1996) observed that murine endothelial cells from peripheral lymph nodes could be specifically activated by IL-7 to induce expression of endogenous lectin adhesion molecule. Further studies showed in the same model a selective induction of addressin (MECA 79 antigen) expression upon IL-7 treatment (Denis *et al.*, 1996). We recently demonstrated the presence of IL-7 receptors in endothelial cells (Dus *et al.*, 2002, in press) but the mechanism of endothelial cells activation by IL-7 is not yet attributed as a receptor-mediated or glycosaminoglycan-presented one, as further described in the case of the chemokines.

Chemokines are secreted molecules that play a capital part in conditioning the endothelial cells. Synthesized by the underlying epithelial tissue cells, they can be taken up by ECs, trancytosed, and finally presented onto the luminal endothelial cell surface thus constituting a chemotactic gradient for the circulating blood borne cells. Because they are synthesized in an organ selective manner as reviewed (Kunkel and Butcher, 2002), they contribute efficiently to the specific character of cell homing. Not only are chemokines preferentially localized in their synthesis, they do stimulate endothelial cells in a selective

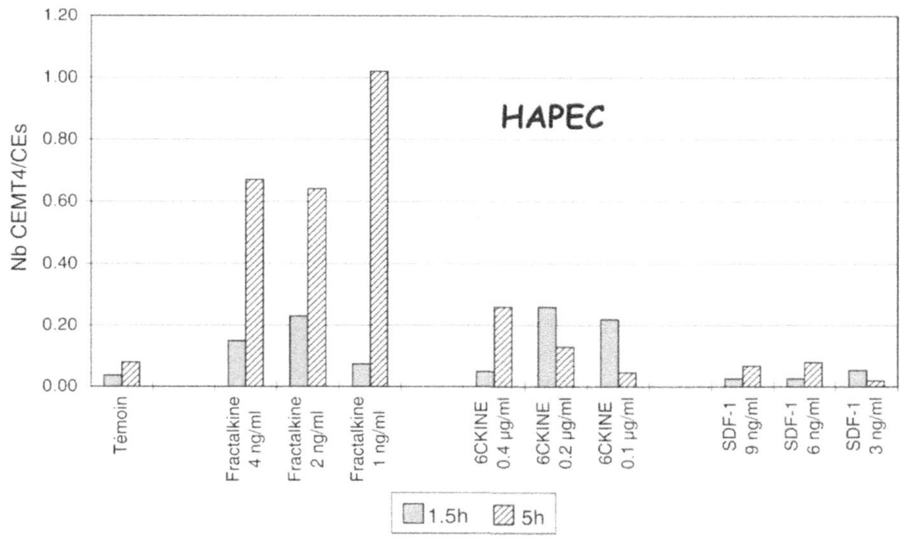

Figure 3. Activation of appendix derived endothelial cells by fractalkine induces increased adhesion properties.

manner to induce endothelial cell adhesion capacity towards lymphocytes. This was mainly illustrated (Crola *et al.*, submitted) in the case of fractalkine (Bazan *et al.*, 1997; Fong *et al.*, 1998; Imai *et al.*, 1997) as shown in Figure 3, and 6Ckine (Kellermann *et al.*, 1999; Sharma *et al.*, 2000; Warnock *et al.*, 2000).

5. SUGAR MODULATION OF ANGIOGENESIS

Angiogenesis is an endothelial cell response to tumor cells factors. Angiogenesis is required for the continual growth of the tumor and provides a gateway for cells to escape the confines of the primary tumor.

This process is a selective reaction (Maehara *et al.*, 2000; Slaton *et al.*, 2001) of the endothelial cells to the tumoral factors. Those factors should be consequently identified to control the endothelial cell response. The genes that are implicated during the endothelial cell response are currently actively studied by DNA arrays and gene display (Kallmann *et al.*, 2002; Slaton *et al.*, 2001; St Croix *et al.*, 2000). New molecules should be identified by these tools as well as by differential gene expression display. By this technique we could identify a new adhesion regulatory molecule expressed in some endothelial cells only (Lamerant and Kieda, data not shown).

A common feature of the adhesion and invasion processes, that is the glycoconjugate-to-protein recognition has to be pointed out. This feature is particularly illustrated during angiogenesis progression by the works aiming for the development of drugs that target the tumor neovasculature in order to inhibit tumor growth. The use of glycoconjugate synthesis inhibitors, like castanospermine which blocks the glucosidases that convert protein N-linked high mannose carbohydrates to complex oligosaccharides, resulted in a significant inhibition of tumor growth *in vivo* in nude mice and of the basic fibroblast growth factor-induced angiogenesis (Pili *et al.*, 1995). Such glucosidase inhibitor prevented the morphological differentiation of endothelial cells *in vitro*. Cell surface oligosaccharides are required for angiogenesis; alterations of these structures on endothelial cells inhibit tumor growth.

Halloran *et al.* (2002), have studied endothelial cells not only as key participants in angiogenic processes that characterize tumor growth, wound repair, and inflammatory diseases, such as human rheumatoid arthritis (RA). They have shown that endothelial cell lectin molecules, such as soluble E-selectin, mediate angiogenesis and described an endothelial cell molecule, Lewis^{Y}-6/H-5-2 glycoconjugate (Le^{Y}/H), that shares some structural features with the soluble E-selectin ligand, sialyl Lewis^{X} (sialyl Lex): Ley/H is rapidly cytokine inducible, up-regulated in RA synovial tissue (where it is cell-bound), and it is up-regulated in a soluble form in angiogenic RA. Soluble Le^{Y}/H is a potent angiogenic mediator suggesting a novel paradigm of soluble blood group antigens as mediators of angiogenic responses and suggest new targets for therapy of diseases, that are characterized by persistent neovascularization (Halloran *et al.*, 2000). This type of reaction can be followed *in vitro* as illustrated in Figure 4.

As reviewed (Nangia-Makker *et al.*, 2000), carbohydrate-binding proteins are highly significantly involved in angiogenesis. The decisive importance of carbohydrate-recognition during angiogenesis stems from the observation that angiogenic factors like the fibroblast growth factor family and vascular endothelial growth factors, bind initially to the extracellular matrix proteoglycans before binding to their cognate receptors; some of the adhesion molecules bind to glycoconjugates present on the surface of the endothelial cells.

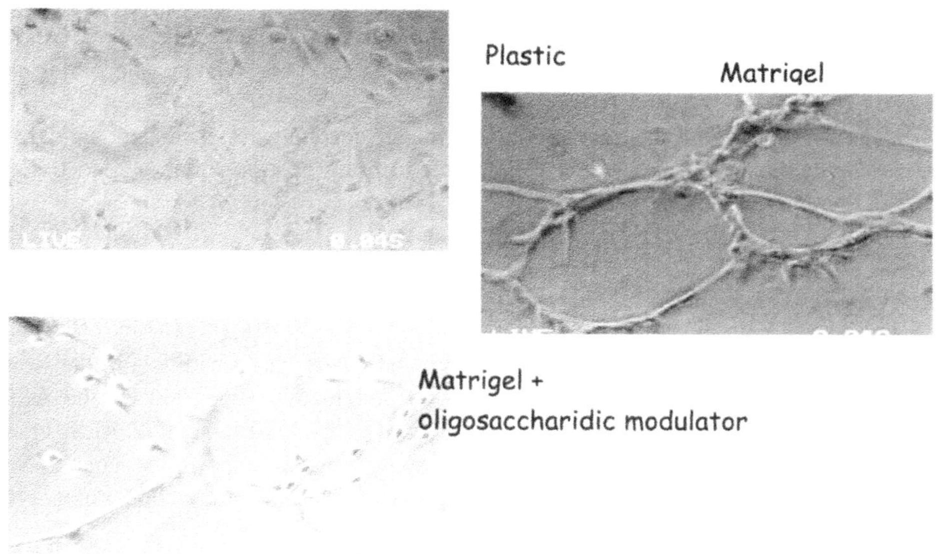

Figure 4. Angiogenesis in matrigel and modulation by oligosaccharides.

Considering the endothelial cell participation into homing specificity, this glycoconjugate-mediated presentation of adhesion molecules and factors seems to be quite a general means used by endothelial cell to handle the molecules that they use for recognition and/or attraction purposes.

6. CONCLUSION AND PERSPECTIVES

Endothelial cell biology is highly directed towards well-defined recognitions and interactions with cells and molecules that modulate the immune responses and namely glycoconjugate and lectins. As portals where tissue entry decisions are taken, endothelial cells are studied to find out the molecules that one has to aim for in order to target a site of the body. Endothelial cells second main characteristic resides in their sensitivity to micro environmental signals by selective activations of adhesion signals that are glycoconjugates dependent.

As such, endothelial cells react differentially to the organ- and tissue-derived factors as chemokines, tumor, and inflammation factors; they also react to the soluble factors in the lumen of the vessel fluids as well as to the cells that get in contact and raises activation.

Indeed this results in the striking organo-specificity of endothelial cell which is a property, thanks to which endothelial cells are able to mediate cell selection by adhesion, recognition, extravasation, and ultimately, antigen-to-lymphocyte meeting thus leading to immune response.

The second level of endothelial cell selectivity appears in the differential activations and responses to microenvironment in pathological situations that, analogically to normal situations, take advantage of the endothelial cell selectivity. Cancer metastases and tumor angiogenesis are examples that were illustrated in this review, which furthermore, pointed out the importance of glycobiology-dependent endothelial cell behavior and biological properties.

The knowledge of the part played by the glycoconjugates mediated biological processes undertaken by endothelial cells seems to bring new potential keys to tissue-targeted therapies.

7. ACKNOWLEDGMENTS

Authors wish to thank Dr Maria Paprocka and Claire Crola with whom the work reported here was largely done in collaboration and mainly supported by the ARC (Association pour la Recherche sur le Cancer N°6515), by INSERM PROGRES (grant N°4P009E France); The State Committee for Scientific Research (KBN, Poland No. 4PO5A14318), the CNRS/PAN PAI "Polonium" French Polish integrated action program N° 99105 (Dr D. DUS, Dr C. KIEDA).

8. REFERENCES

Achen, M.G., Williams, R.A., Baldwin, M.E., Lai, P., Roufail, S., Alitalo, K., and Stacker, S.A., 2002, The angiogenic and lymphangiogenic factor vascular endothelial growth factor-D exhibits a paracrine mode of action in cancer, *Growth Factors.* 20:99–107.

Adachi, S., Yoshida, H., Honda, K., Maki, K., Saijo, K., Ikuta, K., Saito, T., and Nishikawa, S.I., 1998, Essential role of IL-7 receptor alpha in the formation of Peyer's patch anlage, *Int Immunol.* 10:1–6.

Alderson, M.R., Sassenfeld, H.M., and Widmer, M.B., 1990, Interleukin 7 enhances cytolytic T lymphocyte generation and induces lymphokine-activated killer cells from human peripheral blood, *J Exp Med.* 172:577–587.

Alderson, M.R., Tough, T.W., Ziegler, S.F., and Grabstein, K.H., 1991, Interleukin 7 induces cytokine secretion and tumoricidal activity by human peripheral blood monocytes, *J Exp Med.* 173:923–930.

Aoudjit, F., Potworowski, E.F., and St-Pierre, Y., 1998, The metastatic characteristics of murine lymphoma cell lines *in vivo* are manifested after target organ invasion, *Blood.* 91:623–629.

Arvilommi, A.M., Salmi, M., Kalimo, K., and Jalkanen, S., 1996, Lymphocyte binding to vascular endothelium in inflamed skin revisited: a central role for vascular adhesion protein-1 (VAP-1), *Eur J Immunol.* 26:825–833.

Augustin, H.G., Braun, K., Telemenakis, I., Modlich, U., and Kuhn, W., 1995, Ovarian angiogenesis. Phenotypic characterization of endothelial cells in a physiological model of blood vessel growth and regression, *Am J Pathol.* 147:339–351.

Bargatze, R.F., Kurk, S., Butcher, E.C., and Jutila, M.A., 1994, Neutrophils roll on adherent neutrophils bound to cytokine-induced endothelial cells via L-selectin on the rolling cells, *J Exp Med.* 180:1785–1792.

Bastias, J., Wei, M.X., Huynh, R., Chaubet, F., Jozefonvicz, J., and Crepin, M., 2002, Anti-proliferative and anti-tumoral activities of a functionalized dextran (CMDBJ) on the 1205 L-U human tumor melanoma cells, *Anticancer Res.* 22:1603–1613.

Bazan, J.F., Bacon, K.B., Hardiman, G., Wang, W., Soo, K., Rossi, D., Greaves, D.R., Zlotnik, A., and Schall, T.J., 1997, A new class of membrane-bound chemokine with a CX3C motif, *Nature.* 385:640–644.

Beasley, N.J., Prevo, R., Banerji, S., Leek, R.D., Moore, J., van Trappen, P., Cox, G., Harris, A.L., and Jackson, D.G., 2002, Intratumoral lymphangiogenesis and lymph node metastasis in head and neck cancer. *Cancer Res.* 62:1315–1320.

Berg, E.L., Goldstein, L.A., Jutila, M.A., Nakache, M., Picker, L.J., Streeter, P.R., Wu, N.W., Zhou, D., and Butcher, E.C., 1989, Homing receptors and vascular addressins: cell adhesion molecules that direct lymphocyte traffic, *Immunol Rev.* 108:5–18.

Berg, E.L., Magnani, J., Warnock, R.A., Robinson, M.K., and Butcher, E.C., 1992, Comparison of L-selectin and E-selectin ligand specificities: the L-selectin can bind the E-selectin ligands sialyl Le(x) and sialyl Le(a), *Biochem Biophys Res Commun.* 184:1048–1055.

Berg, E.L., McEvoy, L.M., Berlin, C., Bargatze, R.F., and Butcher, E.C., 1993a, L-selectin-mediated lymphocyte rolling on MAdCAM-1, *Nature.* 366:695–698.

Berg, E.L., McEvoy, L.M., Berlin, C., Bargatze, R.F., and Butcher, E.C., 1993b, L-selectin-mediated lymphocyte rolling on MAdCAM-1 [see comments], *Nature.* 366:695–698.

Berg, E.L., Mullowney, A.T., Andrew, D.P., Goldberg, J.E., and Butcher, E.C., 1998, Complexity and differential expression of carbohydrate epitopes associated with L-selectin recognition of high endothelial venules, *Am J Pathol.* 152:469–477.

Berg, E.L., Robinson, M.K., Mansson, O., Butcher, E.C., and Magnani, J.L., 1991a, A carbohydrate domain common to both sialyl Le(a) and sialyl Le(X) is recognized by the endothelial cell leukocyte adhesion molecule ELAM-1, *J Biol Chem.* 266:14869–14872.

Berg, E.L., Yoshino, T., Rott, L.S., Robinson, M.K., Warnock, R.A., Kishimoto, T.K., Picker, L.J., and Butcher, E.C., 1991b, The cutaneous lymphocyte antigen is a skin lymphocyte homing receptor for the vascular lectin endothelial cell-leukocyte adhesion molecule 1, *J Exp Med.* 174:1461–1466.

Berlin, C., Bargatze, R.F., Campbell, J.J., von Andrian, U.H., Szabo, M.C., Hasslen, S.R., Nelson, R.D., Berg, E.L., Erlandsen, S.L., and Butcher, E.C., 1995, alpha 4 integrins mediate lymphocyte attachment and rolling under physiologic flow, *Cell.* 80:413–422.

Bevilacqua, M.P., Nelson, R.M., Mannori, G., and Cecconi, O., 1994, Endothelial-leukocyte adhesion molecules in human disease, *Annu Rev Med.* 45:361–378.

Bizouarne, N., Denis, V., Legrand, A., Monsigny, M., and Kieda, C., 1993a, A SV-40 immortalized murine endothelial cell line from peripheral lymph node high endothelium expresses a new alpha-L-fucose binding protein, *Biol Cell.* 79:209–218.

Bizouarne, N., Mitterrand, M., Monsigny, M., and Kieda, C., 1993b, Characterization of membrane sugar-specific receptors in cultured high endothelial cells from mouse peripheral lymph nodes, *Biol Cell.* 79:27–35.

Borghesi, L.A., Yamashita, Y., and Kincade, P.W., 1999, Heparan sulfate proteoglycans mediate interleukin-7-dependent B lymphopoiesis, *Blood.* 93:140–148.

Briskin, M., Winsor-Hines, D., Shyjan, A., Cochran, N., Bloom, S., Wilson, J., McEvoy, L.M., Butcher, E.C., Kassam, N., Mackay, C.R., Newman, W., and Ringler, D.J., 1997, Human mucosal addressin cell adhesion molecule-1 is preferentially expressed in intestinal tract and associated lymphoid tissue, *Am J Pathol.* 151:97–110.

Butcher, E.C. and Picker, L.J., 1996, Lymphocyte homing and homeostasis, *Science.* 272:60–66.

Butcher, E.C., Williams, M., Youngman, K., Rott, L., and Briskin, M., 1999, Lymphocyte trafficking and regional immunity, *Adv Immunol.* 72:209–253.

Chin, Y.H., Cai, J.P., and Xu, X.M., 1991, Tissue-specific homing receptor mediates lymphocyte adhesion to cytokine-stimulated lymph node high endothelial venule cells, *Immunology.* 74:478–483.

Cooper, C.R., Bhatia, J.K., Muenchen, H.J., McLean, L., Hayasaka, S., Taylor, J., Poncza, P.J., and Pienta, K.J., 2002, The regulation of prostate cancer cell adhesion to human bone marrow endothelial cell monolayers by androgen dihydrotestosterone and cytokines, *Clin Exp Metastasis.* 19:25–33.

Cummings, R.D. and Smith, D.F., 1992, The selectin family of carbohydrate-binding proteins: structure and importance of carbohydrate ligands for cell adhesion, *Bioessays.* 14:849–856.

Davidoff, A.M., Leary, M.A., Ng, C.Y., and Vanin, E.F., 2001, Gene therapy-mediated expression by tumor cells of the angiogenesis inhibitor flk-1 results in inhibition of neuroblastoma growth *in vivo*, *J Pediatr Surg.* 36:30–36.

Deininger, M.H., Trautmann, K., Meyermann, R., and Schluesener, H.J., 2002, Galectin-3 labeling correlates positively in tumor cells and negatively in endothelial cells with malignancy and poor prognosis in oligodendroglioma patients, *Anticancer Res.* 22:1585–1592.

Denis, V., Dupuis, P., Bizouarne, N., de, O.S.S., Hong, L., Lebret, M., Monsigny, M., Nakache, M., and Kieda, C., 1996, Selective induction of peripheral and mucosal endothelial cell addressins with peripheral lymph nodes and Peyer's patch cell-conditioned media, *J Leukoc Biol.* 60:744–752.

Dimitroff, C.J., Lee, J.Y., Rafii, S., Fuhlbrigge, R.C., and Sackstein, R., 2001, CD44 is a major E-selectin ligand on human hematopoietic progenitor cells, *J Cell Biol.* 153:1277–1286.

Duijvestijn, A.M., Kerkhove, M., Bargatze, R.F., and Butcher, E.C., 1987, Lymphoid tissue- and inflammation-specific endothelial cell differentiation defined by monoclonal antibodies, *J Immunol.* 138:713–719.

Ebnet, K., Kaldjian, E.P., Anderson, A.O., and Shaw, S., 1996, Orchestrated information transfer underlying leukocyte endothelial interactions, *Annu Rev Immunol.* 14:155–177.

Fieger, C.B., Emig-Vollmer, S., Petri, T., Grafe, M., Gohlke, M., Debus, N., Semmler, W., Tauber, R., and Volz, B., 2001, The adhesive properties of recombinant soluble L-selectin are modulated by its glycosylation, *Biochim Biophys Acta.* 1524:75–85.

Fong, A.M., Robinson, L.A., Steeber, D.A., Tedder, T.F., Yoshie, O., Imai, T., and Patel, D.D., 1998, Fractalkine and CX3CR1 mediate a novel mechanism of leukocyte capture, firm adhesion, and activation under physiologic flow, *J Exp Med.* 188:1413–1419.

Girard, J.P. and Springer, T.A., 1995, High endothelial venules (HEVs): specialized endothelium for lymphocyte migration, *Immunol Today.* 16:449–457.

Gulubova, M.V., 2002, Expression of cell adhesion molecules, their ligands and tumour necrosis factor alpha in the liver of patients with metastatic gastrointestinal carcinomas, *Histochem J.* 34:67–77.

Halloran, M.M., Carley, W.W., Polverini, P.J., Haskell, C.J., Phan, S., Anderson, B.J., Woods, J.M., Campbell, P.L., Volin, M.V., Backer, A.E., and Koch, A.E., 2000, Ley/H: an endothelial-selective, cytokine-inducible, angiogenic mediator, *J Immunol.* 164:4868–4877.

Haraldsen, G., Kvale, D., Lien, B., Farstad, I.N., and Brandtzaeg, P., 1996, Cytokine-regulated expression of E-selectin, intercellular adhesion molecule-1 (ICAM-1), and vascular cell adhesion molecule-1 (VCAM-1) in human microvascular endothelial cells, *J Immunol.* 156:2558–2565.

Hasslen, S.R., von Andrian, U.H., Butcher, E.C., Nelson, R.D., and Erlandsen, S.L., 1995, Spatial distribution of L-selectin (CD62L) on human lymphocytes and transfected murine L1–2 cells, *Histochem J.* 27:547–554.

Hemmerich, S., Butcher, E.C., and Rosen, S.D., 1994, Sulfation-dependent recognition of high endothelial venules (HEV)-ligands by L-selectin and MECA 79, and adhesion-blocking monoclonal antibody, *J Exp Med.* 180:2219–2226.

Horan, P.K. and Slezak, S.E., 1989, Stable cell membrane labelling, *Nature.* 340:167–168.

Imai, T., Hieshima, K., Haskell, C., Baba, M., Nagira, M., Nishimura, M., Kakizaki, M., Takagi, S., Nomiyama, H., Schall, T.J., and Yoshie, O., 1997, Identification and molecular characterization of fractalkine receptor CX3CR1, which mediates both leukocyte migration and adhesion, *Cell.* 91:521–530.

Imai, Y., Lasky, L.A., and Rosen, S.D., 1992, Further characterization of the interaction between L-selectin and its endothelial ligands, *Glycobiology.* 2:373–381.

Irimura, T. and Nicolson, G.L., 1981, The role of glycoconjugates in metastatic melanoma blood-borne arrest and cell surface properties, *J Supramol Struct Cell Biochem.* 17:325–336.

Jain, R.K. and Padera, T.P., 2002, Prevention and treatment of lymphatic metastasis by antilymphangiogenic therapy, *J Natl Cancer Inst.* 94:785–787.

Jalkanen, S., Steere, A.C., Fox, R.I., and Butcher, E.C., 1986, A distinct endothelial cell recognition system that controls lymphocyte traffic into inflamed synovium, *Science.* 233:556–558.

Kallmann, B.A., Wagner, S., Hummel, V., Buttmann, M., Bayas, A., Tonn, J.C., Rieckmann, P., Sudhoff, T., Sohngen, D., Feugier, P., Jo, D.Y., Shieh, J.H., MacKenzie, K.L., Rafii, S., Crystal, R.G., Moore, M.A., Saigo, K., Sugimoto, T., Matsui, T., Ryo, R., Kumagai, S., Iwaguro, H., Yamaguchi, J., Kalka, C., Murasawa, S., Masuda, H., Hayashi, S., Silver, M., Li, T., Isner, J.M., Asahara, T., Nabel, E.G., Filippi, M.D., Porteu, F., Pesteur, F.L., Schiavon, V., Millot, G.A., Vainchenker, W., de Sauvage, F.J., Dubart Kupperschmitt, A., Sainteny, F., Hendry, J.H., Booth, C., Potten, C.S., Ko, M.K., Kay, E.P., Chekanov, V.S., Nikolaychik, V., Opdenakker, G., Zhang, B., Prendergast, G.C., Fenton, R.G., Nishikawa, S.I., Hirashima, M., Nishikawa, S., Ogawa, M., Brockow, K., Akin, C., Huber, M., Scott, L.M., Schwartz, L.B., Metcalfe, D.D., Voermans, C., van Hennik, P.B., van Der Schoot, C.E., Moehler, T.M., Neben, K., Ho, A.D., Goldschmidt, H., Parati, E.A., Bez, A., Ponti, D., de Grazia, U., Corsini, E., Cova, L., Sala, S., Colombo, A., Alessandri, G., Pagano, S.F., Caprioli, A., Minko, K., Drevon, C., Eichmann, A., Dieterlen-Lievre, F., Jaffredo, T., Podskochy, A., Fagerholm, P., Fiedler, W., Staib, P., Kuse, R., Duhrsen, U., Flasshove, M., Cavalli, F., Hossfeld, D.K., and Berdel, W.E., 2002, Characteristic gene expression profile of primary human cerebral endothelial cells

Circulating Endothelial Adhesion Molecules (sE-selectin, sVCAM-1 and sICAM-1) During rHuG-CSF-Stimulated Stem Cell Mobilization

Ex vivo Expansion of Stem and Progenitor Cells in Co-culture of Mobilized Peripheral Blood CD34+ Cells on Human Endothelium Transfected with Adenovectors Expressing Thrombopoietin, c-kit Ligand, and Flt-3 Ligand

Fluctuations in plasma macrophage colony-stimulating factor levels during autologous peripheral blood stem cell transplantation for haematologic diseases

Endothelial progenitor cell vascular endothelial growth factor gene transfer for vascular regeneration
Stem cells combined with gene transfer for therapeutic vasculogenesis: magic bullets?
Requirement for mitogen-activated protein kinase activation in the response of embryonic stem cell-derived hematopoietic cells to thrombopoietin *in vitro*
Endothelial cells and radiation gastrointestinal syndrome
Differential interaction of molecular chaperones with procollagen I and type IV collagen in corneal endothelial cells
Iron contributes to endothelial dysfunction in acute ischemic syndromes
New insights in the regulation of leukocytosis and the role played by leukocytes in septic shock
Farnesyltransferase inhibitors reverse Ras-mediated inhibition of Fas gene expression
Cell biology of vascular endothelial cells
Levels of mast-cell growth factors in plasma and in suction skin blister fluid in adults with mastocytosis: correlation with dermal mast-cell numbers and mast-cell tryptase
Homing of human hematopoietic stem and progenitor cells: new insights, new challenges?
Angiogenesis in hematologic malignancies
Human neural stem cells express extra-neural markers
Hemangioblast commitment in the avian allantois: cellular and molecular aspects
Repeated UVR exposures cause keratocyte resistance to apoptosis and hyaluronan accumulation in the rabbit cornea
Role of angiogenesis inhibitors in acute myeloid leukemia. *Faseb J.* 12:12.

Karkkainen, M.J., Makinen, T., and Alitalo, K., 2002, Lymphatic endothelium: a new frontier of metastasis research, *Nat Cell Biol.* 4:E2–5.

Kellermann, S.A., Hudak, S., Oldham, E.R., Liu, Y.J., and McEvoy, L.M., 1999, The CC chemokine receptor-7 ligands 6Ckine and macrophage inflammatory protein-3 beta are potent chemoattractants for *in vitro*- and *in vivo*-derived dendritic cells, *J Immunol.* 162:3859–3864.

Kieda, C. and Monsigny, M., 1986, Involvement of membrane sugar receptors and membrane glycoconjugates in the adhesion of 3LL cell subpopulations to cultured pulmonary cells, *Invasion Metastasis.* 6:347–366.

Kieda, C.M., Bowles, D.J., Ravid, A., and Sharon, N., 1978, Lectins in lymphocyte membranes, *FEBS Lett.* 94:391–396.

Kieda, C., Paprocka, M., Krawczenko, A., Zalecki, P., Dupuis, P., Monsigny, M., Radzikowski, C., Dus, D., 2002, Related articles, links abstract new human microvascular endothelial cell lines with specific adhesion molecules phenotypes. *Endothelium.* 9(4):247–261.

Kunkel, E.J. and Butcher, E.C., 2002, Chemokines and the tissue-specific migration of lymphocytes. *Immunity.* 16:1–4.

Ley, K., Zakrzewicz, A., Hanski, C., Stoolman, L.M., and Kansas, G.S., 1995, Sialylated O-glycans and L-selectin sequentially mediate myeloid cell rolling *in vivo*. *Blood.* 85:3727–3735.

Li, L., Short, H.J., Qian, K.X., Elhammer, A.P., and Geng, J.G., 2001, Characterization of glycoprotein ligands for P-selectin on a human small cell lung cancer cell line NCI-H345, *Biochem Biophys Res Commun.* 288:637–644.

Li, Y., Wang, M.N., Li, H., King, K.D., Bassi, R., Sun, H., Santiago, A., Hooper, A.T., Bohlen, P., and Hicklin, D.J., 2002, Active immunization against the vascular endothelial growth factor receptor flk1 inhibits tumor angiogenesis and metastasis, *J Exp Med.* 195:1575–1584.

Lotan, R., Belloni, P.N., Tressler, R.J., Lotan, D., Xu, X.C., and Nicolson, G.L., 1994, Expression of galectins on microvessel endothelial cells and their involvement in tumour cell adhesion, *Glycoconj J.* 11:462–468.

Maehara, Y., Kabashima, A., Koga, T., Tokunaga, E., Takeuchi, H., Kakeji, Y., and Sugimachi, K., 2000, Vascular invasion and potential for tumor angiogenesis and metastasis in gastric carcinoma, *Surgery.* 128:408–416.

Mannori, G., Crottet, P., Cecconi, O., Hanasaki, K., Aruffo, A., Nelson, R.M., Varki, A., and Bevilacqua, M.P., 1995, Differential colon cancer cell adhesion to E-, P-, and L-selectin: role of mucin-type glycoproteins, *Cancer Res.* 55:4425–4431.

Matsushita, Y., Kitajima, S., Goto, M., Tezuka, Y., Sagara, M., Imamura, H., Tanabe, G., Tanaka, S., Aikou, T., and Sato, E., 1998, Selectins induced by interleukin-1beta on the human liver endothelial cells act as ligands for sialyl Lewis X-expressing human colon cancer cell metastasis, *Cancer Lett.* 133:151–160.

McEver, R.P., 1991, GMP-140: a receptor for neutrophils and monocytes on activated platelets and endothelium. *J Cell Biochem.* 45:156–161.

Mehta, P., Cummings, R.D., and McEver, R.P., 1998, Affinity and kinetic analysis of P-selectin binding to P-selectin glycoprotein ligand-1. *J Biol Chem.* 273:32506–32513.

Michie, S.A., Streeter, P.R., Bolt, P.A., Butcher, E.C., and Picker, L.J., 1993, The human peripheral lymph node vascular addressin. An inducible endothelial antigen involved in lymphocyte homing, *Am J Pathol.* 143:1688–1698.

Michie, S.A., Streeter, P.R., Butcher, E.C., and Rouse, R.V., 1995, L-selectin and alpha 4 beta 7 integrin homing receptor pathways mediate peripheral lymphocyte traffic to AKR mouse hyperplastic thymus, *Am J Pathol.* 147:412–421.

Moore, K.L., 1998, Structure and function of P-selectin glycoprotein ligand-1. *Leuk Lymphoma.* 29:1–15.

Morgan, S.M., Samulowitz, U., Darley, L., Simmons, D.L., and Vestweber, D., 1999, Biochemical characterization and molecular cloning of a novel endothelial-specific sialomucin. *Blood.* 93:165–175.

Nakache, M., Berg, E.L., Streeter, P.R., and Butcher, E.C., 1989, The mucosal vascular addressin is a tissue-specific endothelial cell adhesion molecule for circulating lymphocytes, *Nature.* 337:179–181.

Nangia-Makker, P., Baccarini, S., and Raz, A., 2000, Carbohydrate-recognition and angiogenesis, *Cancer Metastasis Rev.* 19:51–57.

Nemoto, Y., Izumi, Y., Tezuka, K., Tamatani, T., and Irimura, T., 1998, Comparison of 16 human colon carcinoma cell lines for their expression of sialyl LeX antigens and their E-selectin-dependent adhesion, *Clin Exp Metastasis.* 16:569–576.

Norman, K.E., Moore, K.L., McEver, R.P., and Ley, K., 1995, Leukocyte rolling *in vivo* is mediated by P-selectin glycoprotein ligand-1, *Blood.* 86:4417–4421.

Ojeifo, J.O., Lee, H.R., Rezza, P., Su, N., and Zwiebel, J.A., 2001, Endothelial cell-based systemic gene therapy of metastatic melanoma, *Cancer Gene Ther.* 8:636–648.

Opolski, A., Wietrzyk, J., Dus, D., Kieda, C., Matejuk, A., Makowska, A., Wojdat, E., Ugorski, M., Laskowska, A., Klopocki, A., Rygaard, J., and Radzikowski, C., 1998, Metastatic potential and saccharide antigens expression of human colon cancer cells xenotransplanted into athymic nude mice, *Folia Microbiol (Praha).* 43:507–510.

Pals, S.T., Horst, E., Scheper, R.J., and Meijer, C.J., 1989, Mechanisms of human lymphocyte migration and their role in the pathogenesis of disease, *Immunol Rev.* 108:111–133.

Pauli, B.U., Augustin-Voss, H.G., el-Sabban, M.E., Johnson, R.C., and Hammer, D.A., 1990, Organ-preference of metastasis. The role of endothelial cell adhesion molecules, *Cancer Metastasis Rev.* 9:175–189.

Picker, L.J. and Butcher, E.C., 1992, Physiological and molecular mechanisms of lymphocyte homing, *Annu Rev Immunol.* 10:561–591.

Picker, L.J., Kishimoto, T.K., Smith, C.W., Warnock, R.A., and Butcher, E.C., 1991, ELAM-1 is an adhesion molecule for skin-homing T cells, *Nature.* 349:796–799.

Picker, L.J., Nakache, M., and Butcher, E.C., 1989, Monoclonal antibodies to human lymphocyte homing receptors define a novel class of adhesion molecules on diverse cell types, *J Cell Biol.* 109:927–937.

Pili, R., Chang, J., Partis, R.A., Mueller, R.A., Chrest, F.J., and Passaniti, A., 1995, The alpha-glucosidase I inhibitor castanospermine alters endothelial cell glycosylation, prevents angiogenesis, and inhibits tumor growth, *Cancer Res.* 55:2920–2926.

Pouyani, T. and Seed, B., 1995, PSGL-1 recognition of P-selectin is controlled by a tyrosine sulfation consensus at the PSGL-1 amino terminus, *Cell.* 83:333–343.

Puri, K.D., Finger, E.B., Gaudernack, G., and Springer, T.A., 1995, Sialomucin CD34 is the major L-selectin ligand in human tonsil high endothelial venules. *J Cell Biol.* 131:261–270.

Saiki, I., Koike, C., Obata, A., Fujii, H., Murata, J., Kiso, M., Hasegawa, A., Komazawa, H., Tsukada, H., Azuma, I., Okada, S., and Oku, N., 1996, Functional role of sialyl Lewis X and fibronectin-derived RGDS peptide analogue on tumor-cell arrest in lungs followed by extravasation, *Int J Cancer.* 65:833–839.

Sassetti, C., Tangemann, K., Singer, M.S., Kershaw, D.B., and Rosen, S.D., 1998, Identification of podocalyxin-like protein as a high endothelial venule ligand for L-selectin: parallels to CD34, *J Exp Med.* 187:1965–1975.

Sassetti, C., Van Zante, A., and Rosen, S.D., 2000, Identification of endoglycan, a member of the CD34/ podocalyxin family of sialomucins, *J Biol Chem.* 275:9001–9010.

Schurmann, G., 1997, Cell adhesion. Molecular principles and initial implications for surgery, *Chirurg.* 68:477–487.

Scott, L.J., Clarke, N.W., George, N.J., Shanks, J.H., Testa, N.G., and Lang, S.H., 2001, Interactions of human prostatic epithelial cells with bone marrow endothelium: binding and invasion, *Br J Cancer.* 84:1417–1423.

Seppo, A., Turunen, J.P., Penttila, L., Keane, A., Renkonen, O., and Renkonen, R., 1996, Synthesis of a tetravalent sialyl Lewis x glycan, a high-affinity inhibitor of L-selectin-mediated lymphocyte binding to endothelium, *Glycobiology.* 6:65–71.

Sharma, S., Stolina, M., Luo, J., Strieter, R.M., Burdick, M., Zhu, L.X., Batra, R.K., and Dubinett, S.M., 2000, Secondary lymphoid tissue chemokine mediates T cell-dependent antitumor responses *in vivo*, *J Immunol.* 164:4558–4563.

Sher, B.T., Bargatze, R., Holzmann, B., Gallatin, W.M., Matthews, D., Wu, N., Picker, L., Butcher, E.C., and Weissman, I.L., 1988, Homing receptors and metastasis. *Adv Cancer Res.* 51:361–390.

Slaton, J.W., Inoue, K., Perrotte, P., El-Naggar, A.K., Swanson, D.A., Fidler, I.J., and Dinney, C.P., 2001, Expression levels of genes that regulate metastasis and angiogenesis correlate with advanced pathological stage of renal cell carcinoma, *Am J Pathol.* 158:735–743.

Springer, T.A., 1990, Adhesion receptors of the immune system, *Nature.* 346:425–434.

Springer, T.A., 1994, Traffic signals for lymphocyte recirculation and leukocyte emigration: the multistep paradigm, *Cell.* 76:301–314.

St Croix, B., Rago, C., Velculescu, V., Traverso, G., Romans, K.E., Montgomery, E., Lal, A., Riggins, G.J., Lengauer, C., Vogelstein, B., and Kinzler, K.W., 2000, Genes expressed in human tumor endothelium, *Science.* 289:1197–1202.

Staroselsky, A.N., Mahlin, T., Savion, N., Klein, O., Nordenberg, J., Donin, N., Michowitz, M., and Leibovici, J., 1996, Metastatic potential and multidrug resistance correlation in the B16 melanoma system, *J Exp Ther Oncol.* 1:251–259.

Steeber, D.A., Tang, M.L., Green, N.E., Zhang, X.Q., Sloane, J.E., and Tedder, T.F., 1999, Leukocyte entry into sites of inflammation requires overlapping interactions between the L-selectin and ICAM-1 pathways, *J Immunol.* 163:2176–2186.

Steegmaier, M., Levinovitz, A., Isenmann, S., Borges, E., Lenter, M., Kocher, H.P., Kleuser, B., and Vestweber, D., 1995, The E-selectin-ligand ESL-1 is a variant of a receptor for fibroblast growth factor, *Nature.* 373:615–620.

Steffen, B.J., Breier, G., Butcher, E.C., Schulz, M., and Engelhardt, B., 1996, ICAM-1, VCAM-1, and MAdCAM-1 are expressed on choroid plexus epithelium but not endothelium and mediate binding of lymphocytes *in vitro*, *Am J Pathol.* 148:1819–1838.

Stoolman, L.M., Yednock, T.A., and Rosen, S.D., 1987, Homing receptors on human and rodent lymphocytes—evidence for a conserved carbohydrate-binding specificity, *Blood.* 70:1842–1850.

St-Pierre, Y., Aoudjit, F., Lalancette, M., and Potworowski, E.F., 1999, Dissemination of T cell lymphoma to target organs: a post-homing event implicating ICAM-1 and matrix metalloproteinases, *Leuk Lymphoma.* 34:53–61.

Streeter, P.R., Berg, E.L., Rouse, B.T., Bargatze, R.F., and Butcher, E.C., 1988, A tissue-specific endothelial cell molecule involved in lymphocyte homing, *Nature.* 331:41–46.

Tanaka, Y., 2000, Integrin activation by chemokines: relevance to inflammatory adhesion cascade during T cell migration, *Histol Histopathol.* 15:1169–1176.

Tu, L., Murphy, P.G., Li, X., and Tedder, T.F., 1999, L-selectin ligands expressed by human leukocytes are HECA-452 antibody-defined carbohydrate epitopes preferentially displayed by P-selectin glycoprotein ligand-1, *J Immunol.* 163:5070–5078.

Vestweber, D. and Blanks, J.E., 1999, Mechanisms that regulate the function of the selectins and their ligands, *Physiol Rev.* 79:181–213.

von Andrian, U.H., Chambers, J.D., Berg, E.L., Michie, S.A., Brown, D.A., Karolak, D., Ramezani, L., Berger, E.M., Arfors, K.E., and Butcher, E.C., 1993, L-selectin mediates neutrophil rolling in inflamed venules through sialyl LewisX-dependent and -independent recognition pathways, *Blood.* 82:182–191.

Wang, F., Wei, L., and Chen, L., 2000, The relationship between vascular endothelial growth factor, microvascular density, lymph node metastasis and prognosis of breast carcinoma, *Zhonghua Bing Li Xue Za Zhi.* 29:172–175.

Wardlaw, A.J., 2001, Eosinophil trafficking in asthma, *Clin Med.* 1:214–218.

Warnock, R.A., Askari, S., Butcher, E.C., and von Andrian, U.H., 1998, Molecular mechanisms of lymphocyte homing to peripheral lymph nodes, *J Exp Med.* 187:205–216.

Warnock, R.A., Campbell, J.J., Dorf, M.E., Matsuzawa, A., McEvoy, L.M., and Butcher, E.C., 2000, The role of chemokines in the microenvironmental control of T versus B cell arrest in Peyer's patch high endothelial venules, *J Exp Med.* 191:77–88.

Wei, Y.Q., Wang, Q.R., Zhao, X., Yang, L., Tian, L., Lu, Y., Kang, B., Lu, C.J., Huang, M.J., Lou, Y.Y., Xiao, F., He, Q.M., Shu, J.M., Xie, X.J., Mao, Y.Q., Lei, S., Luo, F., Zhou, L.Q., Liu, C.E., Zhou, H., Jiang, Y., Peng, F., Yuan, L.P., Li, Q., Wu, Y., and Liu, J.Y., 2000, Immunotherapy of tumors with xenogeneic endothelial cells as a vaccine, *Nat Med.* 6:1160–1166.

Welply, J.K., Abbas, S.Z., Scudder, P., Keene, J.L., Broschat, K., Casnocha, S., Gorka, C., Steininger, C., Howard, S.C., Schmuke, J.J. *et al.*, 1994, Multivalent sialyl-LeX: potent inhibitors of E-selectin-mediated cell adhesion; reagent for staining activated endothelial cells, *Glycobiology*. 4:259–265.

Widmer, M.B., Morrissey, P.J., Goodwin, R.G., Grabstein, K.H., Park, L.S., Watson, J.D., Kincade, P.W., Conlon, P.J., and Namen, A.E., 1990, Lymphopoiesis and IL-7, *Int J Cell Cloning*. 8 Suppl 1:168–170; discussion 171–172.

Woynarowska, B., Dimitroff, C.J., Sharma, M., Matta, K.L., and Bernacki, R.J., 1996, Inhibition of human HT-29 colon carcinoma cell adhesion by a 4-fluoro-glucosamine analogue, *Glycoconj J.* 13:663–674.

Yang, R.B., Ng, C.K., Wasserman, S.M., Colman, S.D., Shenoy, S., Mehraban, F., Komuves, L.G., Tomlinson, J.E., and Topper, J.N., 2002, Identification of a novel family of cell-surface proteins expressed in human vascular endothelium, *J Biol Chem.* 21:21.

Yednock, T.A., Butcher, E.C., Stoolman, L.M., and Rosen, S.D., 1987, Receptors involved in lymphocyte homing: relationship between a carbohydrate-binding receptor and the MEL-14 antigen, *J Cell Biol.* 104:725–731.

Yoshida, H., Honda, K., Shinkura, R., Adachi, S., Nishikawa, S., Maki, K., Ikuta, K., and Nishikawa, S.I., 1999, IL-7 receptor alpha+ CD3(−) cells in the embryonic intestine induces the organizing center of Peyer's patches, *Int Immunol.* 11:643–655.

Zeisig, R., Stahn, R., and Fichtner, I., 2002, Inhibition of carbohydrate mediated cell adhesion by liposomes as a possible [correction of possibile] way to prevent tumour metastasis, *Cell Mol Biol Lett.* 7:270–271.

Zhang, L., Underhill, C.B., and Chen, L., 1995, Hyaluronan on the surface of tumor cells is correlated with metastatic behavior, *Cancer Res.* 55:428–433.

Zhu, D.Z., Cheng, C.F., and Pauli, B.U., 1991, Mediation of lung metastasis of murine melanomas by a lung-specific endothelial cell adhesion molecule, *Proc Natl Acad Sci USA.* 88:9568–9572.

Zoll, B., Lefterova, P., Csipai, M., Finke, S., Trojaneck, B., Ebert, O., Micka, B., Roigk, K., Fehlinger, M., Schmidt-Wolf, G.D., Huhn, D., and Schmidt-Wolf, I.G., 1998, Generation of cytokine-induced killer cells using exogenous interleukin-2, -7 or -12, *Cancer Immunol Immunother.* 47:221–226.

Zollner, O. and Vestweber, D., 1996, The E-selectin ligand-1 is selectively activated in Chinese hamster ovary cells by the alpha(1,3)-fucosyltransferases IV and VII. *J Biol Chem.* 271:33002–33008.

7

INVOLVEMENT OF GAGs IN THE ACTIVITY OF PRO-INFLAMMATORY FACTORS

Fabrice Allain, Christophe Vanpouille, Agnès Denys, Rachel Pakula, Mathieu Carpentier, and Joël Mazurier

Unité de Glycobiologie Structurale et Fonctionnelle
UMR n°8576 du CNRS
Université des Sciences et Technologies de Lille
Villeneuve d'Ascq, France

1. INTRODUCTION

During inflammation, the recruitment of leukocytes is regulated by a cascade of molecular events resulting in adhesion to the endothelium followed by migration into tissue. Initially, selectin-mediated interactions cause leukocytes to roll along the endothelium where they contact factors that trigger strong binding by activating cell integrins. The final step in the exit of leukocytes out of the bloodstream requires cell penetration through the underlying basement membrane (Ebnet and Vestweber, 1999).

Numerous pro-inflammatory proteins, for example chemokines and cytokines, have been identified as attractants and regulatory factors for different types of peripheral blood leukocytes (Baggiolini, 1998; Meager, 1999; Moser *et al.*, 1998). A common feature of some of these soluble mediators is their capacity to interact on one hand, with signaling receptors present on leukocytes and on the other hand, with proteoglycans present in the extracellular matrix (ECM) or expressed on the endothelium or leukocytes themselves. By these interactions, proteoglycans have been shown to be involved in different stages of inflammation, such as the recruitment of inflammatory cells, the release and diffusion of mediators of inflammation by infiltrating leukocytes and neighboring cells, and the turnover of the ECM (Delehedde *et al.*, 2002). For example, proteoglycans may provide tissue-bound reservoirs of inflammatory proteins which can be presented to target cells, as it has been previously described for growth factors. The ability of chemokines/cytokines to bind to and be immobilized by proteoglycans may accentuate their activity by promoting their accumulation in high concentrations at the appropriate location to encounter their

Glycobiology and Medicine, edited by John S. Axford
Kluwer Academic / Plenum Publishers, New York, 2003

target cells. It is believed that proteoglycans capture chemokines in the ECM and on the surface of endothelial cells, leading to preservation, to degradation, and to dilution by the bloodstream. By establishing a local concentration gradient from the source of chemokine secretion, proteoglycans modulate the ability of these factors for triggering integrin-mediated adhesion and migration of leukocyte subsets (Campbell *et al.*, 1998; Tanaka *et al.*, 1993). Finally, proteoglycans may facilitate the assembly of pro-inflammatory factors with their cell surface signaling receptors, and/or contribute to the generation of additional intracellular signals, both events leading to the modulation of signal transduction (Capila and Linhardt, 2002; Kuschert *et al.*, 1999).

2. PROTEOGLYCAN STRUCTURE

Proteoglycans are members of a large family of glycoconjugates formed by core protein decorated with linear chains of repeating disaccharidic units, termed GAGs (Silbert *et al.*, 1997). Based on the nature of these units, four biosynthetically distinct types of GAGs have been defined: heparin/heparan sulfates, chondroitan sulfates/dermatan sulfates, keratan sulfates, and hyaluronic acid. Up to now, most of the inflammatory factors have been described to interact with the heparan sulfate family (Capila and Linhardt, 2002).

2.1. The Heparin/Heparan Sulfate Family

Heparin and heparan sulfates consist of alternating α-D-glucosamine and hexuronate (α-L-iduronate or α-D-glucuronate) residues. There are several fine structural differences among this GAG family, resulting from post-polymerization modifications on the monotonous starting structure of the linear chain (Gallagher *et al.*, 1992; Turnbull *et al.*, 2001). The amino group of glucosamine can be either free or substituted by acetyl or sulfo groups. The N-sulfation often acts as a signal for other modifications. Thus, 6-O sulfation occurs preferentially on N-sulfated glucosamine. Epimerization of glucuronic acid leads to the appearance of iduronic acid, allowing modification in the flexibility of the heparan sulfate chain. Iduronate is often 2-O sulfated while it is very rare for glucuronate to be modified in this way. Finally, 3-O sulfation of glucosamine only occurs rarely as a final modification of highly sulfated disaccharides. In heparin, most of 85% of glucosamines are N-sulfated, whereas in heparan sulfates, only 40–50% of the total potential N-sulfation occur, but importantly these are clustered. The result of this pattern of modifications is that heparan sulfates have high structural diversity, in which domains of low sulfation alternate with highly sulfated domains, that appears essential to the specificity of interactions with heparin-binding proteins (Capila and Linhardt, 2002).

2.2. Cell Surface Proteoglycans

Another degree in the complexity of the proteoglycan family is provided by the nature of the core proteins. For example, cell surface proteoglycans are represented by four families (Silbert *et al.*, 1997).

Syndecan-1 was the first cell surface proteoglycan to be described and is part of a family of four members, syndecan-1 to syndecan-4. They are encoded by distinct genes and all possess a cleavable signal peptide, an extracellular N-terminal domain containing GAG attachment sites, a single transmembrane domain, and a short cytosolic tail. Syndecan-2 and -3 contain only heparan sulfate chains, while syndecan-1 and -4 also carry chondroitane sulfate chains. The most highly conserved domains of the syndecan core proteins are the transmembrane and cytoplasmic domains (more than 50% amino acid identity), that is in contrast to the extracellular domains which vary in length, in the number and placement of GAG chains, and show only limited identity among the four syndecan types. The structural conservation of the cytoplasmic domains strongly suggests that they play a role in an essential intracellular function that may be common to all syndecans. There is growing evidence to suggest that this may be related to their ability to associate with and reorganize cytoskeletal structures (Rapraeger *et al.*, 1998).

Glypican is a second family of six proteoglycans which contain only heparan sulfates. In contrast to syndecans, glypicans are attached to the cell membrane by a glycosylphosphatidyl-inositol lipid anchor.

The third family is represented by a unique member termed betaglycan. This proteoglycan, also termed type III receptor for transforming growth factor β (TGFβ), can bind TGFβ by its core protein as well as the GAG chain and present it to signaling type I and II receptors.

Finally, the last family of cell surface proteoglycans is represented by the CD44 variants. CD44 is described as an ubiquitous cell surface adhesion molecule involved in cell–cell and cell–matrix interactions. The different isoforms are expressed from a single gene by alternative splicing on 10 exons. Most of the CD44 isoforms have chondroitane sulfates, but the splice variants CD44v3, which contain the exon v3, carry also heparan sulfate chains (Jackson *et al.*, 1995). These forms of heparan sulfate proteoglycan (HSPG) act as functional co-receptors that promote cytokine signaling, suggesting a dynamic role of HSPG in cell activation and differentiation (van der Voort *et al.*, 1999).

2.3. Functional Activities of Core Proteins

Interactions of cytosolic domains of syndecans and CD44 variants with cytoskeleton and neighboring signaling molecules have suggested that they may be involved in the regulation of cellular processes related to cell adhesion and migration (Rapraeger *et al.*, 1998).

For example, the cytoplasmic domain of syndecan-1 was demonstrated to be involved in microfilament association (Carey *et al.*, 1996). Syndecan-3 is involved in neurite outgrowth, a mechanism dependent on the association between cytoplasmic domain of the proteoglycan and tyrosine kinases of the Src family (Kinnunen *et al.*, 1998). Syndecan-2 and -4 cooperatively act with integrins in generating Rho-dependent signals for cell spreading on fibronectin and for the assembly of focal adhesions and actin stress fibers (Kusano *et al.*, 2000; Saoncella *et al.*, 1999). In addition, the cytosolic domain of syndecan-4 binds and enhances the activity of PKCα, indicating that it may be involved in signaling evoked by this kinase in cell spreading and adhesion (Oh *et al.*, 1997; Volk *et al.*, 1999). Syntenin, a protein that contains a tandem repeat of PDZ domains, specifically interacts with the cytoplasmic domain of syndecans. Overexpression of this protein induces numerous cell surface extension and cytoskeleton reorganization, making it a putative adaptor that couples syndecans to cytoskeletal proteins or cytosolic signaling effectors (Grootjans *et al.*, 1997).

It is also thought that CD44 may function as a co-signalling receptor in a variety of cell types (Jones *et al.*, 2000; van der Voort *et al.*, 1999). CD44 ligation activates several signaling pathways that culminate in cell proliferation, cytokine secretion, chemokine gene expression, and cytolytic effector functions. One of the earliest events following stimulation via CD44 is tyrosine phosphorylation. The Src family of protein tyrosine kinases such as Lck, Fyn, Lyn, and Hck were shown to be coupled to CD44 in lipid rafts of the plasma membrane, suggesting that CD44 probably generates cellular signals by utilizing the signaling machinery assembled in the plasma membrane microdomains (Ilangumaran *et al.*, 1999).

3. PROTEOGLYCANS AND INFLAMMATION

Growing data are converging to reveal that integral membrane HSPG are critical regulators for growth and inflammatory factors (Delehedde *et al.*, 2002). Infiltrating leukocytes and inflammatory cells express HSPG which show a great structural diversity due to differences in both the core protein and the posttranslational glycosylation. Additionally, changes in proteoglycan structure occur with wound repair and inflammatory stimuli. This diversity in expression and structural features of proteoglycans enhances the potential for presentation and modulation of the activity of inflammatory factors, by determining which ones will bind and by influencing their functional capacity. Tissue- and leukocyte-specific combinations of inflammatory factors will allow the recruitment of cell subsets to be carefully regulated, providing a mechanism for the control of tissue-specific trafficking of leukocytes. Over the past few years, two models have emerged for the functions of proteoglycans in inflammatory response.

3.1. Proteoglycans as Presenting Molecules for Inflammatory Factors

In the first model, heparan sulfates are involved in the specific binding of inflammatory factors and their effective presentation to their cognate signaling receptor. Chemokines have been demonstrated to trigger integrin-mediated adhesion of leukocyte subsets when they were immobilized by binding to GAGs (Campbell *et al.*, 1998; Tanaka *et al.*, 1993). Such a presentation leads to the stabilization of the interactions between inflammatory factors and their receptors and to a consequent sustained signaling (Kuschert *et al.*, 1999).

The mechanism of presentation could involve extracellular proteoglycans as well as cell surface proteoglycans. Indeed, proteoglycans may be visualized as a hand that correctly presents chemokines to their receptors on infiltrating cells (Tanaka *et al.*, 1993). In this way, the interactions with soluble GAGs was demonstrated to enhance the ability of IL-8 to induce neutrophil response (Webb *et al.*, 1993). RANTES and MIP-1α can be bound to proteoglycans in the ECM and modulate T cell adhesion (Gilat *et al.*, 1994). On the other hand, activated cytotoxic T lymphocytes secrete significant amounts of β-chemokines (RANTES, MIP-1α and MIP-1β) together with granular soluble proteoglycans. This association is physiologically relevant and is thought to substitute cell surface proteoglycans in facilitating chemokine activity (Wagner *et al.*, 1998). Intracellular packaging of chemokines bound to proteoglycans has also been described for PF-4 and platelet basic protein, which are stored in platelet alpha granules and are released following activation (Stringer and

Gallagher, 1997). These data suggest that storage of chemokines bound to proteoglycans in specialized granules may be a more general phenomenon for inflammatory effectors.

3.2. Proteoglycans as Co-Receptors for Inflammatory Factors

In the second model, the function of proteoglycans may be that of co-receptors, in which cell surface proteoglycans enable its ligand to activate the cognate signaling receptor. Moreover, simultaneous binding of the inflammatory factor to proteoglycan and signaling receptor allows clustering of the cytosolic domains of both types of cell surface binding sites, that may lead to the activation of co-stimulatory molecules and to the appearance of additional signaling events. This last mechanism implies that both GAGs and core proteins are involved in the binding and co-stimulatory functions of proteoglycans. Therefore, the absence of cell surface proteoglycans cannot be substituted by soluble GAGs, that is a basic difference with the mechanism of presentation.

Several cytokines have been described to bind to heparin and heparan sulfate molecules. However, most of these so-called heparin-binding cytokines, for example IL-1β, IL-2, IL-4, and TNF-α, are not capable of binding to GAGs under physiological conditions, indicating that these interactions are unlikely to take place in a physiological environment. In contrast, growing data are converging to reveal that HSPG are critical regulators of the activity of some cytokines, such as IL-10 or hepatocyte growth factor (HGF). IL-10 is a pleiotropic cytokine that exhibits stimulatory and suppressive effects on the immune and inflammatory systems. For example, IL-10 stimulates the proliferation and differentiation of B cells (Rousset *et al.*, 1992), induces chemotaxis of $CD8^+$ T cells (Jinquan *et al.*, 1993), and enhances cellular toxicity of monocytes (te Velde *et al.*, 1992). In contrast, IL-10 potently inhibits the synthesis of pro-inflammatory factors by macrophages and reduces their ability to serve as antigen-presenting cells for T cells (Moore *et al.*, 1993). Recent studies revealed that cell surface GAGs are important modulators of IL-10 activity (Salek-Ardakani *et al.*, 2000). These data demonstrated that, for optimum stimulatory activity on monocytes, IL-10 requires the presence of HSPG at the cell surface, which may in turn facilitate interaction with high-affinity receptor on these cells or participate in the signaling events. In contrast, soluble GAGs were ineffective to restore the activity of IL-10 on heparinase-treated cells, indicating that they cannot replace the function of cell surface HSPG. HGF is a heparin-binding growth factor that causes epithelial cells to proliferate, differentiate and scatter by activating the tyrosine receptor c-met. Moreover, several findings have suggested a potential role for HGF in leukocyte recruitment (Adams *et al.*, 1994; Delaney *et al.*, 1993; Jiang *et al.*, 1992; Seki *et al.*, 1990). Interestingly, binding and cytokine-like activity of HGF are dependent on the interactions with heparan sulfate chains of CD44v3 variants (van der Voort *et al.*, 1999), providing another example for the role of HSPG in the efficient interaction with high-affinity receptor and participation in signaling events.

4. CYCLOPHILIN B, A MODEL OF HEPARIN-BINDING INFLAMMATORY PROTEIN

Cyclophilins are ubiquitous proteins associated with many molecular mechanisms, that is, signaling, cell growth, protein folding, intracellular transport (Galat, 1999; Marks, 1996).

They were first identified as the main functional binding proteins for the immunosuppressive drug cyclosporine A (CsA) (Handschumacher *et al.*, 1984; Schreiber, 1991). Cyclophilins and the structurally unrelated FK506-binding proteins (FKBPs) exhibit peptidyl-prolyl cis-trans isomerase activity and are able to catalyze *in vitro* the isomerization of peptide bonds and the renaturation of unfolded proteins (Fischer *et al.*, 1998). Both cyclophilins and FKBPs inhibit the phosphatase activity of calcineurin in the presence of their respective immunosuppressive ligands, that is relevant to the inhibition of early T cell activation and constitutes the basis of the prevention of graft rejection (Schreiber, 1991). Cyclophilins A and B (CyPA and CyPB) have been shown to exhibit cytokine-like activity and to participate in cell migration and inflammatory response (Allain *et al.*, 1999; Allain *et al.*, 2002; Bukrinsky, 2002; Gonzalez-Cuadrado *et al.*, 1996; Sherry *et al.*, 1992; Tegeder *et al.*, 1997; Xu *et al.*, 1992). Surface binding sites for CyPA and CyPB have been characterized on leukocytes (Allain *et al.*, 1994; Pushkarsky *et al.*, 2001; Sherry *et al.*, 1998), and interactions with their cell surface signaling receptors appear efficient enough for both cyclophilins to induce chemotaxis of T lymphocytes and neutrophils (Allain *et al.*, 2002; Yurchenko *et al.*, 2001; Yurchenko *et al.*, 2002).

4.1. Binding Properties of CyPB

CyPB has been identified twelve years ago in our laboratory as a secreted component of human milk (Spik *et al.*, 1991). The presence of an extracellular form of this protein led us to analyze its binding properties to cell membrane. In this way, we characterized surface binding sites on T lymphocytes, neutrophils, platelets and endothelial cells (Allain *et al.*, 1994; Allain *et al.*, 1999; Carpentier *et al.*, 1999b; Denys *et al.*, 1997). These binding sites are represented by two types of molecules, a signaling receptor and cell surface proteoglycans (Denys *et al.*, 1998). By site-directed mutagenesis, we identified the binding domains of the protein. We demonstrated that the N-terminal peptide is involved in the interactions with heparan sulfate chains, while the central region of CyPB, which contains catalytic and CsA-binding sites, is involved in the interactions with signaling receptor (Carpentier *et al.*, 1999a). Interestingly, the two binding domains are located on opposite sides of the molecule, indicating that CyPB bound to heparan sulfates may be presented to its signaling receptor. A role for GAGs was reported in HIV-1 attachment through binding to virus-associated CyPA (Saphire *et al.*, 1999), indicating that HSPG may act as binding sites for CyPA. However, we demonstrated that the heparin-binding domain of CyPB, which does not exist in CyPA, is determinant for high-affinity binding to GAGs (Carpentier *et al.*, 1999a).

We then analyzed the structural features of the heparan sulfate motif involved in the interactions with CyPB. This work was performed with heparin-derived oligosaccharides of different sizes, modified by enzymatic or chemical ways. We found that the minimal size of the heparin-binding motif is an octasaccharide and that 6-O sulfation of glucosamine and 2-O sulfation of iduronate are required for the interactions. These last findings are however not surprising, since the involvement of these sulfated groups is often described in the interactions between GAGs and their ligands. More interesting are the findings that the amino group of glucosamine has to be unsubstituted. Indeed, the involvement of free amino group has been only reported for L-selectin (Norgard-Summicht and Varki, 1995) and gD protein of herpes virus (Shukla *et al.*, 1999), but not for the other known heparin-binding proteins.

4.2. Pro-Inflammatory Activities of CyPB

We analyzed the cellular responses induced by CyPB and found that it triggers chemotaxis of peripheral blood T lymphocytes. In our assays, CyPA and CyPB exhibit the same chemotactic activity, indicating that interactions with the same signaling receptor is efficient enough to induce migration of T cells (Allain *et al.*, 2002).

We then demonstrated that CyPB was able to initiate firm adhesion of T lymphocytes to the ECM. The mechanisms by which T cells adhere to the ECM involve functionally active integrins, which have to be activated within seconds for physiological relevance (Hughes and Pfaff, 1998). To understand the underlying mechanism of the pro-adhesive activity of CyPB, we reproduced our experiments in the presence of blocking antibodies to integrin subunits. We demonstrated that only antibodies to α4, β1 and β7 inhibit the adhesion mediated by CyPB, suggesting that CyPB enhances the adhesion of lymphocytes to fibronectin present in the ECM through the activation of α4β1 and α4β7 integrins.

CD147 has been recently described as a signaling receptor for CyPA and CyPB (Yurchenko *et al.*, 2001; Yurchenko *et al.*, 2002). Incubation with blocking antibodies to this receptor inhibits the pro-adhesive activity of CyPB, indicating that CD147 is effectively involved in this cell response (Figure 1A). To check whether HSPGs are involved in the pro-adhesive activity of CyPB, cells were incubated in the presence of free heparin, protamine, a ligand of GAGs or treated with heparinase I, in order to remove cell surface heparan sulfates. In all cases, the effect of CyPB was abolished, indicating that HSPGs are involved in CyPB-mediated adhesion of T lymphocytes (Figure 1A). CyPA, which does not

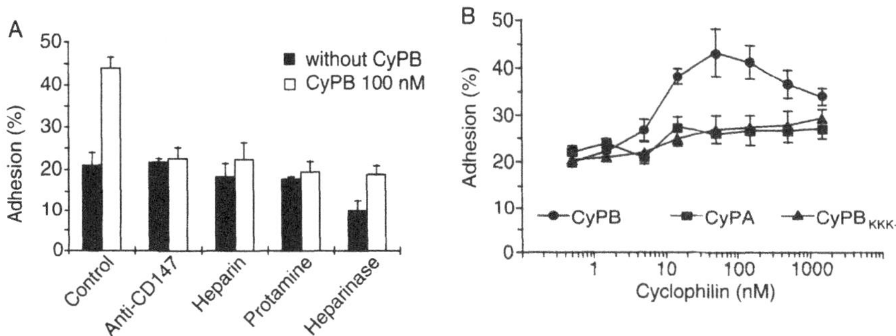

Figure 1. Involvement of cell surface heparan sulfates in CyPB-mediated adhesion of peripheral blood T lymphocytes to the ECM. Freshly isolated T lymphocytes were allowed to adhere into ECM-coated 96-well plates for 30 min at 37 °C. Following extensive washes, remaining firmly attached cells were fixed with 3% paraformaldehyde and then stained with 1% methylene blue in borate buffer. After wash, cells were lysed with 0.1 M HCl and the absorbance of released dye was measured at 595 nm. Cell adhesion was estimated with standard curves where absorbances were related to T cell numbers. (A) Inhibition of CyPB-mediated adhesion of T lymphocytes following incubation with anti-CD147 antibody, heparin (500 μg/ml), protamine (12.5 μM) or cell treatment with heparinase I. T lymphocytes were either pre-incubated with inhibitors for 30 min or treated with heparinase for 1 hr and thereafter used for adhesion assays in the absence (filled bars) or presence of 100 nM CyPB (open bars). (B) Comparison of the effects of CyPB (●), $CyPB_{KKK-}$ (▲) and CyPA (■) on T cell adhesion. T lymphocytes were stimulated in the presence of increasing concentrations of recombinant cyclophilins and allowed to adhere into ECM-coated plates. In both experiments, results are expressed as percentages of initially added T lymphocytes (1×10^6 per well) remaining associated to the ECM-coated plates. Data are means ± SD of quadriplicates and are representative of at least three separate experiments conducted with distinct donors.

bind to the same heparan sulfates, was ineffective for inducing the same cellular response. Moreover, a mutant of CyPB, unable to interact with heparan sulfates, was also unable to induce the adhesion of T lymphocytes (Figure 1B). These data demonstrated that CyPB binding to the signaling receptor is not efficient enough to induce this cell response, while selective interaction with GAGs probably accounts for the unique pro-adhesive activity of this inflammatory factor (Allain *et al.*, 2002).

A specific property of inflammatory proteins is to selectively stimulate different leukocyte subsets. To know whether CyPB shares this property, the phenotype of responsive T cells was analyzed. We found that CyPB was not significantly effective to trigger adhesion of $CD8^+$ T cells. In addition, native $CD4^+$ T lymphocytes were not responsive to CyPB, while the adhesion of memory $CD4^+$ T lymphocytes was strongly increased in the presence of the protein. CD147 is expressed on the whole T cell population (Koch *et al.*, 1999) and cannot explain preferential activity of CyPB on this T cell subset. In contrast, the differential expression of proteoglycans on T cells could account for the cell-specific pro-adhesive effect of CyPB.

4.3. Involvement of Syndecans in Pro-Adhesive Activity of CyPB

Data concerning the expression of proteoglycans on lymphocytes are missing and often controversial. The presence of HSPG on peripheral blood T lymphocytes was not clearly demonstrated (Jackson, 1997; Jackson *et al.*, 1995), but some findings argue for the presence of HSPG on these cells. For example, binding of HIV and activity of β-chemokines are strongly inhibited after cell membrane treatments with heparin lyases (Oravecz *et al.*, 1997; Patel *et al.*, 1993; Saphire *et al.*, 1999). This discrepancy is likely to be due to the low expression of HSPG on T cells and difficulties in detecting them by immunochemical approaches. To check the hypothesis on the role of proteoglycans in the activity of CyPB, we analyzed their expression on freshly isolated peripheral blood lymphocytes. We focused our attention on syndecans and CD44v3 variants. Indeed, the modular structural design of syndecans, that is conserved cytosolic domains but divergent extracellular domains, suggests that they have evolved to carry out specific functions, most likely related to the nature of their interactions with extracellular ligands (Rapraeger and Ott, 1998). This hypothesis is further supported by the highly regulated pattern of expression of specific syndecan types. The expression of CD44 isoforms is also highly regulated upon cell activation and differentiation. A growing interest has focused on the expression of the CD44v3 variants, because of their potential to interact with heparin-binding growth factors and cytokines (Jones *et al.*, 2000; van der Voort *et al.*, 1999).

By RT-PCR and flow cytofluorimetry, we demonstrated that peripheral blood lymphocytes express syndecan-1, -2 and -4, with a higher expression for syndecan-2 on T cells. In addition, the CD44v3 was only found expressed on a sub-population of lymphocytes, mainly corresponding to B cells. To know whether the cell response triggered by CyPB is related to the expression of one of these proteoglycans, T lymphocytes were separated according to their ability to interact with immobilized CyPB. Interestingly, we demonstrated that syndecan-2 is the main proteoglycan expressed by T cells bearing binding sites for CyPB. In the same way, we found that syndecan-2 is the major proteoglycan expressed by $CD4^+/CD45RO^+$ lymphocytes, strongly suggesting that binding and cell-specific activity of CyPB is related to the expression of this HSPG.

To check this hypothesis, we used the cellular model HL-60. Following differentiation in neutrophil-like cells by incubation for 5 days with DMSO, these cells express specific binding sites for CyPB, similar to the ones present on T lymphocytes, and are responsive to the pro-adhesive activity of the protein. Interestingly, we found that the main difference between undifferentiated and neutrophil-like cells is the higher expression of syndecan-1 and -2 on differentiated cells, while other HSPG are not significantly modified. In addition, we demonstrated that the binding of CyPB to cell surface heparan sulfates increases with differentiation and correlates with the overexpression of syndecan-1 and -2, suggesting that both proteoglycans carry the heparin-binding motif recognized by CyPB and play a critical role in the cell responsiveness to this inflammatory factor.

5. CONCLUSION

The findings that proteoglycans can bind and modulate the activity of inflammatory factors suggest that GAG mimetics have possible therapeutic applications as antagonists or agonists of particular activities (Delehedde *et al.*, 2002). Heparin and heparan sulfate chains are however potent anticoagulants, indicating that more minimal structures derived from heparin or other GAGs are required in order to achieve suitable benefits in the patient without side effects. In view of this, further studies are required to identify the structural GAG-binding motifs recognized by inflammatory factors and to make GAG-based therapeutic approaches more common.

6. ACKNOWLEDGMENTS

We are grateful to Maryse Delehedde and David Fernig (School of Biological Sciences, University of Liverpool, U.K.) for their contribution to structural analysis of heparin-binding motifs recognized by pro-inflammatory factors, and to Bernard Haendler (Schering AG, Berlin, Germany) for his help in site-directed mutagenesis of CyPB.

7. REFERENCES

Adams, D.H., Harvath, L., Bottaro, D.P., Interranted, R., Catalano, G., Tanaka, Y., Strain, A., Hubscher, S.G., and Shaw, S., 1994, Hepatocyte growth factor and macrophage inflammatory protein 1 beta: structurally distinct cytokines that induce rapid cytoskeletal changes and subset-preferential migration in T cells, *Proc Natl Acad Sci USA*. 91:7144–7148.

Allain, F., Denys, A., and Spik, G., 1994, Characterization of surface binding sites for cyclophilin B on a human tumor T-cell line, *J Biol Chem*. 269:16537–16540.

Allain, F., Durieux, S., Denys, A., Carpentier, M., and Spik, G., 1999, Cyclophilin B binding to platelets supports calcium-dependent adhesion to collagen, *Blood*. 94:976–983.

Allain, F., Vanpouille, C., Carpentier, M., Slomianny, M.-C., Durieux, S., and Spik, G., 2002, Interaction with glycosaminoglycans is required for cyclophilin B to trigger integrin-mediated adhesion of peripheral blood T lymphocytes to extracellular matrix, *Proc Natl Acad Sci USA*. 99:2714–2719.

Baggiolini, M., 1998, Chemokines and leukocyte traffic, *Nature*. 392:565–568.

Bukrinsky, M.I., 2002, Cyclophilins: unexpected messengers in intercellular communications, *Trends Immunol*. 23:323–325.

Campbell, J.J., Hedrick, J., Zlotnik, A., Siani, M.A., Thompson, D.A., and Butcher, E.C., 1998, Chemokines and the arrest of lymphocytes rolling under flow conditions, *Science*. 279:381–384.

Capila, I. and Linhardt, R.J., 2002, Heparin-protein interactions, *Angew Chem Int Ed.* 41:390–412.

Carey, D.J., Bendt, K.M., and Stahl, R.C., 1996, The cytoplasmic domain of syndecan-1 is required for cytoskeleton association but not detergent insolubility. Identification of essential cytoplasmic domain residues, *J Biol Chem.* 271:15253–15260.

Carpentier, M., Allain, F., Haendler, B., Denys, A., Mariller, C., Benaïssa, M., and Spik, G., 1999a, Two distinct regions of cyclophilin B are involved in the recognition of a functional receptor and of glycosaminoglycans on T lymphocytes, *J Biol Chem.* 274:10990–10998.

Carpentier, M., Descamps, L., Allain, F., Denys, A., Durieux, S., Kieda, C., Cecchelli, R., and Spik, G., 1999b, Receptor-mediated transcytosis of cyclophilin B through the blood-brain barrier, *J Neurochem.* 73:260–270.

Delaney, B., Koh, W.S., Yang, K.H., Strom, S.C., and Kaminski, N.E., 1993, Hepatocyte growth factor enhances B-cell activity, *Life Sci.* 53:89–93.

Delehedde, M., Allain, F., Payne, S.J., Borgo, R., Vanpouille, C., Fernig, D.G., and Deudon, E., 2002, Proteoglycans in inflammation, *Curr Med Chem.* 1:89–102.

Denys, A., Allain, F., Foxwell, B., and Spik, G., 1997, Distribution of cyclophilin B-binding sites in the subsets of human peripheral blood lymphocytes, *Immunology.* 91:609–617.

Denys, A., Allain, F., Carpentier, M., and Spik, G., 1998, Involvement of two classes of binding sites in the interactions of cyclophilin B with peripheral blood T-lymphocytes, *Biochem J.* 336:689–697.

Ebnet, K. and Vestweber, D., 1999, Molecular mechanisms that control leukocyte extravasation: the selectins and the chemokines, *Histochem Cell Biol.* 112:1–23.

Fischer, G., Tradler, T., and Zarnt, T., 1998, The mode of action of peptidyl prolyl cis/trans isomerase *in vivo*: binding vs. catalysis, *FEBS Lett.* 426:17–20.

Galat, A., 1999, Variations of sequences and amino acid compositions of proteins that sustain their biological functions: an analysis of the cyclophilin family of proteins, *Arch Biochem Biophys.* 371:149–162.

Gallagher, J.T., Turnbull, J.E., and Lyon, M., 1992, Patterns of sulphation in heparan sulphate: polymorphism based on a common structural theme, *Int J Biochem.* 24:553–560.

Gilat, D., Hershkoviz, R., Mekori, Y.A., Vlodavsky, I., and Lider, O., 1994, Regulation of adhesion of CD4+ T lymphocytes to intact or heparinase- treated subendothelial extracellular matrix by diffusible or anchored RANTES and MIP-1 beta, *J Immunol.* 153:4899–4906.

Gonzalez-Cuadrado, S., Bustos, C., Ruiz-Ortega, M., Ortiz, A., Guijarro, C., Plaza, J.J., and Egido, J., 1996, Expression of leucocyte chemoattractants by interstitial renal fibroblasts: up-regulation by drugs associated with interstitial fibrosis, *Clin Exp Immunol.* 106:518–522.

Grootjans, J.J., Zimmermann, P., Reekmans, G., Smets, A., Degeest, G., Durr, J., and David, G., 1997, Syntenin, a PDZ protein that binds syndecan cytoplasmic domains, *Proc Natl Acad Sci USA.* 94:13683–13688.

Handschumacher, R.E., Harding, M.W., Rice, J., Drugge, R.J., and Speicher, D.W., 1984, Cyclophilin: a specific cytosolic binding protein for cyclosporin A, *Science.* 226:544–547.

Hughes, P.E. and Pfaff, M., 1998, Integrin affinity modulation, *Trends Cell Biol.* 8:359–364.

Ilangumaran, S., Borisch, B., and Hoessli, D.C., 1999, Signal transduction via CD44: role of plasma membrane microdomains, *Leuk Lymphoma.* 35:455–469.

Jackson, D.G., Bell, J.I., Dickinson, R., Timans, J., Shields, J., and Whittle, N., 1995, Proteoglycan forms of the lymphocyte homing receptor CD44 are alternatively spliced variants containing the v3 exon, *J Cell Biol.* 128:673–685.

Jackson, D.G., 1997, Human leucocyte heparan sulphate proteoglycans and their roles in inflammation, *Biochem Soc Transactions.* 25:220–224.

Jiang, W., Puntis, M.C., Nakamura, T., and Hallett, M.B., 1992, Neutrophil priming by hepatocyte growth factor, a novel cytokine, *Immunology.* 77:147–149.

Jinquan, T., Larsen, C.G., Gesser, B., Matsushima, K., and Thestrup-Pedersen, K., 1993, Human IL-10 is a chemoattractant for CD8+ T lymphocytes and an inhibitor of IL8-induced CD4+ T lymphocyte migration, *J Immunol.* 151:4545–4551.

Jones, M., Tussey, L., Athanasou, N., and Jackson, D.G., 2000, Heparan sulfate proteoglycan isoforms of the CD44 hyaluronan receptor induced in human inflammatory macrophages can function as paracrine regulators of fibroblast growth factor action, *J Biol Chem.* 275:7964–7974.

Kinnunen, A., Kinnunen, T., Kaksonen, M., Nolo, R., Panula, P., and Rauvala, H., 1998, N-syndecan and HB-GAM (heparin-binding growth-associated molecule) associate with early axonal tracts in the rat brain, *Eur J Neurosci.* 10:635–648.

Koch, C., Staffler, G., Huttinger, R., Hilgert, I., Prager, E., Cerny, J., Steinlein, P., Majdic, O., Horejsi, V., and Stockinger, H., 1999, T cell activation-associated epitopes of CD147 in regulation of the T cell response, and their definition by antibody affinity and antigen density, *Int Immunol.* 11:777–786.

Kusano, Y., Oguri, K., Nagayasu, Y., Munesue, S., Ishihara, M., Saiki, I., Yonekura, H., Yamamoto, H., and Okayama, M., 2000, Participation of syndecan-2 in the induction of stress fiber formation in cooperation with integrin alpha5beta1: structural characteristics of heparan sulfate chains with avidity to COOH-terminal heparin-binding domain of fibronectin, *Exp Cell Res.* 256:434–444.

Kuschert, G.S., Coulin, F., Power, C.A., Proudfoot, A.E., Hubbard, R.E., Hoogewerf, A.J., and Wells, T.N., 1999, Glycosaminoglycans interact selectively with chemokines and modulate receptor binding and cellular responses, *Biochemistry.* 38:12959–12968.

Marks, A.R., 1996, Cellular functions of immunophilins, *Physiol Rev.* 76:631–649.

Meager, A., 1999, Cytokine regulation of cellular adhesion molecule expression in inflammation, *Cytokine Growth Factor Rev.* 10:27–39.

Moore, K.W., O'Garra, A., de Waal Malefyt, R., Vieira, P., and Mosmann, T.R., 1993, Interleukin-10, *Annu Rev Immunol.* 11:165–190.

Moser, B., Loetscher, M., Piali, L., and Loetscher, P., 1998, Lymphocyte responses to chemokines, *Int Rev Immunol.* 16:323–344.

Norgard-Summicht, K. and Varki, A., 1995, Endothelial heparan sulfate proteoglycans that bind to L-Selectin have glucosamine residues with unsubstituted amino groups, *J Biol Chem.* 270:12012–12024.

Oh, E.S., Couchman, J.R., and Woods, A., 1997, Serine phosphorylation of syndecan-2 proteoglycan cytoplasmic domain, *Arch Biochem Biophys.* 344:67–74.

Oravecz, T., Pall, M., Wang, J., Roderiquez, G., Ditto, M., and Norcross, M.A., 1997, Regulation of anti-HIV-1 activity of RANTES by heparan sulfate proteoglycans, *J Immunol.* 159:4587–4592.

Patel, M., Yanagishita, M., Roderiquez, G., Bou-Habib, D.C., Oravecz, T., Hascall, V.C., and Norcross, M., 1993, Cell-surface heparan sulfate proteoglycan mediates HIV-1 infection of T-cell lines, *AIDS Res Hum Retrov.* 9:167–174.

Pushkarsky, T., Zybarth, G., Dubrovsky, L., Yurchenko, V., Tang, H., Guo, H., Toole, B., Sherry, B., and Bukrinsky, M., 2001, CD147 facilitates HIV-1 infection by interacting with virus-associated cyclophilin A, *Proc Natl Acad Sci USA.* 98:6360–6365.

Rapraeger, A.C. and Ott, V.L., 1998, Molecular interactions of the syndecan core proteins, *Curr Opin Cell Biol.* 10:620–628.

Rousset, F., Garcia, E., Defrance, T., Peronne, C., Vezzio, N., Hsu, D.H., Kastelein, R., Moore, K.W., and Banchereau, J., 1992, Interleukin 10 is a potent growth and diffentiation factor for activated human B lymphocytes, *Proc Natl Acad Sci USA.* 89:1890–1893.

Salek-Ardakani, S., Arrand, J.R., Shaw, D., and Mackett, M., 2000, Heparin and heparan sulfate bind interleukin-10 and modulate its activity, *Blood.* 96:1879–1888.

Saoncella, S., Echtermeyer, F., Denhez, F., Nowlen, J.K., Mosher, D.F., Robinson, S.D., Hynes, R.O., and Goetinck, P.F., 1999, Syndecan-4 signals cooperatively with integrins in a Rho-dependent manner in the assembly of focal adhesions and actin stress fibers, *Proc Natl Acad Sci USA.* 96:2805–2810.

Saphire, A.C., Bobardt, M.D., and Gallay, P.A., 1999, Host cyclophilin A mediates HIV-1 attachment to target cells via heparans, *EMBO J.* 18:6771–6785.

Schreiber, S.L., 1991, Chemistry and biology of the immunophilins and their immunosuppressive ligands, *Science.* 251:283–287.

Seki, T., Ihara, I., Sugimura, A., Shimonishi, M., Nishizawa, T., Asami, O., Hagiya, M., Nakamura, T., and Shimizu, S., 1990, Isolation and expression of cDNA for different forms of hepatocyte growth factor from human leukocyte, *Biochem Biophys Res Commun.* 172:321–327.

Sherry, B., Yarlett, N., Strupp, A., and Cerami, A., 1992, Identification of cyclophilin as a proinflammatory secretory product of lipopolysaccharide-activated macrophages. *Proc Natl Acad Sci USA.* 89:3511–3515.

Sherry, B., Zybarth, G., Alfano, M., Dubrovsky, L., Mitchell, R., Rich, D., Ulrich, P., Bucala, R., Cerami, A., and Bukrinsky, M., 1998, Role of cyclophilin A in the uptake of HIV-1 by macrophages and T lymphocytes, *Proc Natl Acad Sci USA.* 95:1758–1763.

Shukla, D., Liu, J., Blaiklock, P., Shworal, N.W., Bai, X., Esko, J.D., Cohen, G.H., Eisenberg, R.J., Rosenberg, R.D., and Spear, P.G., 1999, A novel role for 3-O-sulphated heparan sulfate in Herpes simplex virus 1 entry, *Cell.* 99:13–22.

Silbert, J.E., Bernfield, M., and Konenyesi, R., 1997, Proteoglycans: a special class of glycoproteins, Chapter 1, *Glycoproteins II*, Elsevier Science BV.

Spik, G., Haendler, B., Delmas, O., Mariller, C., Chamoux, M., Maes, P., Tartar, A., Montreuil, J., Stedman, K., Kocher, H.P., Keller, R., Hiestand, P.C., and Movva, N.R., 1991, A novel secreted cyclophilin-like protein (SCYLP), *J Biol Chem.* 266:10735–10738.

Stringer, S.E. and Gallagher, J.T., 1997, Specific binding of the chemokine platelet factor 4 to heparan sulfate, *J Biol Chem.* 272:20508–20514.

Tanaka, Y., Adams, D.H., and Shaw, S., 1993, Proteoglycans on endothelial cells present adhesion-inducing cytokines to leukocytes, *Immunol Today.* 14:111–115.

Tegeder, I., Schumacher, A., John, S., Geiger, H., Geisslinger, G., Bang, H., and Brune, K., 1997, Elevated serum cyclophilins levels in patients with severe sepsis, *J Clinical Immunol.* 17:380–386.

te Velde, A.A., de Waal Malefijt, R., Huijbens, R.J., de Vries, J.E., and Figdor, C.G., 1992, IL-10 stimulates monocyte Fc gamma R surface expression and cytotoxic activity. Distinct regulation of antibody-dependent cellular cytotoxicity by IFN-gamma, IL-4 and IL-10, *J Immunol.* 149:4048–4052.

Turnbull, J., Powell, A., and Guimond, S., 2001, Heparan sulfate: decoding a dynamic multifunctional cell regulator, *Trends Cell Biol.* 11:75–82.

Van der Voort, R., Taher, T.E., Wielenga, V.J., Spaargaren, M., Prevo, R., Smit, L., David, G., Hartmann, G., Gherardi, E., and Pals, S.T., 1999, Heparan sulfate-modified CD44 promotes hepatocyte growth factor/scatter factor-induced signal transduction through the receptor tyrosine kinase c-Met, *J Biol Chem.* 274:6499–6506.

Volk, R., Schwartz, J.J., Li, J., Rosenberg, R.D., and Simons, M., 1999, The role of syndecan cytoplasmic domain in basic fibroblast growth factor dependent signal transduction, *J Biol Chem.* 274:24417–24424.

Wagner, L., Yang, O.O., Garcia-Zepeda, E.A., Ge, Y., Kalams, S.A., Walker, B.D., Pasternack, M.S., and Luster, A.D., 1998, b-chemokines are released from HIV-1-specific cytolitic T-cell granules complexed to proteoglycans, *Nature* 391:908–911.

Webb, L.M., Ehrengruber, M.U., Clark-Lewis, I., Baggiolini, M., and Rot, A., 1993, Binding to heparan sulphate or heparin enhances neutrophil responses to interleukin 8, *Proc Natl Acad Sci USA.* 90:7158–7162.

Xu, Q., Leiva, M.C., Fischkoff, S.A., Handschumacher, R.E., and Lyttle, C.R., 1992, Leukocyte chemotactic activity of cyclophilin, *J Biol Chem.* 267:11968–11971.

Yurchenko, V., O'Connor, M., Dai, W.W., Guo, H., Toole, B., Sherry, B., and Bukrinsky, M., 2001, CD147 is a signaling receptor for cyclophilin B. *Biochem Biophys Res Commun.* 288:786–788.

Yurchenko, V., Zybarth, G., O'Connor, M., Dai, W.W., Franchin, G., Hao, T., Guo, H., Hung, H.C., Toole, B., Gallay, P., Sherry, B., and Bukrinsky, M., 2002, Active site residues of cyclophilin A are crucial for its signaling activity via CD147, *J Biol Chem.* 277:22959–22965.

8

LECTIN DOMAINS ON CYTOKINES

Jean-Pierre Zanetta and Gérard Vergoten

CNRS Unité Mixte de Recherche 8576
Laboratoire de Glycobiologie Structurale et Fonctionnelle
Université des Sciences et Technologies de Lille
Bâtiment C9, 59655 Villeneuve d'Ascq Cedex (France)

1. INTRODUCTION

There is increasing evidence that carbohydrate-binding proteins (lectins) are widespread in mammals (Kilpatrick, 2002), and especially in cells of the immune system. Several tenths calcium-dependent lectins (C-type lectins; Dodd and Drickamer, 2001; Drickamer, 1997) have been identified each endowed with a different carbohydrate-binding specificity. For example, L-selectin (Foxall *et al.*, 1992; Lasky, 1992) recognizes the 6'-sulphated sialyl-Lewis^x epitope. Mannan-binding lectins constitute a family of related calcium-dependent lectins with important implications in the field of human pathology (Kawasaki, 1999; Kilpatrick, 2002). Likewise the calcium-independent lectins comprise the lactose-binding galectins (Cooper, 2002), the I-type lectins (Angata and Brinkman-van der Linden, 2002; Kelm *et al.*, 1994; Nath *et al.*, 1995; Powell and Varki, 1994) and CSL, which is specific for oligomannosidic N-glycans with 6 mannose residues (Zanetta *et al.*, 1987). The MR60-ERGIC-53 mannose-binding lectin (Arar *et al.*, 1995; Fiedler and Simmons, 1994) is related to plant lectins. Besides these molecules involved either in homotypic or heterotypic cell adhesion, or in intracellular traffic of glycoconjugates, those that are polyvalent (having two carbohydrate-recognition domains (CRD) or organized as oligomers) may have clustering effects on their ligands. This may represent the endogenous mechanism mimicked by specific polyvalent antibodies or plant lectins (Feizi and Childs, 1987). For example, the lectin CSL, which is rapidly over-expressed and externalized after stimulation of human cells with phorbol esters and binds to its surface ligands (including the glycosylated forms of CD3 on T cells and CD24 on B cells) is responsible for the major tyrosine-phosphorylation changes occurring in the early stages of cell activation (Zanetta *et al.*, 1995).

Such a role of soluble extracellular lectins in signal transduction raises the question of how small soluble proteins, through their binding to a specific receptor, may generate

Glycobiology and Medicine, edited by John S. Axford
Kluwer Academic / Plenum Publishers, New York, 2003

specific intracellular signals. It has been frequently observed that when a cytokine binds to its receptor, the receptor intracytoplasmic domain becomes phosphorylated/dephosphorylated. This change in phosphorylation could not be explained simply by the putative oligomerization of the cytokine since these intracytoplasmic domains are frequently devoid of kinase/phosphatase activities. Indeed, the intracellular change in phosphorylations is due to enzymes associated with another surface receptor complex. When phosphorylation studies are performed on an intact cell, the change in phosphorylations is due to a unique kinase/phosphatase, suggesting that, rather than unspecific conformational changes of the receptor intracytoplasmic domain upon cytokine binding, the cytokine itself would be responsible for the specific extracellular association of its receptor to another surface complex.

Cytokines are soluble factors mediating communication between cells of both immune and non immune systems. Biological responses to cytokines involve specific cellular receptors associated with other transducing molecules transmitting the biological signal induced by the cytokine to the nucleus (Kishimoto *et al.*, 1992a). Phosphorylated receptor then initiates a phosphorylation/dephosphorylation cascade leading to the biological response. The common view that the cytokine binding to its receptor induces a conformational change of the intracytoplasmic domain of the receptor can be successfully challenged considering several major points. One cytokine can act differently depending on the target cell. Interleukin-6 is very typical of this view, since this cytokine can exert its biological functions on a wide range of target cells (Kishimoto *et al.*, 1992b). In contrast, different cytokines can act on same cells to induce similar effects. Understanding of these features of cytokines can be only allowed by considering cytokines as bi-functional molecules, able to extracellularly associate their receptors with other shared surface molecular complexes bearing a ligand of the cytokine. This view on the bi-functionality of cytokines is supported by numerous site-directed mutagenesis studies and domain-specific antibodies, indicating that besides a receptor-binding domain (RBD), the cytokines possess another domain essential for the cytokine-dependent signalling (Somers *et al.*, 1997). Subtle changes in this second domain did not modify the receptor-binding domain, but suppressed the biological activity. In the three-dimensional structure of cytokines, this second domain was localized at the opposite of the receptor-binding site (Wells *et al.*, 1994). Several authors observed that inhibitors of glycosylation inhibit ligand binding and/or signal transduction induced by cytokines, thus defining glycosylation as a key factor in the signal transduction induced by cytokines (Ding *et al.*, 1995; Mancilla *et al.*, 1992; Niu *et al.*, 2000; Schaaf-Lafontaine *et al.*, 1985; Shibuya *et al.*, 1991; Wall *et al.*, 1988). The biological significance of these observations could be easily explained considering the second domain of the cytokines as a carbohydrate-recognition domain (CRD).

Several authors already described such lectin-like activities for cytokines, including IL-1α and IL-1β, IL-2, TNF-α and TNF-β (Brody and Durum, 1989; Fukushima *et al.*, 1993 and 1997; Hession *et al.*, 1987; Muchmore and Decker, 1987; Sherblom *et al.*, 1988; Winkelstein *et al.*, 1990; Zanetta *et al.*, 1996). However, in most cases, the cytokines presented low affinities for the identified oligosaccharide ligands, which were generally common oligosaccharide structures in human. This low affinity and the tissue abundance of the identified ligands did not defend the concept of an important role of these lectin-like activities. Since most of the previous studies on the lectin activities of cytokines were performed by measuring the binding of ^{125}I-labelled cytokines in solid-phase binding assays, and since cytokines could lose their lectin activity upon chemical or radiochemical labelling (Kaplan *et al.*, 1995), previous results were re-examined using a new method allowing the screening of lectin activities of cytokines with proteins having a fully preserved biological activity. Using this method (Cebo *et al.*, 2001), it was possible to attribute specific carbohydrate-binding properties and identify high affinity ligands for several cytokines, including IL-1α, IL-1β, IL-2, IL-4, IL-6 and IL-7 (Cebo *et al.*, 2001) and more recently, to IL-3 (Zanetta *et al.*, 2002).

2. BIOLOGICAL FUNCTION OF THE LECTIN ACTIVITY OF IL-2

Interleukin 2 (IL-2) is a molecule produced essentially by activated CD4+ T cells, which can stimulate the proliferation of T cells, induce the cytotoxicity of CD8+ cells and the activation of B cells. Three IL-2 receptors (IL-2R) have been identified (α, β, γ), only the IL-2Rβ being constitutively expressed on resting cells (Zola *et al.*, 1991) and receiving first the IL-2 signal. Studies on the signal transduction pathway resulting from IL-2 binding (Taniguchi and Minami, 1993; Waldmann, 1991) showed that, although IL-2Rβ has no kinase activities, its intracytoplasmic domain is tyrosine-phosphorylated upon IL-2 binding by the $p56^{lck}$ kinase (Farrar *et al.*, 1990; Hatakeyama *et al.*, 1991; Shibuya *et al.*, 1994). Since this kinase is associated with the CD3/TCR complex, it is suggested that when IL-2 binds to its receptor, IL-2Rβ and the CD3/TCR complex become associated. After tyrosine phosphorylation, $p56^{lck}$ binds through an SH2 domain specific of the *src* family kinases to a short sequence of the IL-2Rβ receptor containing the phosphotyrosine residue and remains firmly attached to the IL-2Rβ.

The common hypothesis to account for the interaction between $p56^{lck}$ and the IL-2Rβ receptor is that IL-2 binding initiates a conformational change of the intracytoplasmic domain of IL-2Rβ, providing a site for tyrosine phosphorylation. However, this could neither explain why only $p56^{lck}$ recognizes the intracytoplasmic domain of IL-2Rβ when experiments are performed on an intact cell nor data acquired by site-directed mutagenesis of IL-2. Indeed, it was demonstrated (Cohen *et al.*, 1986; Ju *et al.*, 1987) that IL-2 needs two domains for expressing its full biological activity: one involved in the binding to its receptors, and the other necessary for the expression of the biological activity. Furthermore, the use of domain-specific antibodies reinforced the concept of two domains, suggesting that *in vivo*, IL-2 behaves as a bi-functional molecule. However, the nature of the second site indispensable for the biological activity remained uncertain.

IL-2 was first described as a lectin specific for oligomannosidic N-glycans with 5 and 6 mannose residues but not oligomannosides with a higher number of mannose residues or glucosyl-oligomannosides (Fukushima and Yamashita, 2001; Sherblom *et al.*, 1989; Zanetta *et al.*, 1998a). Using a complex experimental design (Zanetta *et al.*, 1996), it was shown, on resting human lymphocytes, that in the presence of IL-2, and only in its presence, an anti-TCR antibody co-immunoprecipitated IL-2Rβ. IL-2Rβ was released from the complex in a mechanism independent of the oligomannoside with 9 mannose residues, but dependent on oligomannosides with 5 and 6 mannose residues. Moreover, this specifically released IL-2Rβ was found to be associated with the $p56^{lck}$ kinase, verifying the strong association between IL-2Rβ and $p56^{lck}$. This confirmed that the lectin

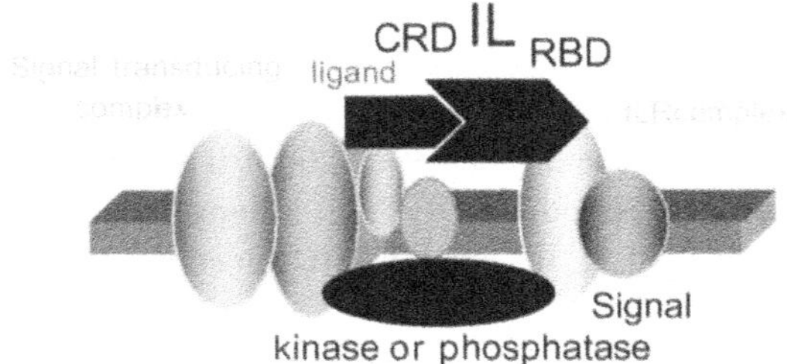

Figure 1. Hypothetical scheme on the biological function of the bi-functional interleukins. The Interleukin (IL) bound to its receptor complex (ILRcomplex) through its receptor-binding domain (RBD) associates extracellularly with the latter through the interaction of its carbohydrate-recognition domain (CRD) with its oligosaccharide ligand (ligand) with another surface complex (Signal-transducing complex). The action of specific intracellular kinases or phosphatases generates a signal specific of the interleukin.

activity of IL-2 provokes an extracellular association between IL-2Rβ and glycoprotein constituents of the TCR complex. One of the ligands of IL-2 in T cells was a glycosylated form of CD3. Because it is actually bi-functional, IL-2 is able to recognize specific oligomannosides at the cell surface leading to specific intracytoplasmic associations. The specific phosphorylation of IL-2Rβ by $p56^{lck}$ can be explained because of the specific carbohydrate-dependent association of IL-2Rβ with CD3/TCR. Thus, these experiments demonstrate that an extracellular glycobiological interaction, which occurs *in vivo*, can modify specifically the intracellular organization of molecules. Therefore, it was hypothesized that this mechanism evidenced for IL-2 could be generalized to other interleukins (Figure 1).

The carbohydrate-binding properties of IL-2 for oligomannosidic N-glycans with 5 and 6 mannose residues allowed proposing a new understanding in the field of immunodeficiency. Based on knockout experiments of the gene of IL-2 (Horak, 1995), the pattern of an IL-2-dependent immunodeficiency was characterized. These symptoms resembled those found in human or animal diseases, which include α-mannosidosis, candidiasis, cancer and AIDS (Zanetta *et al.*, 1998a, b). For example, although the most abundant structures of microorganism cell wall were not potential IL-2 ligands, the *Candida albicans* cell wall glycans bond (specifically compared to other yeast strains) huge amounts of IL-2 in a mechanism dependent of oligomannosides with five and six mannose residues and independent of the oligomannoside with nine mannose residues.

3. IL-1α AND IL-1β EXHIBIT DISTINCT CARBOHYDRATE-BINDING PROPERTIES

IL-1 is a pro-inflammatory cytokine exhibiting a wide range of biological effects including fever induction, prostaglandins synthesis and T-lymphocyte activation (Dinarello, 1994). In fact, the IL-1 family consists of three different members, named IL-1 alpha (IL-1α), IL-1 beta (IL-1β) and IL-1 receptor antagonist (IL-1Rα) (Carter *et al.*, 1990). Two IL-1 receptors have been identified. All three IL-1 can bind to the two IL-1 receptors but produce distinct biological responses, thus asking again the question of how an unspecific conformational change in the intracellular domain of the receptor upon ligand binding can lead to such divergent signalling pathways. In contrast, these observations can be explained considering cytokines as bi-functional molecules, the specific lectin activity of each IL-1 family member associating the common receptor with a particular transducing complex, thus leading to distinct signalling pathways. Evidence that IL-1α was a carbohydrate-binding molecule was clearly demonstrated by Muchmore and Decker (1986). Indeed, the oligosaccharides isolated by pronase digestion of uromodulin were able to inhibit T-cell proliferation induced by the cytokine, indicating that the biological function of IL-1α was supported by its lectin-like activity (Muchmore and Decker, 1987). However, the identified putative ligand of IL-1α, the tri-antennary N-glycan of fetuin, was only effective at the 0.1 mM range. In a recent study (Cebo *et al.*, 2001), it was demonstrated that the IL-1α ligand was not the previous compound (ineffective), but the disialylated di-antennary N-glycan bearing two Neu5Acα2,3 residues. The corresponding mono-antennary N-glycan as well as the di-antennary bearing either one or two Neu5Acα2,6 residues was also ineffective, as were shorter oligosaccharides like sialyllactoses. Therefore, IL-1α appeared as a lectin specific for a di-antennary glycan very rare in human tissues.

IL-1β was suggested to bind mannose 6-phosphate diester present in some GPI-anchored proteins (Fukushima *et al.*, 1997). However, this was not verified and, in contrast (Cebo *et al.*, 2001), it was demonstrated that IL-1β interacted specifically with the GM_4 glycolipid isolated from rat brain myelin. The binding of IL-1β to GM_4 could not be reversed using sialyllactoses, in such a way that the identification of an oligosaccharide inhibitor was still pending. However, it may be speculated that the IL-1β ligand includes also the polar head of the long-chain bases of this GM4, rich in C22 sphingenine and phytosphingosine.

Although the biological function of the lectin-like activities of these two cytokines have not been yet studied, the fact that the two IL-1 have different oligosaccharide ligands could explain why these two compounds, which have the same receptors, have different signalling profiles and different target cells. The specificity of IL-1β for GM_4 may open new fields in the understanding of its function. Indeed, based on immunohistochemical studies on the rat cerebellum (Ozawa *et al.*, 1993), GM_4 was found only as a constituent of astrocytes. Based on its localization in the cerebellum, it may be suggested that these astrocytes are type-II astrocytes, the ones surrounding the Ranvier's nodes. Interestingly, astrocytes are the only cells in the central nervous system (CNS) producing IL-1β and possessing the IL-1β receptor. IL-1β could act in an autocrine as well as in a paracrine way. Therefore, it may be suggested that the lectin activity of IL-1β induces the association of its receptor with other surface molecular complexes containing GM_4. It is noteworthy that IL-1β is the cytokine responsible for fever (mouse with a knock-out of the gene for IL-1 did not produce fever (Licinio and Wong, 1996), a mechanism directed by the CNS. The question remains to know if GM_4 is actually present at the surface of specific cells of the immune system.

4. LECTIN ACTIVITY OF IL-4

IL-4 is a cytokine regulating major functions of B lymphocytes, including proliferation and production of IgE antibodies by activated cells, expression of class II MHC antigens and low affinity IgE receptor (CD23) on resting B cells. IL-4 also induces the proliferation of activated T cells and differentiation of naive T lymphocytes into Th2 cells (Paul, 1991). The high affinity IL-4 receptor is an heterodimer of a ligand binding subunit, named IL-4Rα/CD124, and a transducing molecule, the common gamma chain (CD132) shared by other cytokine signalling complexes including those for IL-2, IL-7, IL-9 and IL-15 (Takeshita *et al.*, 1992). Upon ligand binding, the Jak-Stat and the PI3 kinase pathways are initiated thus leading to the biological response to the cytokine (Nelms *et al.*, 1999).

No evidence was previously provided for a lectin activity of IL-4. Nevertheless, several authors showed IL-4 binding to a highly glycosylated protein called IL-4BP (IL-4 Binding Protein) present in human urine and biological fluids of mice. IL-4BP inhibits IL-4- but not IL-2- dependent T-cell proliferation, thus demonstrating that the transduction pathways initiated by IL-4 were specifically concerned (Christie *et al.*, 1995; Fernandez Botran and Vitetta, 1990). Similarly, IL-4 bound to some gangliosides and, especially a GM_3 ganglioside isolated from melanoma (Chu and Sharom, 1995). However, authors did not provide evidence that this binding was carbohydrate-dependent. Our recent studies defined the 1,7 lactone of N-acetyl-neuraminic acid as a high-affinity ligand of IL-4 (Cebo *et al.*, 2001). This compound was never detected before in biological fluids or tissues and was evidenced using a new GC/MS method of analysis of sialic acid (Zanetta *et al.*, 2001). It was found to be present as a very minor sialic acid in normal human tissues (including normal human lymphocytes) but was the dominant sialic acid found in some malignant tumors or cells (unpublished data). This 1,7 lactone had a very strong stability to alkaline and acidic treatment as compared with the other lactones, and could have been confused with N-glycolyl-neuraminic acid in some occasions. The abundance of this lactone in mucins from the eggs of *Bufo bufo* (Zanetta *et al.*, 2001) allowed isolating this compound for testing the biological activity of this lectin activity of IL-4.

The addition of the lactone inhibited the tyrosine dephosphorylation and the serine/threonine phosphorylation of a 63 kDa protein (p63; Cebo *et al.*, submitted), indicating that the early signal transduction events initiated by IL-4 on resting lymphocytes (Huang and Paul, 2000; Imani *et al.*, 1997; Kolb and Abadie, 1993) were dependent on its lectin activity.

5. FUNCTION OF THE LECTIN ACTIVITY OF IL-6

Interleukin 6 (IL-6) is a pleiotropic cytokine showing essential roles in immunity, haematopoiesis and inflammation (Akira *et al.*, 1990; Fukada *et al.*, 1999; Hirano *et al.*, 1997). The detailed mechanism by which the IL-6 binding to its receptor (IL-6Rα) generates a signal remains partially understood, although this signal takes place through a "signal-transducing molecule," the gp130 glycoprotein in healthy human (Fukada *et al.*, 1999). This molecule is considered as a second receptor (IL-6Rβ) by several authors (Savino *et al.*, 1994; Somers *et al.*, 1997), the binding of IL-6 to IL-6Rα provoking the association of the latter with gp130. Putative domains of direct interactions between these molecules have been described based on X-ray crystallographic data of IL-6 and molecular modelling of the interaction of IL-6 with the extracellular domains of IL-6Rα and of gp130 (Somers *et al.*, 1997) and on site directed mutagenesis experiments in various sub-domains of the molecule (Fontaine *et al.*, 1993, 1994; Leebeck *et al.*, 1992; Li *et al.*, 1993;

Paonessa *et al.*, 1995; Savino *et al.*, 1993, 1994; Yasueda *et al.*, 1992). These data did not take into account the possibility that IL-6 could be, as other interleukins (Cebo *et al.*, 2001), a bi-functional molecule having, beside a receptor-binding domain, a carbohydrate-recognition domain (CRD). As demonstrated (Cebo *et al.*, 2001), IL-6 is a lectin specific for O- and N-glycans possessing the HNK-1 epitope HO_3S-3-GlcAβ1,3Galβ1,4. Using very high affinity oligosaccharide ligands isolated from the eggs of *Rana temporaria*, we were able to demonstrate that the lectin activity of IL-6 is necessary for the initial signal transduction due to IL-6. Indeed, the oligosaccharide ligands (at less than the nanomolar range for the higher affinity ligand) were able to completely inhibit the dephosphorylation of phospho-tyrosine residues of a few proteins induced by IL-6 on resting human lymphocytes (Cebo *et al.*, 2002). Furthermore, gp130, the signal-transducing molecule of the IL-6 system was bearing itself a N-glycan endowed with a HNK-1 epitope (Cebo *et al.*, 2002). Consequently, as observed for IL-2, the lectin activity of IL-6 makes this molecule bi-functional, this bi-functionality allowing the specific extracellular association of IL-6Rα with the signal-transducing complex containing gp130.

6. LECTIN ACTIVITIES OF OTHER CYTOKINES

Several other cytokines were described as lectins. Cebo *et al.* (2001) demonstrated that IL-7 binds specifically to the ovine submandibular mucin (OSM), a mucin considered as rich in the sialyl-Tn antigen, but not to glycosaminoglycans as suggested previously (Clarke *et al.*, 1995). In fact the binding of IL-7 to OSM was specifically inhibited by a small size glycopeptide isolated by extensive pronase digestion of OSM. The overall composition of this glycopeptide showed the presence of 1 Neu5Ac, 1 GalNAc and 1 Ser residue, suggesting that it corresponded to the sialyl-Tn antigen, an onco-fetal antigen extremely rare in healthy human tissues. Recently (Zanetta *et al.*, 2002), we observed that IL-3 recognized specifically a proteoglycan isolated from adult rat brains termed as PGS3 (Normand *et al.*, 1988), but did not recognize either the common commercially available glycosaminoglycans, or other proteoglycans isolated from the same source. The chemical analyses of this compound indicated an important over-sulfatation of the glycosaminoglycan part that was not observed in the other compounds. 1H-NMR analysis indicated that the high affinity ligand of IL-3 was the glycosaminoglycan composed of the disaccharide unit GlcA(2S)β1,3GalNAc(4S)β1.

The biological relevance of our findings is sustained by the fact that cells sensitive to IL-3 possess a relatively high level of di-sulfated chondroitin sulfate chains (although the higher affinity ligand proposed here has not been so far identified in these cells). A further argument comes from the brain immuno-localization of PGS3. Indeed, this compound presents a preferential localization on a few neurons of the cerebellum and of the forebrain. This localization has to be related to the RT-PCR and immunohistochemical studies showing the presence of IL-3Rα in the central nervous system, with a preferential localization in large cholinergic neurons (Kamegai *et al.*, 1990; Konishi *et al.*, 1995). Since it has been demonstrated that IL-3 has a neurotrophic effect on these neurons (increases in acetylcholinesterase and choline-acetyltransferase), it may be speculated that the IL-3 signalling occurs through a similar mechanism as that described for the other interleukins, that is the extracellular association of the IL-3Rα with its chondroitin sulfate proteoglycan ligand due to the bi-functional IL-3, allowing specific phosphorylation/dephosphorylation mechanisms.

Throughout 13 different cytokines tested for their lectin activities in our system, seven showed high affinity bindings to specific glycoconjugates. This did not necessarily mean that the others are not endowed with such activities. It should be considered that the number of immobilized glycoconjugates used in the initial study (Cebo *et al.*, 2001) was relatively reduced as compared with the extreme possibilities of oligosaccharide structures found in nature, and that the proper high affinity ligand was not present in the population of immobilized glycoconjugates. Because of the absence of precise data in literature, the methodology used for these studies was to examine the binding of different cytokines to heterogeneous immobilized glycoconjugates: fetuin, a mixture of bovine ribonuclease B and of bovine lactotransferrin, ovalbumin, a mixture of mucins (ovine and the equine submaxillary and the mucins from the eggs of *Bufo bufo*), a glycosaminoglycan mixture, a sciatic nerve SDS extract, a mixture of gangliosides and a mixture of neutral lipids from the human meconium.

The lectin activity of IL-4 for the 1,7 lactone of Neu5Ac would have been missed, if the mucins of the eggs of *B. bufo* were not used as immobilized putative ligands. Similarly, the specificity of IL-3 would have been missed in the absence of a specific brain proteoglycan as an immobilized ligand. Therefore, before concluding to the absence of carbohydrate-binding properties for cytokines, it should be considered, as observed in all cases described above, that the high affinity ligands were always rare structures in humans and in mammals.

7. DEFINITION OF THE CRD OF CYTOKINES

The CRD is defined as a domain of the cytokines different from that of the receptor-binding domain (RBD), generally localized at the opposite of the molecule, not involved in the receptor binding but necessary for the signalling. This might be evident considering the site-directed mutagenesis experiments published in literature. Unfortunately, most of the site-directed mutagenesis experiments published in literature did not take into account the possibility of a CRD for cytokines, in such a way that amino acids, generally important for defining a CRD were not taken as essential targets for mutagenesis. For example, the CRD of most lectins so far co-crystallized with oligosaccharide ligands involve aromatic amino acids localized in an external water-accessible position. Unfortunately, these amino acids were generally considered as internal hydrophobic constituents involved in the stabilization of the three-dimensional structure of the molecule. Nevertheless, the increasing number of three-dimensional structures of cytokines and of lectins, the latter being co-crystallized with their ligands emphasizes the role of these amino acids (Trp, Tyr, Phe). The importance of these aromatic amino acids as a template for carbohydrate binding has been clearly demonstrated for galectin-2 (co-crystallized with lactose), the pyranic ring of Gal interacting by strong van der Waals interactions with the aromatic ring of Trp65 (Lobsanov *et al.*, 1993; Siebert *et al.*, 1997). Site directed mutagenesis indicated that the CRD is preserved only if Trp is replaced by another aromatic amino acid, although its replacement by Tyr or Phe decreases the affinity by a factor of ten. In *pertussis* toxin co-crystallized with its ligand Neu5Acα2,6Gal, the aromatic amino acids involved as template of the binding are two water-exposed Tyr residues (Stein *et al.*, 1994) providing a cradle for the pyranic ring of Neu5Ac. This scheme of the CRD of calcium-independent lectins (Kilpatrick, 2002) seems also preserved in viruses, as in the *influenza* virus hemagglutinins (Sauter *et al.*, 1992) and plant lectins (Lescar *et al.*, 2002). The specificity of binding is due to the formation of several hydrogen bonds concerned with the monosaccharide defining the crude specificity of the lectin (Gal for galectins, Neu5Ac for *pertussis* toxin) and a few (one or two) with the non-reducing monosaccharide. Furthermore, several studies reported that the fixation of a monosaccharide or oligosaccharide ligand in a lectin does not significantly change either the conformation of the CRD of the lectin or the more thermodynamically stable conformation of the oligosaccharide ligand. Therefore, the determination of the structure of the lower energy conformers of interleukin high-affinity ligands was an essential point. Unfortunately, NMR determinations of oligosaccharide conformations and computational calculations still give evasive results, although, the comparisons of predicted structures with X-ray-crystallography data of the interactions between a lectin with its ligand can be of essential importance. However, none of the X-ray data were concerned with very high affinity glycan ligands (Kd in the range of 1 mM for lactose and galectin-2) as we determined for interleukins (between 10^{-7} and 10^{-10} for interleukins). Therefore, although the essential features of CRDs could be the same, the precise docking of the endogenous ligand still remains unknown. Affino-blotting analyses of the endogenous ligands of endogenous lectins seem to be indicative of this problem. For example, the human MBP recognized four bands in human liver, although several tenths of liver glycoproteins have terminal mannose-residues (Mori *et al.*, 1988). Based on these criteria, it was possible to determine the localization of the CRD of a cytokine and to propose coherent docking pictures of the ligand in a cytokine for which the three-dimensional structure has been determined, provided the knowledge of the conformation of the lower energy conformers of the oligosaccharide ligands. Such studies may be helped by the knowledge of the structures of different ligands and non-ligands of the cytokines.

7.1. CRD of Human IL-6

IL-6 recognizes oligosaccharides having the HNK-1 epitope (see above) present on mammalian N-glycans and on O-glycans from the mucins of the eggs of *Rana temporaria*, SO3H-3-GlcAβ1,3Galβ1,4-. Comparison of the binding to IL-6 of different oligosaccharides from eggs of different amphibians indicated that the presence of sulfate and of GlcA bound to a Gal residue were an absolute requirement for the binding. The substitution of Gal in position 4 by another Galβ residue inhibits the binding. Analyses of the conformations of the *R. temporaria* and N-glycan HNK-1 epitope indicated that the lower energy conformers possessed an identical conformation (Cebo *et al.*, 2002). Computational docking of these conformers in the IL-6 structure was performed using the initial hypothesis that the sulfate part of the ligand should replace a sulfate group identified in the three-dimensional structure of IL-6, this sulfate group being strongly hydrogen-bonded to Asn155. This analysis allowed docking the ligand in a very stable structure as shown in Figure 2 (Cebo *et al.*, 2002).

Besides the three H-bonds involving sulfate, a H-bond is formed with the 2-OH group of GlcA, and two H-bonds are formed with the 4-OH and 6-OH group of Gal. The Trp157 residue of IL-6 is able to interact by van der Waals forces with the two cycles of GlcNAc and Man of the glycoprotein HNK-1, ensuring the stability of the binding. This allows understanding why the presence of a substituent on one of the hydroxyl of the $C_{(4)}$ carbon atom of Gal will destroy the affinity of the oligosaccharide for IL-6. This site was

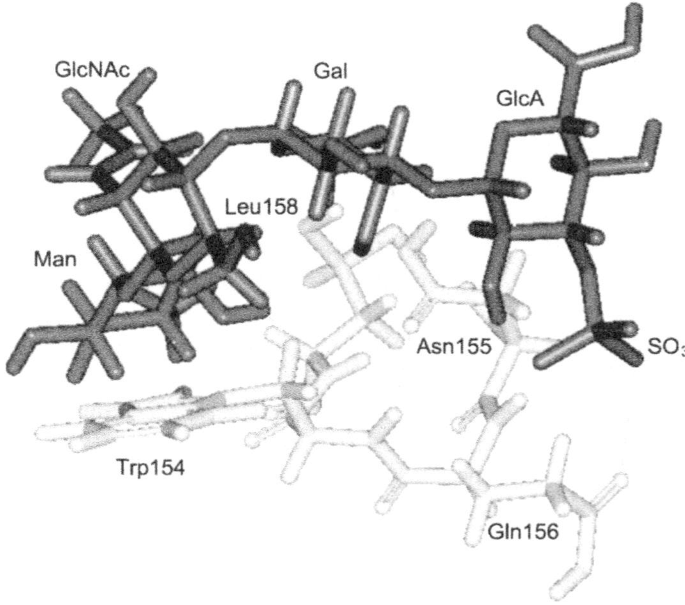

Figure 2. Representation of the calculated CRD of human IL-6. The HNK-1 oligosaccharide (dark structure) interacts by hydrogen bonds with Trp104, Asn155, Gln156, Leu158 and with van der Waals interactions with Trp154.

already defined (Somers *et al.*, 1997) as a site of interaction between IL-6 and the signal transducing molecule gp130.

7.2. CRD of IL-3

Based on site directed mutagenesis experiments, two different sites were identified in the structure of IL-3. The receptor-binding domain was defined as a large area. The second site was likely localized around the water-exposed Trp104 since the replacement of this amino acid eliminated all biological activity of IL-3 without affecting its binding to its receptor. Molecular modelling of the different disaccharides of chondroitin sulphates and *in silico* docking experiments (Zanetta *et al.*, 2002) indicated that: (i) the lower energy conformers of the biochemically identified disaccharide ligand showed a preferential binding in this area in contrast with the others (Figure 3); (ii) refinement of the docking indicated that the best docking was obtained for a conformer having an intermediate conformation between the two lower energy conformers. This involved a strong van der Waals interaction of the pyranic ring of GalNAc with Trp104 and four strong hydrogen bonds.

Interestingly, a mutation P33G of a proline residue close to Trp104 was shown to significantly increase the biological activity of IL-3 (14 fold) without affecting the binding to the receptor (Lokker *et al.*, 1991). In order to have an idea of the mechanism of the increased biological activity, we performed a computational modelling of the P33G mutation. In fact, our data indicated that the conformation of the putative CRD was completely changed with an opening of the cavity around Trp104, which became completely exposed.

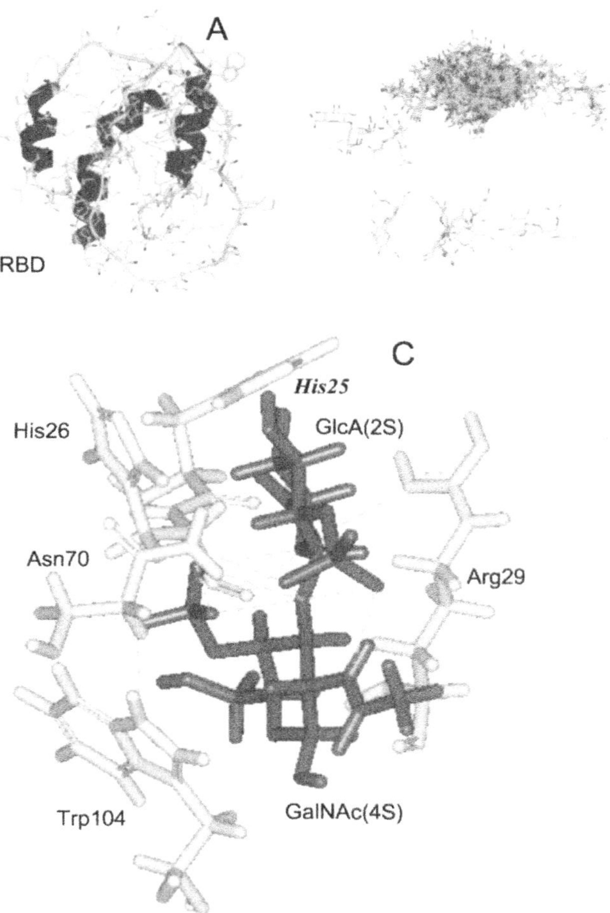

Figure 3. Computational docking of the lower energy conformers of the IL-3 ligand onto the IL-3 molecule. Note that more than 80% of the ligand (B) were docking in the same cavity of the IL-3 molecule (A) at the bottom of which Trp104 is localized. (C) Magnification of the optimized CRD of IL-3. The ligand (dark structure) forms hydrogen bonds with His26, Arg29, Asn70 and Trp104. Strong van der Waals interactions occur with His26 and Trp104, a weaker interaction occurring with His25.

The RBD region was not significantly changed, in agreement with the absence of significant modification of receptor binding. This would suggest that the increased activity was due to an increased affinity of IL-3 mutein for its ligand.

In fact, this was not true. Indeed, *in silico* docking experiments of the high affinity ligand of the native IL-3 showed an absence of significant binding of the 100 lower energy conformers of this ligand, in contrast with the hypothesis of the increased affinity mentioned above. However, when performing the same experiments with the other chondroitin sulfate disaccharides, it appeared that the classical disaccharide of chondroitin-4-S showed a good docking in this area, but with a complete change of the pattern of hydrogen bonds. Therefore, the idea emerged that the P33G mutation, affecting only the initial CRD of

IL-3, changed the carbohydrate-specificity of IL-3 from a rare chondroitin sulfate to a common and abundant chondroitin sulfate: chondroitin-4-S. Therefore, the increased activity of the mutein could be explained in terms of increased IL-3-dependent association between the IL-3 receptor and the signal-transducing complex containing the new IL-3 ligand, chondroitin-4-sulfate. This would result in an increased signalling, thus increased biological function, but induce a loss of the cell specificity of the action of IL-3, because the new ligand has a wide distribution in human (Zanetta *et al.*, 2002). Although this hypothesis is still not sustained by experimental data, this new concept would be extremely easy to confirm or infer by specialists of molecular biology.

7.3. CRD of other Cytokines

Our data on IL-6 (Cebo *et al.*, 2002), on IL-3 (Zanetta *et al.*, 2002) and on IL-4 (Cebo *et al.*, submitted) allowed to define a putative CRD for these molecules, the general scheme of the CRD being conserved: a water-exposed aromatic residue localized in a cavity surrounded by amino acids able to form easily hydrogen bonds (Asn, Gln, Arg, His), a situation very similar to that found for the three examples shown above (galectin 2; *pertussis* toxin and *influenza* hemagglutinin). Independent *in silico* docking of the high affinity ligand in the three-dimensional structure of the cytokines and studies of the site-directed mutagenesis experiments published in the literature allowed a very good reliability of the results. Therefore, in the few examples shown here, the identification of a CRD seemed to be quite easy when both the three-dimensional structure of the cytokine and reliable mutagenesis experiments are accessible in protein data banks and in literature. In fact, we found that the problem is easier to solve with the later discovered cytokines, because the mutagenesis experiments were not randomly performed, but based on the knowledge of the three-dimensional structure and of the discrimination between internal amino acids important for stabilizing the three-dimensional structure of the molecule and those important for the CRD. This is evidenced by the data from site-directed mutagenesis experiments performed on the first discovered interleukins (IL-2, IL-1α and IL-1β respectively), ignoring the presence of a CRD.

Recent studies performed on the putative CRD of IL-2 gave unexpected results. Indeed, docking experiments performed with the lower energy conformers of $Man_{5-6}GlcNac_2$ showed a systematic interaction of the reducing end of the glycans with IL-2, a situation corresponding to a non-sense since this reducing end is attached to a protein. Such a situation was also observed for high size glycans. In order to overcome this aberration, we artificially attached the oligomannoside to a lure, consisting in a 24-carbon atom ball, all carbon atoms bearing diol groups. In these conditions, 93% of the $Man_6GlcNAc_2$ ligand of IL-2 were correctly oriented. Moreover, more than 50% of the lower energy complexes between IL-2 and its ligand were localized in a large cavity of IL-2, different from the site defined as the IL-2Rβ-binding domain (Figure 4).

The interaction involved 10 hydrogen bonds of the ligand with Thr-3, Lys-9, Thr-10, Gln-13, Glu-95, Leu-96 and Lys-97, but no significant van der Waals interactions. But strong van der Waals interactions with the pyranic rings of the ligands can be obtained after a rotation of 180° of the six N-terminal amino acids. This suggested that the carbohydrate-recognition domain of interleukin-2, involving a two-step process, was essentially different from all known lectins. This peculiarity of the CRD of IL-2 was previously suggested

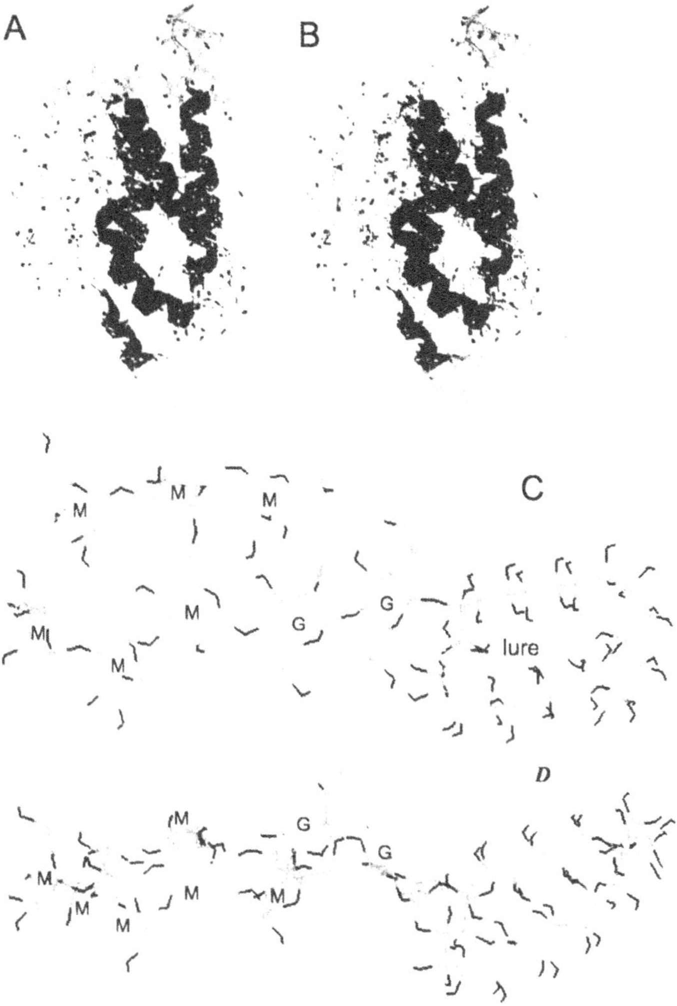

Figure 4. Representation of the docking experiments of $Man_6GlcNAc_2$ onto the IL-2 structure. IL-2 (A) presents a large cavity in which the lure-attached ligand is included (B). The reason why oligomannosides are specifically docking into this cavity is almost planar of the lower energy conformers as shown in C and D, the two pictures representing view after a rotation of 90°.

(Fukushima and Yamashita, 2001) considering the fact that once IL-2 has bound its oligomannoside ligand, it was very difficult to displace it. In fact, the entrapping of the oligomannosides with five and six mannose residues by the rotation of Pro-2 decreased the energy of the ligand/IL2 complex by 115 $kCal.mole^{-1}$ (Zanetta *et al.*, submitted), explaining the difficulty to dissociate the complex once formed. The strict specificity of IL-2 for the previous ligands and the absence of affinity for oligomannosides with a higher number of mannose residues can be easily explained since the latter are too big and the branches are not distributed in a single plane as for $Man_5GlcNAc_2$ and $Man_6GlcNAc_2$ in such a way that they cannot penetrate into the cavity of the IL-2 CRD.

8. CONCLUSION AND PERSPECTIVES

The concept that cytokines could have a lectin activity is emerging, especially since the demonstration of the biological function of this lectin activity. In cases so far examined, cytokines recognize with higher affinity always glycans having a rare and specific expression in normal tissues. This implicates that cells responding to a particular cytokine should have both the receptor and the ligand at their surface. The outcome of the initial receptor change in phosphorylations results in the commitment of specific pathways, followed by the expression of new molecules involved in cell proliferation or differentiation. However, due to the cell-specific association of the cytokine receptor with a peculiar surface molecular complex, the cascade of signal transduction would be cell-specific and would explain the divergence between the cascades, which are engaged. The actual relevance to an *in vivo* situation of experiments of gene transfection of cytokine receptors in malignant or immortalized cells, which can express cytokine ligands entirely different from the endogenous ones is questioned. This is the case for IL-2. Indeed, although the carbohydrate-binding properties of IL-2 for oligomannosides with 5 and 6 mannose residue is well established (Fukushima and Yamashita, 2001; Zanetta *et al.*, 1996), the CRD can associate its receptor beta with different molecules depending on the cell type and on the degree of stimulation of one single cell type. Indeed, in the CTLL-2 mouse cell line, one of the ligands of IL-2 was demonstrated to be the molecule known as IL-2Rα. This carbohydrate-dependent association between IL-2Rβ and IL-2Rα triggers the formation of the high affinity complex between the IL-2 receptors leading to cellular signalling. However, this mechanism observed on a cell line actually possessing IL-2Rα at the cell surface is very different from that observed in the initial steps of antigen-specific or general activation of human lymphocytes. Indeed, IL-2Rα is not expressed at the surface of resting human lymphocytes (Zola *et al.*, 1991). In contrast, TCRα and TCRβ subunits as well as CD3 in T cells are present. Upon the initial association between IL-2Rβ and the TCR complex due to the lectin activity of IL-2, this complex gave a signal allowing the expression of IL-2Rα and IL-2Rβ as well as IL-2 at the cell surface together with the internalization of the previous complex (Luto *et al.*, 1997). The same situation was found for IL-6 (Cebo *et al.*, 2002). Indeed, human resting lymphocytes have a small number and amount of HNK-1 glycans. This completely differed in the HepG2 cell line (one of the most frequent model for studying the IL-6-dependent response) in which the quantity and number of IL-6 ligands is increased by a factor close to 20 fold. This could induce a multiplicity of signals and multiple signal transduction pathways, probably different from that observed in normal cells.

The increased number of three-dimensional structures of the cytokines, the site-directed mutagenesis experiments (when taking into account the possibilities of multiple different functional sites in cytokine molecules) and the progresses in the computational software used for glycan conformations and docking experiments allow proposing the localization of the putative carbohydrate-recognition domains. As calculated for IL-6 and IL-3, relatively water-exposed aromatic residues (Trp, Tyr, Phe, His) seem to be of essential importance, as well as hydrogen bonds. In our hands, ionic interactions did not play a significant role in the binding of an oligosaccharide in the CRD (no ionic interactions were observed between IL-4 and its disaccharide ligand and IL-6 with its ligands). This is in perfect agreement with the structures of all CRD of calcium-independent lectins co-crystallized with their ligands.

9. ACKNOWLEDGMENTS

This work was supported by grants "Glycotrains" and "Euraman" from the European Community. The authors thank Dr. Christelle Cebo, Thierry Dambrouck, Emmanuel Maes, Christine Laden, Gérard Strecker, Jean-Claude Michalski and Alexandre Pons for their collaboration and Mr. Philippe Timmerman and Yves Leroy for assistance.

10. REFERENCES

Akira, S., Hirano, T., Taga, T., and Kishimoto, T., 1990, Expression of interleukin-2 receptor gamma chain on human neutrophils, *FASEB J.* 4:2860–2867.

Angata, T. and Brinkman-van der Linden, 2002, E.C.M., I-type lectins, *Biochim Biophys Acta.* 1572:294–316.

Arar, C., Carpentier, V., Le Caer, J.P., Monsigny, M., Legrand, A., and Roche, A.C., 1995, ERGIC-53, a membrane protein of the endoplasmic reticulum-Golgi intermediate compartment, is identical to MR60, an intracellular mannose-specific lectin of myelomonocytic cells, *J Biol Chem.* 270:3551–3553.

Brody, D.T. and Durum, S.K., 1989, Membrane IL-1: IL-1 alpha precursor binds to the plasma membrane via a lectin-like interaction, *J Immunol.* 143:1183–1187.

Carter, D.B., Deibel Jr, M.R., Dunn, C.J., Tomich, C.S., Laborde, A.L., Slightom, J.L., Berger, A.E., Bienkowski, M.J., Sun, F.F., McEwan, R.N. *et al.*, 1990, Purification, cloning, expression and biological characterization of an interleukin-1 receptor antagonist protein, *Nature.* 344:633–638.

Cebo, C., Dambrouck, T., Maes, E., Laden, C., Strecker, G., Michalski, J.C., and Zanetta, J.P., 2001, Recombinant human interleukins IL-1alpha, IL-1beta, IL-4, IL-6, and IL-7 show different and specific calcium-independent carbohydrate-binding properties, *J Biol Chem.* 276:5685–5691.

Cebo, C., Durier, V., Lagant, P., Maes, E., Florea, D., Lefebvre, T., Strecker, G., Vergoten, G., and Zanetta, J.P., 2002, Function and molecular modelling of the interaction between human interleukin 6 and its HNK-1 oligosaccharide ligands, *J Biol Chem.* 277:12246–12252.

Christie, G., Dacey, I., and Weston, J., 1995, Identification of a soluble, high affinity human interleukin 4 binding protein in normal human urine, *Cytokine.* 7:305–310.

Chu, J.W. and Sharom, F., 1995, Gangliosides interact with interleukin-4 and inhibit interleukin-4-stimulated helper T-cell proliferation, *Immunology.* 84: 396–403.

Clarke, D., Katoh, O., Gibbs, R.V., Griffiths, S.D., and Gordon, M.Y., 1995, Interaction of interleukin 7 (IL-7) with glycosaminoglycans and its biological relevance, *Cytokine.* 7:325–330.

Cohen, F.E., Kosen, P.A., Kuntz, I.D., Epstein, L.B., Ciardelli, T.L., and Smith, K.A., 1986, Structure-activity studies of interleukin-2, *Science.* 234:349–352.

Cooper, D.N.W., 2002, Galectinomics: finding themes in complexity, *Biochim Biophys Acta.* 1572: 209–231.

Dinarello, C.A., 1994, The interleukin-1 family: 10 years of discovery, *FASEB J.* 8:1314–1325.

Ding, D.X., Vera, J.C., Heaney, M.L., and Golde, D.W., 1995, N-glycosylation of the human granulocyte-macrophage colony-stimulating factor receptor alpha subunit is essential for ligand binding and signal transduction, *J Biol Chem.* 270:24580–24584.

Dodd, R.B. and Drickamer, K., 2001, Lectin-like proteins in model organisms: implications for evolution of carbohydrate-binding activity, *Glycobiology.* 11:71–79.

Drickamer, K., 1997, Making a fitting choice: common aspects of sugar-binding sites in plant and animal lectins, *Structure.* 5:465–468.

Farrar, W.L., Garcia Garcia, G., Evans, G., Michiel, D., and Linnekin, D., 1990, Cytokine regulation of protein phosphorylation, *Cytokine.* 2:77–91.

Feizi, T. and Childs, R.A., 1987, Carbohydrates as antigenic determinants of glycoproteins, *Biochem J.* 245:1–11.

Fernandez-Botran, R. and Vitetta, E.S., 1990, A soluble, high-affinity, interleukin-4-binding protein is present in the biological fluids of mice, *Proc Natl Acad Sci USA.* 87:4202–4206.

Fiedler, K. and Simons, K., 1994, The role of N-glycans in the secretory pathway, *Cell.* 77:625–626.

Fontaine, V., Ooms, J., and Content, J., 1994, Mutagenesis of the human interleukin-6 fourth predicted alpha-helix: involvement of the Arg168 in the binding site, *Eur J Immunol.* 24:1041–1045.

Fontaine, V., Savino, R., Arcone, R., de Wit, L., Brakenhoff, J.P., Content, J., and Ciliberto, G., 1993, Involvement of the Arg179 in the active site of human IL-6, *Eur J Biochem.* 211:749–755.

Foxall, C., Watson, S.R., Dowbenko, D., Fennie, C., Lasky, L.A., Kiso, M., Hasegawa, A., Asa, A., and Brandley, B.K., 1992, The three members of the selectin receptor family recognize a common carbohydrate epitope, the sialyl Lewis(x) oligosaccharide, *J Cell Biol.* 117:895–902.

Fukada, T., Yoshida, Y., Nishida, K., Ohtani, T., Shirogane, T., Hibi, M., and Hirano, T., 1999, Signalling through Gp130: toward a general scenario of cytokine action, *Growth Factors.* 17:81–91.

Fukushima, K. and Yamashita K., 2001, Carbohydrate recognition site of interleukin-2 in relation to cell proliferation, *J Biol Chem.* 276:7351–7356.

Fukushima, K., Hara-Kuge, S., Ohkura, T., Seko, A., Ideo, H., Inazu, T., and Yamashita, K., 1997, Lectin-like characteristics of recombinant human interleukin-1beta recognizing glycans of the glycosylphosphatidylinositol anchor, *J Biol Chem.* 272:10579–10584.

Fukushima, K., Watanabe, H., Takeo, K., Nomura, M., Asahi, T., and Yamashita, K., 1993, N-linked sugar chain structure of recombinant human lymphotoxin produced by CHO cells: the functional role of carbohydrate as to its lectin-like character and clearance velocity, *Arch Biochem Biophys.* 304:144–153.

Hatakeyama, M., Kono, T., Kobayashi, N., Kawahara, A., Levin, S.D., Perlmutter, R.M., and Taniguchi, T., 1991, Interleukin-2 receptor beta chain gene: generation of three receptor forms by cloned human alpha and beta chain cDNA's, *Science.* 252:1523–1528.

Hession, C., Decker, J.M., Sherblom, A.P., Kumar, S., Yue, C.C., Mattallano, R.J., Tizard, R., Kawashima, E., Schmeissner, U., Heletky, S., Chow, E.P., Burne, C.A, Shaw, A., and Muchmore, A.V., 1987, Uromodulin (Tamm-Horsfall glycoprotein): a renal ligand for lymphokines, *Science.* 237:1479–1484.

Hirano, T., 1998, Interleukin 6 and its receptor: ten years later, *Int Rev Immunol.* 16:249–284.

Hirano, T., Nakajima, K., and Hibi, M., 1997, Signalling mechanisms through gp130: a model of the cytokine system, *Cytokine Growth Factor Rev.* 8:241–252.

Horak, I., 1995, Immunodeficiency in IL-2-knockout mice, *Clin Immunol Immunopathol.* 76:172–173.

Huang, H. and Paul, W.E., 2000, Protein tyrosine phosphatase activity is required for IL-4 induction of IL-4 receptor alpha-chain, *J Immunol.* 164:1211–1215.

Imani, F., Rager, K.J., Catipovic, B., and Marsh, D.G., 1997, Interleukin-4 (IL-4) induces phosphatidylinositol 3-kinase (p85) dephosphorylation. Implications for the role of SHP-1 in the IL-4-induced signals in human B cells, *J Biol Chem.* 272:7927–7931.

Ju, G., Collins, L., Kaffka, K.L., Tsien, W.H., Chizzonite, R., Crowl, R., Bhatt, R., and Kilian, P.L., 1987, Structure-function analysis of human interleukin-2. Identification of amino acid residues required for biological activity, *J Biol Chem.* 262:5723–5731.

Kamegai, M., Niijima, K., Kunishita, T., Nishizawa, M., Ogawa, M., Araki, M., Ueki, A., Konishi, Y., and Tabira, T., 1990, Interleukin 3 as a trophic factor for central cholinergic neurons *in vitro* and *in vivo*, *Neuron.* 4:429–436.

Kaplan, D., Smith, D., Huang, R., and Yildirim, Z., 1995, Self-association of interleukin 2 bound to its receptor, *FASEB J.* 9:1096–1102.

Kawasaki T., 1999, Structure and biology of mannan-binding protein, MBP, an important component of innate immunity, *Biochim Biophys Acta.* 1473:186–195.

Kelm, S., Schauer, R., Manuguerra, J.C., Gross H.J., and Crocker, P.R., 1994, Modifications of cell surface sialic acids modulate cell adhesion mediated by sialoadhesin and CD22, *Glycoconjugate J.* 11:576–585.

Kilpatrick, D., 2002, Animal lectins: a historical introduction and overview, *Biochim Biophys Acta.* 1572:187–197.

Kilpatrick, D.C., 2002, Mannan-binding lectin: clinical significance and applications, *Biochim Biophys Acta.* 1572,401–413.

Kishimoto, T., Taga, T., and Akira S., 1992a, Cytokine signal transduction, *Cell.* 76: 253–262.

Kishimoto, T., Akira, S., and Taga, T., 1992b, Interleukin-6 and its receptor: a paradigm for cytokines, *Science.* 258:593–597.

Kolb, J.P. and Abadie, A., 1993, Inhibitors of protein tyrosine kinases and protein tyrosine phosphatases suppress IL-4-induced CD23 expression and release by human B lymphocytes, *Eur Cytokine Netw.* 4:429–438.

Konishi, Y., Chui, D.H., Kunishita, T., Yamamura, T., Higashi, Y., and Tabira, T., 1995, Demonstration of interleukin-3 receptor-associated antigen in the central nervous system, *J Neurosci Res.* 41:572–582.

Lasky, L.A., 1992, Selectins: interpreters of cell-specific carbohydrate information during inflammation, *Science.* 258:964–969.

Leebeck, F.W., Kariya, K., Schwabe, M., and Fowlkes, D.M., 1992, Identification of a receptor binding site in the carboxyl terminus of human interleukin-6, *J Biol Chem.* 267:14832–14838.

Lescar, J., Loris, R., Mitchell, E., Gautier, C., Chazalet, V., Cox, V., Wyns, L., Perez, S., Breton, C., and Imberty, A., 2002, Isolectins I-A and I-B of Griffonia (Bandeiraea) simplicifolia. Crystal structure of metal-free GS I-B(4) and molecular basis for metal binding and monosaccharide specificity, *J Biol Chem.* 277:6608–6614.

Li, X., Rock, F., Chong, P., Cockle, S., Keating, A., Ziltener, H., and Klein, M., 1993, Structure-function analysis of the C-terminal segment of human interleukin-6, *J Biol Chem.* 268:22377–22384.

Licinio, J. and Wong, M.L., 1996, Interleukin 1 beta and fever, *Nat Med.* 2:1314–1315.

Lobsanov, Y.D., Gitt, M.A., Leffler, H., Barondes, S.H., and Rini, J.M., 1993, X-ray crystal structure of the human dimeric S-Lac lectin, L-14-II, in complex with lactose at 2.9-A resolution, *J Biol Chem.* 268:27034–27038.

Lokker, N.A., Movva, N.R., Strittmatter, U., Fagg, B., and Zenke, G., 1991, Structure-activity relationship study of human interleukin-3. Identification of residues required for biological activity by site-directed mutagenesis, *J Biol Chem.* 266:10624–10631.

Luton, F., Legendre, V., Gorvel, J.P., Schmitt-Verhulst, A.M., and Boyer, C., 1997, Tyrosine and serine protein kinase activities associated with ligand-induced internalized TCR/CD3 complexes, *J Immunol.* 158:3140–3147.

Mancilla, J., Ikejima, T., and Dinarello, C.A., 1992, Glycosylation of the interleukin-1 receptor type I is required for optimal binding of interleukin-1, *Lymphokine Cytokine Res.* 11:197–205.

Mori, K., Kawasaki, T., Yamashina, I., 1988, Isolation and characterization of endogenous ligands for liver mannan-binding protein, *Arch Biochem Biophys.* 264:647–656.

Muchmore, A.V. and Decker, J.M., 1986, Uromodulin. An immunosuppressive 85-kilodalton glycoprotein isolated from human pregnancy urine is a high affinity ligand for recombinant interleukin 1 alpha, *J Biol Chem.* 261:13404–13407.

Muchmore, A.V. and Decker, J.M., 1987, The lectin-like interaction between recombinant tumor necrosis factor and uromodulin, *J Immunol.* 138:2541–2546.

Nath, D., van der Merwe, P.A., Kelm, S., Bradfield, P., and Crocker, P.R., 1995, The amino-terminal immunoglobulin-like domain of sialoadhesin contains the sialic acid binding site. Comparison with CD22, *J Biol Chem.* 270:26184–26191.

Nelms, K., Keegan, A.D., Zamorano, J., Ryan, J.J., and Paul, W.E., 1999, The IL-4 receptor: signalling mechanisms and biologic functions, *Annu Rev Immunol.* 17:701–738.

Niu, L., Heaney, M.L., Vera, J.C., and Golde, D.W., 2000, High-affinity binding to the GM-CSF receptor requires intact N-glycosylation sites in the extracellular domain of the beta subunit, *Blood.* 95:3357–3362.

Noguchi, N., Nakamura, Y., Russell, S.M., Ziegler, S.F., Tsang, M., Cao, X., and Leonard, W.J., 1993, Interleukin-2 receptor gamma chain: a functional component of the interleukin-7 receptor, *Science.* 262:1877–1880.

Normand, G., Kuchler, S., Meyer, A., Vincendon, G., and Zanetta, J.P., 1988, Isolation and immunohistochemical localization of a chondroitin sulfate proteoglycan from adult rat brain, *J Neurochem.* 51:665–676.

Ozawa, H., Kotani, M., Kawashima, I., Numata, M., Ogawa, T., Terashima, T., and Tai, T., 1993, Generation of a monoclonal antibody specific for ganglioside GM4: evidence for GM4 expression on astrocytes in chicken cerebellum, *J Biochem.* 114:5–8.

Paonessa, G., Graziani, R., de Serio, A., Savino, R., Ciapponi, L., Lahm, A., Salvati, A.L., Toniatti, C., and Ciliberto, G., 1995, Definition of a composite binding site for gp130 in human interleukin-6, *EMBO J.* 14:1942–1951.

Paul, W.E., 1991, Interleukin-4: a prototypic immunoregulatory lymphokine, *Blood.* 77:1859–1870.

Powell, L.D. and Varki, A., 1994, CD22-mediated cell adhesion to cytokine-activated human endothelial cells. Positive and negative regulation by alpha 2–6-sialylation of cellular glycoproteins, *J Biol Chem.* 269:10628–10636.

Sauter, N.K., Glick, G.D., Crowther, R.L., Park, S.J., Eisen, M.B., Skehel, J.J., Knowles, J.R., and Wiley, D.C., 1992, Crystallographic detection of a second ligand binding site in influenza virus hemagglutinin, *Proc Natl Acad Sci USA.* 89:324–328.

Savino, R., Ciapponi, L., Lahm, A., Demartis, A., Cabibbo, A., Toniatti, C., Delmastro, P., Altamura, S., and Ciliberto, G., 1994, Rational design of a receptor super-antagonist of human interleukin-6, *EMBO J.* 13:5863–5870.

Savino, R., Lahm, A., Giorgio, A., Cabiddo, A., Tramontano, A., and Ciliberto, G., 1993, Saturation mutagenesis of the human interleukin 6 receptor-binding site: implications for its three-dimensional structure, *Proc Natl Acad Sci USA.* 9:4067–4071.

Schaaf-Lafontaine, N., Balthazart, C., and Hooghe, R.J., 1985, Membrane carbohydrates of lymphoid cells: the receptor for interleukin 2, *Immunobiology.* 170:249–255.

Sherblom, A.P., Decker, J.M., and Muchmore, A.V., 1988, The lectin-like interaction between recombinant tumor necrosis factor and uromodulin, *J Biol Chem.* 263:5418–5424.

Sherblom, A.P., Sathyamoorthy, N., Decker, J.M., and Muchmore, A.V., 1989, IL-2, a lectin with specificity for high mannose glycopeptides, *J Immunol.* 143:939–944.

Shibuya, H., Kohu, K., Yamada, K., Barsoumian, E.L., Perlmutter, R.M., and Taniguchi, T., 1994, Functional dissection of p56lck, a protein tyrosine kinase which mediates interleukin-2-induced activation of the c-fos gene, *Mol Cell Biol.* 14:5812–5819.

Shibuya, K., Chiba, S., Miyagawa, K., Kitamura, T., Miyazono, K., and Takaku, F., 1991, Structural and functional analyses of glycosylation on the distinct molecules of human GM-CSF receptors, *Eur J Biochem.* 198:659–666.

Siebert, H.C., Adar, R., Arango, R., Burchert, M., Kaltner, H., Kayser, G., Tajkhorshid, E., von der Lieth, C.W., Kaptein, R., Sharon, N., Vliegenthart, J.F.G., and Gabius, H.J., 1997, Involvement of laser photo-CIDNP (chemically induced dynamic nuclear polarization)-reactive amino acid side chains in ligand binding by galactoside-specific lectins in solution, *Eur J Biochem.* 249:27–38.

Somers, W., Stahl, M., and Seehra, J.S., 1997, 1.9 A crystal structure of interleukin 6: implications for a novel mode of receptor dimerization and signalling, *EMBO J.* 16:989–997.

Stein, P.E., Boodhoo, A., Armstrong, G.D., Heerze, L.D., Cockle, S.A., Klein, M.H., and Read, R.J., 1994, Structure of a pertussis toxin-sugar complex as a model for receptor binding, *Nat Struct Biol.* 1:591–596.

Takeshita, T., Asao, H., Ohtani, K., Ishii, N., Kumaki, S., Tanaka, N., Munakata, H., Nakamura, M., and Sugamura, K., 1992, Cloning of the gamma chain of the human IL-2 receptor, *Science.* 257:379–382.

Taniguchi, T. and Minami, Y., 1993, The IL-2/IL-2 receptor system: a current overview, *Cell.* 73:5–8.

Waldmann, T.A., 1991, The interleukin-2 receptor, *J Biol Chem.* 266:2681–2684.

Wall, K.A., Pierce, J.D., and Elbein, A.D. Inhibitors of glycoprotein processing alter T-cell proliferative responses to antigen and to interleukin 2, *Proc Natl Acad Sci USA.* 85:5644–5648.

Wells, T.N.C., Graber, P., Proudfoot, A.E.I., Arod, C.Y., Jordan, S.R., Lambert, M.H., Hassel, A.M., and Milburn, M.V., 1994, The three-dimensional structure of human interleukin-5 at 2.4-angstroms resolution: implication for the structures of other cytokines, *Ann N Y Acad Sci.* 725:118–127.

Winkelstein, A., Muchmore, A.V., Decker, J.M., and Blaese, R.M., 1990, Uromodulin: a specific inhibitor of IL-1-initiated human T cell colony formation, *Immunopharmacol.* 20:201–205.

Yasueda, H., Miyasaka, Y., Shimamura, T., and Matsui, H., 1992, Effect of semi-random mutagenesis at the C-terminal 4 amino acids of human interleukin-6 on its biological activity, *Biochem Biophys Res Commun.* 187:18–25.

Zanetta, J.P., Alonso, C., and Michalski, J.C., 1996, Interleukin 2 is a lectin that associates its receptor with the T-cell receptor complex, *Biochem J.* 318:49–53.

Zanetta, J.P., Bindeus, R., Normand, G., Durier, V., Lagant, P., Maes, E., and Vergoten, G., 2002, Evidence for a lectin activity for human interleukin-3 and modelling of its carbohydrate-recognition domain, *J Biol Chem.* 277:38764–38771.

Zanetta, J.P., Bonaly, R., Maschke, S., Strecker, G., and Michalski, J.C., 1988, Differential binding of lectins IL-2 and CSL to *Candida albicans* and cancer cells, *Glycobiology.* 8:221–225.

Zanetta, J.P., Bonaly, R., Maschke, S., Strecker, G., and Michalski, J.C., 1988, Hypothesis: Immunodeficiencies in α-mannosidosis, mycosis, aids and cancer: a common mechanism of inhibition of the function of interleukin 2 by oligomannosides, *Glycobiology.* 8:v–xi.

Zanetta, J.P., Meyer, A., Kuchler, S., and Vincendon, G., 1987, Isolation and immunochemical study of a soluble cerebellar lectin delineating its structure and function, *J Neurochem.* 49:1250–1257.

Zanetta, J.P., Pons, A., Iwersen, M., Mariller, C., Leroy, Y., Timmerman, P., and Schauer, R., 2001, Diversity of sialic acids revealed using gas chromatography/mass spectrometry of heptafluorobutyrate derivatives, *Glycobiology.* 11:663–676.

Zanetta, J.P., Wantyghem, J., Kuchler-Bopp, S., Badache, A., and Aubery, M., 1995, Human lymphocyte activation is associated with the early and high-level expression of the endogenous lectin CSL at the cell surface, *Biochem J.* 311:629–636.

Zola, H., Weedon, H., Thompson, G.R., Fung, M.C., Ingley, E., and Hapel, A.J., 1991, Expression of IL-2 receptor p55 and p75 chains by human B lymphocytes: effects of activation and differentiation, *Immunology.* 72:167–173.

9

CYTOKINES AND GLYCOSAMINOGLYCANS (GAGs)

Roslyn V. Gibbs

School of Pharmacy & Biomedical Sciences
University of Portsmouth
St Michael's Building, White Swan Road
Portsmouth, Hampshire, PO1 2DT, UK

1. INTRODUCTION

Glycosaminoglycans (GAGs), formerly known as mucopolysaccharides, are anionic polysaccharide molecules that are widely distributed among animal tissues and produced by most cell types. Several key immune cytokines have been shown to bind strongly and selectively to these molecules resulting in modulation of their bioactivity and/or tissue distribution. Thus a greater understanding of these interactions is required to provide insight into the role of GAGs as potential regulators of immune responses.

2. GLYCOSAMINOGLYCANS (GAGs)

2.1. Structure of Glycosaminoglycans

GAGs are linear polymers of repeating disaccharide units which are substituted with carboxylate ester and/or sulfate ester groups. The constituent disaccharide units consist of an amino sugar, usually D-glucosamine (GlcN) or D-galactosamine (GalN), which is linked typically to a uronic acid residue of either D-glucuronic acid (GlcUA) or L-iduronic acid (IdUA). This unit can be variably sulfated at both N- and O-positions, which endows these molecules with considerable structural heterogeneity. GAGs are usually classified into four main groups according to the structure of their disaccharide repeat unit (Table 1) (Scott, 1993).

Between one and approximately one hundred GAG chains are usually found covalently bound to a protein core to form a macromolecular structure known as a proteoglycan

Glycobiology and Medicine, edited by John S. Axford
Kluwer Academic / Plenum Publishers, New York, 2003

Table 1. The constituent disaccharide repeat units of GAGs that form the basis of their classification, where Gal—galactose; GalNAc—N-acetyl galactosamine; GlcNAc—N-acetyl glucosamine; GlcN—glucosamine; GlcUA—glucuronic acid; and IdUA—iduronic acid.

GAG	Disaccharide repeat	
Hyaluronic Acid (HA)	GlcNAc β1-4 GlcUA	
Keratan Sulfate (KS)	GlcNAc β1-4 Gal	
Chondroitin Sulfate (CS)	GalNAc-4S β1-4 GlcUA	Type A
	GalNAc-6S β1-4 GlcUA	Type C
	GalNAc-4S6S β1-4 GlcUA	Type E
Dermatan Sulfate (DS)	GalNAc-4S β1-4 IdUA	CS Type B
	GalNAc-4S β1-4 IdUA-2S	
Heparan Sulfate (HS)	GlcNAc α1-4 GlcUA	
	GlcNS α1-4 IdUA-2S	
Heparin (hep)	GlcNS-6S α1-4 IdUA-2S	

(Yanagishita, 1994). GAG chains can be attached through both O- and N-glycosidic linkages, although they are most commonly found O- linked to serine residues (Hascall *et al.*, 1994).

2.2. Properties and Tissue Distribution of GAGs

Proteoglycans can be found intracellularly, usually within secretory granules, on the surface of cells or within the extracellular matrix. Hyaluronic acid (HA), however, is the only GAG that is both unsulfated and synthesized in free form. Although it is found ubiquitously within the extracellular space of higher animals, it is present in highest concentrations in soft connective tissues such as embryonic tissues, synovial fluid, vitreous humor of the eye, and umbilical cord. The structure, function and potential therapeutic applications of HA have been reviewed by Laurent and Fraser (1992). Keratan sulfate (KS) also has a wide tissue distribution but is particularly abundant in cartilage, invertebral discs and the cornea. In the cornea, KS abundance has been related to the maintenance of the high level of hydration required for corneal transparency (Funderburgh, 2000).

Chondroitin sulfate (CS) is the most commonly occurring GAG where it is particularly abundant in the cornea, cartilage and adult bone. The GalNAc unit of the constituent disaccharide can be substituted with sulfate groups at the C4 or C6 positions, to form types A (CSA) and C (CSC) respectively, or at both positions to form type E (CSE) (Table 1) (Yamada *et al.*, 1992). CS polysaccharide chains normally consist of both type A and C disaccharides, with CSE the predominant form in mast cells and activated macrophages and CSA in hematopoietic cells (Kolset and Gallagher, 1990). The disaccharide repeat forming Dermatan sulfate (DS) is a structural isomer of that found in the CSs, in which most of the GlcUA residues have been replaced by IdUA residues as a result of epimerization of the carboxyl group at C5 of GlcUA. In addition to sulfation at C4 and C6 of the GalNAc unit, the C2 position of IdUA unit can also be sulfated in DS. DS therefore usually presents as a copolymer comprising disaccharides containing both IdUA and GlcUA and therefore is also referred to as CSB (Hascall *et al.*, 1994; Otsu *et al.*, 1985).

Heparan sulfate proteoglycans (HSPGs) are widely distributed throughout animal tissues where they are localized to two main areas: either the plasma membrane of cells or the basement membranes (Yanagishita and Hascall, 1992). In contrast, heparin (hep) is mainly a product of mast cells where it is stored intracellularly, complexed to potent inflammatory mediators and released upon degranulation of these cells during the inflammatory response. Hep and heparan sulfate (HS) are structurally related and comprise the same monosaccharide residues (Table 1). The amino sugar is an α1-4 linked glucosamine (GlcN), which can be N-acetylated or N-sulfated and the hexuronate either GlcUA or IdUA. During synthesis modifications such as 6-O- and/or 3-O-sulfation of GlcN and 2-O-sulfation of IdUA can occur. HS contains approximately equal amount of N-acetylated and N-sulfated GlcN, in contrast to hep in which 70% or more of GlcN residues are sulfated. Furthermore, HS contains a higher proportion of GlcUA residues than hep, which is rich in IdUA (Casu, 1985; Gallagher *et al.*, 1986; Vives *et al.*, 1999). HS polymers characteristically display a high level of structural heterogeneity mainly due to variations in levels of sulfation. N-Sulfation of GlcN residues occurs in defined regions of the HS polymer, which are termed S-domains, and HS polymers frequently comprise alternating S-domains separated by regions of low sulfation (unmodified GlcNAc-GlcUA repeats) (Turnbull and Gallagher, 1990a, b). Hep polymers, which are extensively N-sulfated and rich in IdUA residues and O-sulfates, resemble S-domains along the entire length of their chains. Thus, varying patterns of sulfation and epimerization of hexuronate residues endows great variability to the structures of hep and HS, a property which is not found in other members of the GAG family whose structures are relatively constant and well-defined. On this basis, Toida *et al.* (1997) have suggested that the simple classification of these heterogenous molecules as HS is not sufficient in defining their structural characteristics. Furthermore, since structure is closely associated with activity, the structural differences between HSs found in different tissues and cell types may have a role in defining the biological activities of these molecules within specific tissues.

2.3. Biological Functions of GAGs

GAGs have a diverse range of biological roles, from the straightforward mechanical support functions required for maintaining the structural integrity of tissues to a contribution to the regulation of cellular proliferation, differentiation, and communication (Kjellen and Lindahl, 1991). Most of these biological functions are defined by the ability of GAGs to bind to a number of regulatory proteins thereby influencing their tissue distribution, stability, and biological activity. A large number of proteins have been found to bind to hep/HS and other GAGs, including enzyme inhibitors (Lindahl *et al.*, 1984; Maimone and Tollefsen, 1990), matrix proteins (San Antonio *et al.*, 1993), viral proteins (Rider *et al.*, 1994), growth factors and cytokines (Tanaka *et al.*, 1993; Yayon *et al.*, 1991). Despite this diversity in protein binding, studies have shown that these interactions are specific with proteins recognizing defined saccharide sequences in the GAG chains. For example, antithrombin III has been shown to recognize a unique pentasaccharide sequence in heparin that contains an unusual 3-O-sulfated glucosamine unit that is essential for binding (Lindahl *et al.*, 1984).

3. CYTOKINE–GAG INTERACTIONS

The term cytokine is an umbrella term for several groups of related molecules including the Interleukins (ILs), Colony Stimulating Factors (CSFs), Interferons (IFNs),

Chemokines, and the Tumor Necrosis Factor (TNF) family of molecules. The members of these protein families play pivotal roles in the immune response, acting as soluble communicators that regulate the function of immune cells. Cytokines exert their functions through interactions with specific high affinity receptors on the surface of their target cells. In addition, an increasing number of these cytokines have also been shown to bind specifically to hep or other GAGs with these interactions having a range of effects on the cytokines involved. These cytokines include the interleukins IL-1 and IL-2 (Ramsden and Rider, 1992), IL-3 (Roberts *et al.*, 1988), IL-4 (Lortat-Jacob *et al.*, 1997), IL-5 (Lipscombe *et al.*, 1998), IL-6 (Ramsden and Rider, 1992), IL-7 (Clarke *et al.*, 1995), IL-8 (Webb *et al.*, 1993), IL-10 (Salek-Ardakani *et al.*, 2000), IL-12 (Hasan *et al.*, 1999), GM-CSF (Gordon *et al.*, 1987), gamma interferon (IFN-γ) (Lortat-Jacob and Grimaud, 1991) and RANTES (Martin *et al.*, 2001). This section will focus on examples of those cytokines whose GAG-properties have been relatively well characterized.

3.1. Interleukin-2 (IL-2)

Initially termed T Cell Growth Factor, IL-2 is produced by activated T cells and its major activity is the autocrine and paracrine stimulation of these cells, triggering their proliferation and differentiation into effector cells (Coutinho *et al.*, 1979; Meuer *et al.*, 1984) and the maintenance of T cell homeostasis (Wrenshall and Platt, 1999). This action is mediated through binding to the high affinity trimeric cell surface IL-2 receptor (IL-2R$\alpha\beta\gamma_c$) which is expressed on activated T cells. In addition IL-2 can also stimulate B cell proliferation and differentiation (Farrar *et al.*, 1982), promote the generation of lymphokine activated cells (LAK cells) (Grimm *et al.*, 1983) and augment NK cell activity (Heney *et al.*, 1981). IL-2 is a member of the 4α-helix bundle family of cytokines and its structure is stabilized by a single disulfide bond which is important in defining biological activity (Bazan and McKay, 1992; McKay, 1992). In contrast, its single O-glycan linked to threonine at position 3 is not necessary for function (Robb and Smith, 1981). Ramsden and Rider (1992) first demonstrated the ability of IL-2 to bind to hep and fucoidan, a sulfated polysaccharide from brown algae, but not CS, which suggested these interactions were specific. Furthermore, inhibition of IL-2 bioactivity could also be demonstrated, but only in the presence of high concentrations of fucoidan: hep was found to have no effect. Further studies on the selective binding of GAGs by IL-2 revealed that highly sulfated preparations of HS were also recognized by IL-2, but HSs of low sulfation were inactive, as was DS (CSB) (Najjam *et al.*, 1997). IL-2–hep binding was also found to be size dependent, with chain lengths of 15 residues retaining binding activity. Estimates of the binding affinity showed IL-2–hep interactions to be relatively weak (estimated $K_d \sim 500$ nM) and an order of magnitude less than the affinity to which hep has been shown to bind other cytokines. In further studies, *in vitro* bioassays confirmed earlier reports that hep had no effect on the biological activity of IL-2 (Najjam *et al.*, 1998). These observations were supported by other workers who demonstrated the *in vivo* localization of IL-2 to lymphoid organs through interactions with cell surface HS (Wrenshall and Platt, 1999). IL-2 retained in this way was also shown to be fully biologically active, such that it could induce proliferation of and contribute to the activation-induced cell death of T cells. It has therefore been proposed that GAGs may serve to retain IL-2 in a biologically active form close to its site of secretion, favoring a paracrine role for this cytokine. Structural studies on IL-2 appear to support this hypothesis. Najjam *et al.* (1998) have identified a putative hep binding site on IL-2, comprising the basic residues K32, K76, R81, and R83, that is distinct from its receptor binding sites. Modeling studies have shown that this site lies in a shallow groove formed by the three IL-2R subunits which is sufficiently wide to accommodate a hep molecule.

3.2. Interleukin-7 (IL-7)

IL-7 is produced by bone marrow and thymic stromal cells, intestinal epithelial cells and keratinocytes (He and Malek, 1998). The first reported biological activity of IL-7 was that of a pre-B cell growth factor (Billips *et al.*, 1992) but other functions have since been assigned to this cytokine. IL-7, alone or in conjunction with other factors, can stimulate the proliferation of thymocytes (Costello *et al.*, 1993), peripheral blood T cells and NK cells (Chazen *et al.*, 1989; Naume and Espevik, 1991), induce the generation of LAK cells (Lynch and Miller, 1992) and activate the tumoricidal and antimicrobial activities of macrophages (Appasamy, 1999). Like IL-2, IL-7 has been predicted to assume a 4α-helix bundle conformation stabilized by three disulfide bonds, which are essential for activity (Cosenza *et al.*, 1997; Kroemer *et al.*, 1996). IL-7 also possesses three potential N-glycan sites, however glycosylation is not required for biological function. In 1995, IL-7 was shown to bind to the GAGs hep and HS and, to a lesser extent, DS (CSB), but not CSA (Clarke *et al.*, 1995). Furthermore, affinity studies reported equilibrium constants (K_d) of 25 nM and 82 nM for binding to hep and HS, respectively. These observations have been both confirmed and extended by other workers who showed that IL-7 also bound fucoidan strongly but did not recognize KS or HA (Guelle *et al.*, 2000). Furthermore, using chemically modified heps it was also suggested that 6-O-sulfation of hep/HS is important for IL-7 binding, thus supporting the suggestion that these interactions are specific. The effects of GAG binding on IL-7 bioactivity were first investigated using *in vitro* cell bioassays where it was clearly demonstrated that hep, but not HS, could inhibit the IL-7 induced proliferation of a murine pre-B cell line. This observation has since been confirmed by other workers who also demonstrated the inhibitory effects of exogenous free hep on IL-7 induced proliferation of B cell precursors (Borghesi *et al.*, 1999). In addition, they reported that pro-B cells lacking cell surface HSPG responded poorly to stimulation with IL-7. This lead to the proposal that HSPG can directly regulate the bioactivity and bioavailablity of IL-7 to B cell precursors through its role as a component of the IL-7 receptor. The ability of exogenous hep but not HS to inhibit IL-7 activity, as observed by Clarke *et al.* (1995), could be accounted for in the relative affinities of these GAGs for IL-7. Thus, free hep could readily inhibit IL-7 binding to cell surface HSPG as it binds to IL-7 at least three times more strongly than HS. As a result, exogenous HS may not compete as effectively as hep, allowing binding of IL-7 to the HSPG receptor and subsequent proliferation of the B cell precursors. This effect may also be influenced by the species of HS used. In this study, HS from bovine kidney was used and failed to inhibit IL-7 bioactivity. Due to the high degree of structural heterogeneity exhibited by HSs, it is possible that this fraction may not be enriched in sequences recognized by IL-7 with high affinity, such as those expressed on the surface of the B cell progenitor, thereby accounting for its inability to serve as an efficient competitor. The selective recognition of different HS species has been demonstrated by Guelle *et al.* (personal communication) who showed that IL-7, like IL-2, binds more strongly to highly sulfated HSs that more closely resemble hep molecules.

3.3. Interleukin-8 (IL-8)

IL-8 is an 8 kDa protein which belongs to the structurally related C-X-C family of chemokines. IL-8 mediates acute inflammatory reactions through its actions as a chemoattractant and activator of leucocytes. Its effects are exerted through binding to specific receptors which are members of the G-protein coupled seven transmembrane helix family. The principal target cells of this cytokine are neutrophils but T cells also respond to IL-8 stimulation (McFadden and Kelvin, 1997). Chemokines, including IL-8, are key mediators in the transmigration of leucocytes across the endothelium. This is a complex multistep process, the first stage of which is the initial contact between leucocytes and the endothelium, a process mediated through interactions between selectins and their sugar ligands. As these interactions are relatively weak the leucocytes do not firmly adhere to endothelial cells but roll along the endothelial cell surfaces under the influence of blood flow. In the next stage, rolling is attenuated and leucocytes adhere firmly to the endothelium, mediated by strong interactions between leucocyte integrins and immunoglobulin-like endothelial cell adhesion molecules. This second step is induced by chemotactic molecules, such as IL-8, which activate the integrins constitutively expressed on the leucocyte surface (Rot *et al.*, 1996). As this process must occur in the correct sequence it has been proposed that the chemokines mediating integrin activation are immobilized on the endothelial cell surface. Leucocytes whose integrins are activated before initial contact with the endothelium lose their ability to adhere and transmigrate. Thus, only IL-8 bound to endothelial cell surfaces can promote neutrophil adhesion and emigration while, in its soluble form, IL-8 inhibits this process (Rot, 1992). IL-8 is now known to be sequestered on endothelial cell surfaces by GAGs (Hoogewerf *et al.*, 1997). The exact GAG target that mediates this interaction is unknown but it is presumed to be GAG molecules of the HS class. *In vitro*, IL-8 has been shown to bind to GAGs of the hep/HS class and HS, but not hep, has been shown to increase neutrophil chemotaxis four fold across artificial membranes (Webb *et al.*, 1993). As a result, much work has been conducted on IL-8–GAG interactions in order to understand further their immunoregulatory effects. Studies have demonstrated that IL-8 at low concentrations exist in monomeric form, whilst at higher concentrations form dimers or higher multimers. Furthermore, this oligomerization can be mediated by hep/HS and endothelial cell surface GAGs which bind IL-8 thereby increasing its local concentration (Hoogewerf *et al.*, 1997). This is supported by other workers who identified the active binding sequence in hep/HS as approximately 22–24 residues long and comprising two N-sulfated regions separated by a fully N-acetylated region. The two N-sulfated domains each bound an IL-8 monomer resulting in the formation and stabilization of an IL-8 dimer (Spillmann *et al.*, 1998). This study also revealed that the affinity of monomeric IL-8 for hep/HS oligosaccharides was too weak to allow binding at physiological ionic strength whereas the affinity of the dimer for hep/HS was of adequate strength. This has recently been contradicted by other workers who reported that IL-8 monomers bound HSs with affinities several orders of magnitude greater than IL-8 dimers (Goger *et al.*, 2002). To this end they identified a HS octamer that bound to monomeric IL-8 with a $K_d < 5$ nM, compared to binding constants in the μM range for binding of HS oligosaccharides to dimeric IL-8. This observation supports previous reports that IL-8 dimers bound to hep and HS with similar affinities ($K_d \sim 6$ μM) (Witt and Lander, 1994). The differences observed in GAG binding affinities between monomeric and dimeric IL-8 were thought to be due to different modes of binding to the GAG chains. Thus it was proposed that IL-8 monomer–GAG interactions involved the dimer interface thereby enabling high affinity binding but as these sites were not accessible in the dimers they bound hep/HS with much lower affinity (Goger *et al.*, 2002). The biological significance of this is as yet unclear although it has been hypothesized that monomeric IL-8 is the biologically active form of the chemokine and dimerization is a concentration-dependent means of reducing IL-8 affinity to induce diffusion of the chemokine away from the endothelium. This in effect represents a self-regulatory mechanism for the dispersal of IL-8 chemotactic gradients established through high affinity binding of monomeric IL-8 to GAGs expressed on the endothelium. Studies on the distribution of IL-8 binding sites on endothelial cells revealed different patterns of IL-8 binding to blood vessels of similar type in different organs and histological sites within organs (Rot *et al.*, 1996). These differences in binding were attributed to differences in the GAG type and structure expressed by these tissues. Thus GAGs may play key roles in defining the tissue specific expression of IL-8 and other chemokines which, in turn, regulate the selective recruitment of different leucocyte populations necessary for effective immune responses.

NMR and X-ray crystallographic studies have shown IL-8 to comprise three antiparallel β-strands connected by loop regions and one long α-helix located at the C-terminus. Two disulfide bridges contribute to the integrity of the structure (Baldwin *et al.*, 1991). IL-8 binds to hep/HS through a specific site involving the C-terminal α-helix (Webb *et al.*, 1993) and residues in the proximal loop. This loop is also implicated in receptor binding although the GAG- and receptor-sites are believed to form distinct surfaces on the IL-8 molecule (Kuschert *et al.*, 1998).

3.4. Interleukin-12 (IL-12)

IL-12 is a 70 kDa heterodimer comprising p35 and p40 subunits (Wolf *et al.*, 1991). The smaller p35 subunit has a 4α-helical bundle structure and the larger subunit shows homology to the sIL-6Rα chain. Structurally, IL-12 represents a cytokine:cytokine soluble receptor complex such as the IL-6 : sIL-6Rα complex (Meager, 1998). IL-12 is secreted largely by macrophages and antigen presenting cells (APCs) and has important functions in the early stages of the immune response, in particular in the stimulation of Th1 responses (Trinchieri, 1995). This cytokine stimulates the secretion of interferon-γ (IFN-γ) from NK cells and IFN-γ then directs the differentiation of naive Th cells into Th1 cells. Th1 cells then produce IFN-γ which stimulates further production of IL-12 from APC generating a positive feedback loop (Frucht *et al.*, 2001). Thus, IL-12 and IFN-γ function at the interface between the innate and adaptive responses, where they initiate the activation of a cell-mediated immune response. IL-12 has been shown to bind strongly to the GAGs hep and HS and, to a lesser extent, DS (CSB). The interaction was specific as IL-12 failed to recognize CSA or CSC (Hasan *et al.*, 1999). In keeping with several other cytokines, IL-12 was found to bind strongly to highly sulfated HSs and weakly or not at all to other HS species. This study also implicated the disaccharide GlcNS α1-4 IdUA-2S as contributing to the hep/HS sequence recognized by IL-12. Furthermore, the p40 subunit was also shown to bind heparin and the measured affinity, in the order of 10 nM, was found to be similar to that of the IL-12 heterodimer, suggesting that the heparin binding site may be located on this subunit. Studies on the biological effects of GAG-binding using a murine NK cell line (KY-1) showed that exogenous GAGs had no effect on the ability of IL-12 to induce IFN-γ secretion in these cells (Garnier *et al.*, 2002). However, treatment of the cells with chondroitin ABCase lead to a significant reduction in IFN-γ secretion in response to IL-12, an effect that could not be reproduced with other GAGases including chondroitinases ACI and ACII. Inhibition of IFN-γ secretion in response to IL-12 was also observed following treatment of cells with 4-methylumbelliferyl-7-β-D-xyloside, an inhibitor of intact proteoglycan synthesis. This lead to the proposal that cell surface CSB is an important mediator in the IL-12 stimulated IFN-γ secretion from NK cells, although the precise role of CSB proteoglycans in this process remains to be determined.

3.5. Gamma Interferon (IFN-γ)

X-ray studies have shown IFN-γ to be an α-helical protein comprising six helices (Ealick *et al.*, 1991). It has two potential glycosylation sites and two major forms of the cytokine exist, with one form being glycosylated at both sites (25 kDa) and the other at only one site (20 kDa). In its native state, IFN-γ is a dimer composed of both the 20 and 25 kDa forms and, although glycosylation is not required for biological activity, it is necessary for dimerization (Rinderknecht *et al.*, 1984; Yip *et al.*, 1982). The main producers of IFN-γ or immune IFN are activated T cells, NK cells and LAK cells. IFN-γ is highly pleiotropic and can be implicated in nearly all phases of the inflammatory and immune response. For example, IFN-γ induces the differentiation of naive Th cells into Th1 cells, in combination with IL-12 (Frucht *et al.*, 2001). IFN-γ also promotes the expression of surface markers or receptors on target cells. Thus, it enhances expression of MHC proteins on a range of cell types (Giacomini *et al.*, 1988; Trinchieri and Perussia, 1985), T cell Activating Protein (TAP) on T cells (Dumont *et al.*, 1988) and immunoglobulin receptors on phagocytes (Petroni *et al.*, 1988). In addition it stimulates the expression of the immunoglobulin secretory component thereby increasing the exocrine secretion of IgA and IgM antibodies (Solid *et al.*, 1987). IFN-γ can also activate macrophages and neutrophils (Pace *et al.*, 1983) as well as inducing the cytolytic activity of CTLs and LAK cells (Giovarelli *et al.*, 1988). The GAG binding properties of IFN-γ were first reported in 1991 and have subsequently been well characterized (Lortat-Jacob *et al.*, 1991). IFN-γ binds to GAGs of the hep/HS class through two domains (C1 and C2) rich in basic residues located at the C-terminus of the protein (Lortat-Jacob and Grimaud, 1991). Studies on the active HS binding sequence have shown that it consists of two N-sulfated domains of 3–4 disaccharide residues separated by an extended (15–16 disaccharide) N-acetylated region. The two N-sulfated domains interact with the C-termini of two IFN-γ monomers whilst the intervening N-acetylated sequence forms a bridge between the monomers (Lortat-Jacob *et al.*, 1995). IFN-γ–hep/HS interactions have been shown to affect the proteolytic processing of IFN-γ and hence regulate its activity. IFN-γ, in the absence of hep/HS, is cleaved at the C-terminus to remove 16 or more amino acids, with the effect of reducing the specific activity of the cytokine. When bound to hep/HS, a degree of protection is afforded and only the C-terminal 8 residues are cleaved, which surprisingly increases the specific activity of the cytokine by a factor of ten (Lortat-Jacob *et al.*, 1996a). In addition, HS may also influence IFN-γ bioactivity by regulating its interaction with its specific receptor. Surface plasmon resonance studies have shown that the affinity of IFN-γ for its receptor is reduced in the presence of hep/HS due to receptor and hep/HS binding to the same C-terminal C1 domain (Sadir *et al.*, 1998). Thus, it has been proposed that HS may regulate binding of interferon-γ to its receptor via competitive binding to the C1 domain. These studies have been supported by other workers who have demonstrated that free hep is able to inhibit the

IFN-γ stimulated anti-parasitic activity of glioblastoma cells (Daubener *et al.*, 1995) and the expression of MHC Class II proteins on endothelial cells (Fritchley *et al.*, 2000). Hep/HS has also been shown to influence the tissue distribution of IFN-γ. *In vivo* studies have shown that the accumulation and localization of IFN-γ to tissues, via binding to HS, can be inhibited by co-administration of IFN-γ with hep (Lortat-Jacob *et al.*, 1996b). Furthermore, hep increases the circulatory half-life of IFN-γ from 1.1 to 99 min by preventing its accumulation in the tissues (Lortat-Jacob *et al.*, 1996a). Thus, it appears that hep/HS can regulate IFN-γ activity in three ways. Firstly by influencing its accumulation and distribution in the tissues, secondly by regulating its specific activity through modifying the proteolytic processing of its carboxyl terminus and finally, by regulating its biological activity through competition with its specific cell surface receptor.

3.6. Summary

A number of key immune cytokines show strong and selective binding to GAG molecules. There is a large body of evidence that confirms that these interactions are not merely the result of non-specific electrostatic attraction, but that cytokines recognize specific sequences in their target GAGs. For example, GAG oligosaccharides that bind IFN-γ and IL-8 with high affinity have been isolated and the specific GAG ligand for IL-12 appears to be a sequence found in DS (CSB). The biological effect of these interactions also varies between different cytokines. Thus, for IL-2 and IL-8, GAGs serve to increase their local concentrations and present them in a biologically active form to target cells. For other cytokines, binding to GAGs directly regulates their bioactivity, as exemplified by IFN-γ, IL-7 and IL-12. Thus, exogenous hep/HS has an inhibitory effect on IL-7 and IFN-γ, and treatment of target cells with GAGases reduces the biological activities of all three cytokines. Interactions with GAGs can also protect cytokines from proteolytic degradation (IL-7 and IFN-γ) and influence the specific activity of the cytokine (IFN-γ).

Although some heparin-binding cytokines are structurally similar, others share no structural homology implying there is no single common binding mechanism by which they all recognize their GAG ligands. For example, IL-2 and IL-7 are both 4α-helix bundle proteins and also share a common receptor subunit (γc). However, the biological effects of hep/HS binding on the activity of these two cytokines are very different and the affinities with which they bind hep differ by an order of magnitude. This suggests that, despite their structural homology, they recognize GAGs via different mechanisms. IFN-γ and IL-8 share no structural homology but both interact with GAGs through binding sites located at their C-termini. Furthermore, GAG binding does appear to promote dimerization of these cytokines, however, their GAG ligands exert very different regulatory effects on their bioactivities. In order to understand more fully the functional significance of cytokine–GAG binding and its influence on immune responses, an in-depth analysis of specific GAG ligands and the mechanisms by which cytokines recognize them is required.

4. TECHNIQUES FOR EVALUATING CYTOKINE–GAG INTERACTIONS

A range of techniques has been used to study the interactions between cytokines and GAGs. These center on identifying the different classes of GAG recognised by cytokines and, more specifically, the high affinity GAG oligosaccharide that represents the *in vivo* target of the cytokine. Other studies serve to characterize GAG binding sites on the cytokine molecules, identifying the amino acid residues that contribute to GAG recognition. Finally, a spectrum of *in vitro* and *in vivo* studies have been performed to understand the biological functions of GAG–cytokine interactions and their roles in regulating cytokine activity. The following section will provide an overview of some of the experimental approaches used by workers to study cytokine–GAG interactions.

4.1. Identification of GAG-Binding Cytokines and Evaluation of GAG-Binding Properties

4.1.1. Affinity Chromatography. This technique has been used widely in the study of GAG binding to several cytokines, including IL-1α, IL-2, IL-6 (Ramsden and Rider, 1992) and IL-7 (Clarke *et al.*, 1995). It involves covalently binding hep or other GAGs to a solid support matrix and passing solutions of the cytokine over the GAG-matrix support. Binding to GAGs is determined by retention of the cytokine on the matrix. The strength of the interaction can also be ascertained in terms of the molar salt concentration required for eluting the cytokine from the immobilized GAG. Using this technique, Clarke *et al.* (1995) determined that IL-7 bound hep and HS but not CS or DS. Furthermore, the interactions of IL-7 with hep were shown to be stronger than those to HS, since IL-7 was eluted from a hep-sepharose column in buffer containing 0.6 M NaCl and from HS-sepharose columns in the presence of 0.3 M NaCl. Although this technique is simple to perform and allows strength of GAG binding to be investigated, affinities in terms of M NaCl cannot be readily compared to equilibrium constants determined for cytokine interactions with other molecules, such as their specific receptors. A comparison of the relative affinities of both is significant when evaluating the biological role of cytokine–GAG binding. Techniques that will enable the estimation of equilibrium constants (K_d) for cytokine–GAG interactions are affinity co-electrophoresis (ACE) and, more recently, surface plasmon resonance (SPR), both of which will be discussed in Sections 4.1.3. and 4.1.4.

4.1.2. Elisa. ELISA-based techniques have proven to be highly informative in the identification of hep-binding cytokines and the structural requirements for these interactions. One such technique was reported in 1997 and involves the synthesis of a heparin–bovine serum albumin (hep–BSA) conjugate as a means of immobilising heparin to the surface of an ELISA plate well (Najjam *et al.*, 1997). A solution of the cytokine of increasing concentration is then added and bound cytokine detected using appropriate anti-cytokine antibodies. Furthermore, the interaction of the cytokine with the hep–BSA complex can be inhibited by pre-incubation of the cytokine with increasing concentrations of free hep. The concentration of hep giving 50% inhibition of binding (IC_{50}) thus provides an estimate of the affinity of the interaction. This technique was validated using two well-characterized hep-binding proteins, basic fibroblast growth factor (bFGF) and antithrombin III (ATIII). For both proteins the IC_{50} values obtained, 80 nM for ATIII and 12 nM for bFGF, were shown to correlate well with their previously determined equilibrium constants (50 nM and 6 nM, respectively). Dose dependent binding to hep has subsequently been demonstrated for IL-2 (Najjam *et al.*, 1997), IL-7 (Guelle *et al.*, 2000) and IL-12 (Hasan *et al.*, 1999) using this ELISA technique and the IC_{50} values estimated for all three cytokines were within the nM range. Inhibition studies have also been used to evaluate the full GAG binding properties of specific cytokines. These involved pre-incubating a solution of the cytokine with a fixed concentration of glycosaminoglycan before applying it to the immobilized hep–BSA complex. Interaction with free GAG thus inhibits cytokine binding to the complex and the extent of the observed inhibition is a reflection of how strongly the cytokine binds to a particular class of GAG. This technique has been used to demonstrate selective IL-7 binding to highly sulfated GAG species such as hep, HS and fucoidan (Guelle *et al.*, 2000). Such inhibition studies have also been used to identify the structural requirements for GAG recognition by cytokines. Using chemically modified heparins that have been selectively N-, 2-O-, or 6-O-desulfated as the inhibitor species, the requirement for particular sulfate groups on the hep molecule for high affinity binding can be ascertained. Using this approach, hep-IL-7 binding was shown to be strongly influenced by the presence of 6-O-sulfate groups and, to a lesser extent, N- and 2-O-sulfates (Guelle *et al.*, personal communication). Heparin oligosaccharides of defined size, generated by both chemical and enzymatic cleavage, can also be used to identify those that bind with high affinity. Such information provides insight into the nature of the GAG ligands specifically recognized by individual cytokines. For example, IL-12 was reported to bind more strongly to decasaccharides and octasaccharides generated by chemical cleavage than to equivalent size counterparts generated by cleavage with heparinase I. Since heparinase I cleaves selectively between N-sulfated glucosamines and 2-O-sulfated iduronates, the presence of the disaccharide GlcNS α1-4 IdUA-2S was strongly implicated in IL-12 binding (Hasan *et al.*, 1999).

4.1.3. Affinity Co-electrophoresis (ACE). This method measures the interaction of growth factors or cytokines with ^{125}I-GAGs and has been used to estimate the dissociation constants (K_d) of these interactions (Lee and Lander, 1991). Essentially the technique involves incorporating varying concentrations of a cytokine, prepared in an agarose support, into large wells running lengthways through an agarose gel. Radioiodinated GAGs are then applied to a well at the top of the gel and electrophoresed through the cytokine samples. The electrophoretic migration of the radiolabelled GAG can be visualized by autoradiography and the extent to which migration has been retarded through growth factor binding measured. From this the retardation coefficient (R) can be calculated and, since R varies with protein concentration according to the Scatchard equation, the K_d for cytokine–GAG binding can be readily determined. Detailed methodology for this technique has been described by Gibbs and Gordon (1996). This technique has been used to estimate the affinity constants for the interaction of several proteins, including IL-7, with their GAG ligands (Clarke *et al.*, 1995; Lee and Lander, 1991; San Antonio *et al.*, 1993). One disadvantage of this technique, however, is that it relies on the use of radioiodinated GAGs and hence modification of the GAG chains, which could potentially interfere with active binding sequences along the GAG chain. Despite this, the K_d values determined have been shown to be in broad agreement with estimations of binding affinities generated by other techniques. Thus, for IL-7-hep binding, a K_d value of 25 nM was determined by ACE (Clarke *et al.*, 1995) and an IC_{50} value of 100 nM estimated by ELISA (Guelle *et al.*, 2000).

4.1.4. Surface Plasmon Resonance. More recently, the technique of surface plasmon resonance (SPR) has found value in examining the binding kinetics of cytokine–heparin interactions. In this technique, biomolecular interactions take place on a sensor chip surface, which consists of a glass slide coated on one side with a thin layer of gold. The gold surface is covered with a covalently bound matrix, usually carboxymethylated dextran, onto which biomolecules are immobilized. The immobilized ligand is then exposed to an interacting molecule (analyte) which is passed over the sensor chip surface under conditions of continuous flow. As biomolecular interactions occur, the refractive index of the aqueous medium changes in direct relation to the mass quantity of analyte that is bound to the sensor chip surface. A biosensor instrument measures changes in the refractive index of the aqueous medium, through a SPR phenomenon, and records these changes as a sensogram of changes in SPR signal with time. The advantage of SPR technology for evaluating biomolecular interactions is that it allows for the study of real-time binding events which is ideal for obtaining kinetic data. Furthermore, it is a rapid, automated technique that does not necessarily require that the molecules under study are labelled or otherwise modified. More detailed information on the theory of SPR and use of biosensors has been reported by Nice and Catimel (1999), Rich and Myszaka (2000), and Salamon *et al.* (1997).

Biosensors have been used to evaluate successfully the interaction of GAGs with a number of cytokines, including acidic fibroblast growth factor (aFGF) (Mach *et al.*, 1993), basic fibroblast growth factor (bFGF) (Stearns *et al.*, 1997) and platelet derived growth factor (PDGF) (Lustig *et al.*, 1996). In these studies, hep was biotinylated and immobilized onto the surface of sensor chips coated with streptavidin. Other workers have avoided the need for chemical modification of hep prior to immobilization by conjugating hep via its reducing terminus to either biotinylated bovine serum albumin, which subsequently interacts with immobilized streptavidin (Guelle *et al.*, 2000, Salek-Ardakani *et al.*, 2000), or by direct conjugation onto carboxymethylated sensor chip surfaces (Nika, K., personal communication). Using this approach, Salek-Ardakani *et al.* (2000) calculated an equilibrium constant (K_d) of 54 nM for the interaction of IL-10 with hep. Preliminary studies on IL-7–hep binding generated similar K_d values of 50 nM and 40 nM for the interaction of human and murine IL-7 with unfractionated heparin (Guelle *et al.*, 2000), however, subsequent refinement of the experimental procedure indicated that they were nearer to 70 nM and 160 nM for these cytokines (Nika K., personal communication). Nevertheless, these values broadly agree with previous IC_{50} values of ~80 nM and 50 nM, determined for human and murine IL-7 using a modified ELISA approach (Guelle *et al.*, 2000), but are higher than the dissociation constant (K_d) of 25 nM, determined by ACE (Clarke *et al.*, 1995). Biosensors have also been used to evaluate the role of hep/HS in the recognition, by IFN-γ, of its specific receptor (Sadir *et al.*, 1998). These studies revealed that HS reduced the affinity of IFN-γ for its receptor as the residues forming the HS binding site, the C-terminal C1 domain, also functioned to increase the on-rate of the IFN-γ : IFN-γR binding reaction. Thus SPR studies may provide valuable information on the role of GAGs in mediating cytokine receptor recognition.

4.2. Identification of GAG Binding Sites on Cytokine Molecules

GAG binding sites are known to involve clusters of basic residues that interact with the negatively charged sulfate groups on GAG molecules. These basic residue clusters are located approximately 20 Å apart on the cytokine surface, a distance that corresponds to the spatial arrangement of sulfate groups on a hep molecule (Spillmann and Lindahl, 1994). Residues contributing to a particular GAG site can all originate from the same part of the polypeptide chain, forming a continuous site, or from different positions within the primary sequence which then come together on folding of the polypeptide chain (discontinuous site). From the study of many heparin-binding proteins, several workers have published consensus sequences for the basic residues forming continuous sites (Cardin and Weintraub, 1989; Fromm *et al.*, 1997; Hileman *et al.*, 1998), although these do not occur in all proteins that bind heparin. Thus, the lack of a known consensus sequence in the primary structure of a cytokine does not automatically infer that it will not bind to GAGs. The basic residues contributing to GAG binding sites have only been identified for a small number of immune cytokines, including IFN-γ, IL-8 and IL-2 and, in each study, several approaches were employed to identify these residues.

In a study conducted by Lortat-Jacob and Grimaud (1991), the hep/HS binding domain on IFN-γ was localized to the C-terminus by domain mapping with epitope-mapped monoclonal anti-IFN-γ antibodies, in the presence and absence of HS. Found within this region of the IFN-γ protein were two clusters of basic residues that were referred to as Domain 1 (124–131) and Domain 2 (137–140). Peptides representing these two domains were then synthesized, but only the peptide representing Domain 1 was shown to compete with hep/HS for binding to IFN-γ. However, HS binding was shown to inhibit digestion of IFN-γ by both chymotrypsin, which cleaves between basic residues (K-R) in Domain 1 and Carboxypeptidase Y, which removed C-terminal residues including Domain 2. Thus, it was proposed that basic residues in Domain 1 (^{128}KRKR^{131}) and, to a lesser extent, Domain 2 (^{137}RGRR^{140}) both contributed to the hep/HS binding site on IFN-γ.

The C-terminal region of IL-8 was also implicated in hep/HS binding. This was accomplished by assessing the ability of N- and C-terminally truncated IL-8 mutants to bind to hep-sepharose columns. Binding was not observed in those mutants with truncated C-termini, suggesting that the hep binding site had been disrupted through loss of the C-terminus (Webb *et al.*, 1993). In another study, IL-8 muteins were produced containing single point mutations at the positions of each basic residue. The ability of mutants to bind to hep-sepharose columns was then assessed leading to the proposal that residue K20 in addition to the C-terminal residues R60, K64, K67, and R68 were involved in hep/HS binding (Kuschert *et al.*, 1998), thus confirming previous work. Also in this study, the binding of an active hep dissacharide to ^{15}N-IL-8 was examined using NMR spectroscopy. Results of these experiments implicated residues 18–23, which form an N-terminal loop region, in addition to the C-terminal α-helix as contributing to a discontinuous hep/HS binding surface on IL-8.

In 1998, two putative heparin binding sites were identified on the surface of IL-2. The first of these was a continuous site comprising residues K48, K49, K54, and H55 and the second a discontinuous site formed from residues K32, K76, R81, and R83 (Najjam *et al.*, 1998). These sites were identified by a combination of molecular modelling techniques, domain mapping with epitope mapped monclonal anti-IL-2 antibodies and assessment of the heparin binding properties of a series of IL-2 muteins using a modified ELISA technique. Molecular modelling revealed that both of these proposed sites did not interfere with IL-2-IL-2R binding, which was consistent with the hypothesis that hep/HS GAGs sequester IL-2 in an active form close to its site of production (Najjam *et al.*, 1997). This was further supported by inhibition studies by the same workers who showed that soluble recombinant IL-2Rα and IL-2Rβ polypeptides did not interfere with IL-2 heparin binding.

4.3. Evaluation of the Biological Role of Cytokine–GAG Interactions

To date, studies investigating the biological role of cytokine–GAG interactions have centred on *in vitro* cell culture assays. These frequently employ cell lines that produce a measurable biological response as a result of exposure to a specific cytokine. For example, studies on IL-2 (Ramsden and Rider, 1992), IL-5 (Lipscombe *et al.*, 1998), IL-6 (Mummery and Rider, 2000) and IL-7 (Clarke *et al.*, 1995) have used cell lines that proliferate in response to cytokine exposure and the proliferative response can be readily measured by, for example, 3H-thymidine uptake. In other cell lines used, cytokines induce the expression of other molecules. For example, studies on IL-12 have used a NK cell line KY-1 that secretes IFN-γ in response to IL-12 stimulation; the IFN-γ produced is subsequently quantitated using an ELISA technique (Garnier *et al.*, 2002). Investigations into the biological role of IFN-γ-hep/HS interactions have used a number of different cell lines including endothelial (Fritchley *et al.*, 1998), glioblastoma (Daubener *et al.*, 1995) and adenocarcinoma (Fernandez-Botran *et al.*, 1999) cell lines. In all instances these can be induced to express MHC class II and other proteins in response to stimulation with IFN-γ. Expression of these molecules can then be assessed by immunofluorescence or RT-PCR methods. Other workers have used primary cultures rather than cell lines. For example, IL-10-hep biological activity was examined using peripheral blood monocytes which express the markers CD16 and CD64 in response to IL-10 stimulation (Salek-Ardakani *et al.*, 2000). Similarly, IL-7 bioactivity has been studied using murine long term bone marrow cultures (LTBMC) cultured under conditions to support both lymphopoiesis and myelopoiesis (Borghesi *et al.*, 1999). In all of these investigations, the biological role of GAG–cytokine interactions was assessed in two ways. Firstly, by examining the effect of exogenous free GAG on the bioactivity of a specific cytokine and, secondly, by removing GAGs from the surface of target cells using specific GAGases. Through measurements of cell proliferation or induction of protein expression, the ability of GAGs to regulate cytokine bioactivity has been ascertained.

Only a small number of *in vivo* studies have been performed and these have focused on evaluating the role of GAGs in regulating the *in vivo* distribution and tissue localization of cytokines. Investigations have been conducted in mice on the immune cytokines IL-2 (Wrenshall and Platt, 1999) and IFN-γ (Lortat-Jacob *et al.*, 1996a; Lortat-Jacob *et al.*, 1996b). In both instances, the cytokines were shown to localise to specific tissues through interactions with cell surface or extracellular matrix HS. Furthermore, this effect could be blocked by the administration of free GAG (hep) that effectively acted as a competitive inhibitor of these interactions. Thus GAGs have been shown to have a central role in regulating the bioavailability of immune cytokines within different tissues and, in this way, having the potential to make significant contributions to the regulation of immune responses.

5. CONCLUSIONS

A number of immune cytokines have been shown to bind to GAGs, recognising specific saccharide sequences in the GAG chains with high affinity. Structural studies indicate that GAG recognition does not occur via a common binding mechanism and this is reflected in the broad range of biological effects reported for these interactions. For some cytokines, binding to GAGs appears to be a requirement for subsequent interaction with specific receptors on target cells. For other cytokines, GAG-binding has no measurable effect on cytokine bioactivity but instead is a potentially important means for regulating cytokine bioavailability. Thus, in conjunction with their tissue specific expression, GAGs may provide an important mechanism by which immune responses can be regulated. A thorough understanding of these interactions will therefore contribute to our understanding of the immune system and its regulation, which is essential for the successful manipulation of the immune response for therapeutic gain.

6. REFERENCES

Appasamy, P., 1999, Biological and clinical implications of interleukin-7 and lymphopoiesis, *Cytokines Cell Mol Ther.* 5:25–39.

Baldwin, E.T., Weber, I.T., St. Charles, R., Xuan, J.-C., Appella, E., Yamada, M., Matsushima, K., Edwards, B.F.P., Clore, G.M., Gronenborn, A.M., and Wlodawer, A., 1991, Crystal structure of interleukin 8: symbiosis of NMR and crystallography, *Proc Natl Acad Sci.* 88:502–506.

Bazan, F.J. and McKay, D.B., 1992, Unraveling the structure of IL-2, *Science.* 257:410–412.

Billips, L.G., Petitte, D., Dorshkind, K., Narayanan, R., Chiu, C., and Landreth, K.S., 1992, Differential roles of stromal cells, interleukin-7 and *kit*-ligand in the regulation of B lymphopoiesis, *Blood.* 79:1185–1192.

Borghesi, L.A., Yamashita, Y., and Kincade, P.W., 1999, Heparan sulfate proteoglycans mediate interleukin-7-dependent B lymphopoiesis, *Blood.* 93:140–148.

Cardin, A.D. and Weintraub, H.J.R., 1989, Molecular modeling of protein-glycosaminoglycan interactions, *Arteriosclerosis.* 9:21–32.

Casu, B., 1985, Structure and biological activity of heparin, *Adv Carbohydr Chem Biochem.* 43:51–135.

Chazen, G.D., Pereira, G.M.B., LeGros, G., Gillis, S., and Shevach, E.M., 1989, Interleukin-7 is a T cell growth factor, *Proc Natl Acad Sci.* 86:5923–5927.

Clarke, D., Katoh, O., Gibbs, R.V., Griffiths, S.D., and Gordon, M.Y., 1995, Interaction of interleukin 7 (IL-7) with glycosaminoglycans and its biological relevance, *Cytokine.* 7:325–330.

Cosenza, L.L., Sweeney, E.E., and Murphy, J.R., 1997, Disulphide bond assignment in human interleukin-7 by matrix assisted laser desorption/ionization mass spectroscopy and site-directed cysteine to serine mutational analysis, *J Biol Chem.* 272:32995–33000.

Costello, R., Imbert, J., and Olive, D., 1993, Interleukin-7, a major T-lymphocyte cytokine, *Eur Cytokine Netw.* 4:253–262.

Coutino, A., Lotta, E., Larsson, E., Gronvik, K., and Andersson, J., 1979, Studies on T cell activation II. The targets for concanavalin A-induced growth factors, *Eur J Immunol.* 9:587–592.

Daubener, W., Nockemann, S., Gutsche, M., and Hadding, U., 1995, Heparin inhibits the antiparasitic and immune modulatory effects of human recombinant interferon-γ, *Eur J Immunol.* 25:688–692.

Dumont, F.J., Palfree, R.G.E., and Fisher, P.A., 1988, The T-cell activating protein (TAP) is up-regulated by endogenous IFN-γ in activated T cells, *Immunol.* 64:267–271.

Ealick, S.E., Cook, W.J., Vijay-Kumar, S., Carson, M., Nagabhushan, T.L., Trotta, P.P., and Bugg, C.E., 1991, Three-dimensional structure of recombinant human interferon-γ, *Science.* 252:698–702.

Farrar, J.J., Benjamin, W.R., Hilfiker, M.L., Howard, M., Farrar, W.L., and Farrar-Fuller, J., 1982, The biochemistry, biology and role of interleukin 2 in the induction of cytotoxic T cell and antibody-forming B cell responses, *Immunol Rev.* 63:129–166.

Fernandez-Botran, R., Yan, J., and Justus, D.E., 1999, Binding of interferon γ by glycosaminoglycans: a strategy for localization and/or inhibition of its activity, *Cytokine.* 11:313–325.

Fritchley, S.J., Kirby, J.A., and Ali, S., 2000, The antagonism of interferon-gamma (IFN-γ) by heparin: examination of the blockade of class II MHC antigen and heat shock protein-70 expression, *Clin Exp Immunol.* 120:247–252.

Fromm, J.R., Hileman, R.E., Caldwell, E.E.O., Weiler, J.M, and Linhardt, R.J., 1997, Pattern and spacing of basic amino acids in heparin binding sites, *Arch Biochem Biophys.* 343:92–100.

Frucht, D.M., Fukao, T., Bogdan, C., Schindler, H., O'Shea, J.J., and Koyasu, S., 2001, IFN-γ production by antigen presenting cells: mechanisms emerge, *Trends Immunol.* 22:556–560.

Funderburgh, J.L., 2000, Keratan sulphate: structure, biosynthesis and function, *Glycobiology.* 10:951–958.

Gallagher, J.F., Lyon, M., and Steward, W.P., 1986, Structure and function of heparan sulphate proteoglycans, *Biochem J.* 236:313–325.

Garnier, P., Gibbs, R.V., and Rider, C.C., 2003, A role for chondroitin sulphate B in the activation of IL-12 in stimulating γ-IFN secretion, *Immunol Lett.* 85:53–58.

Giacomini, P., Tecce, R., Gambari, R., Sacchi, A., Fisher, P.B., and Natali, P.G., 1988, Recombinant human IFN-γ, but not IFN-α or IFN-β, enhances MHC- and non-MHC-encoded glycoproteins by a protein-synthesis-dependent mechanism, *J Immunol.* 140:3073–3081.

Gibbs, R.V. and Gordon, M.Y., 1996, Binding of growth factors to extracellular matrix proteins and glycoproteins, in: *Cell and tissue culture: laboratory procedures.* Ch 8E 1.1–1.19. eds. J.B. Griffiths, A. Doyle, D.G. Newell, John Wiley and Sons Ltd, W. Sussex, UK.

Giovarelli, M., Santoni, A., Jemma, C., Musso, T., Giuffrida, A., Cavallo, G., Landolfo, S., and Forni, G., 1988, Obligatory role of IFN-γ in induction of lymphokine-activated and T lymphocyte killer cell activity, but not in boosting of natural cytotoxicity, *J Immunol.* 141:2831–2836.

Goger, B., Halden, Y., Rek, A., Mosi, R., Pye, D., Gallagher, J., and Kungl, A.J., 2002, Different affinities of glycosaminoglycan oligosaccharides for monomeric and dimeric interleukin-8: a model for chemokine regulation at inflammatory sites, *Biochemistry.* 41:1640–1646.

Gordon, M.Y., Riley, G.P., Watt, S.M., and Greaves, M.F., 1987, Compartmentalization of a growth factor (GM-CSF) by glycosaminoglycans in the bone marrow microenvironment, *Nature.* 326:403–405.

Grimm, E.A., Robb, R.J., Roth, L.M., Neckers, L.M., Lachman, L.B., Wilson, D.J., and Rosenburg, S.A., 1983, Lymphokine-activated killer cell phenomenon. III. Evidence that IL-2 is sufficient for direct activation of peripheral blood lymphocytes, *J Exp Med.* 158:1356–1361.

Guelle, M., Nika, K., Mernagh, D., Mulloy, B., Forster M., and Gibbs, R., 2000, Characterisation of interleukin-7 (IL-7)-glycosaminoglycan interactions, *Minerva Biotecnologica.* 12:121–122.

Hasan, M., Najjam, S., Gordon, M.Y., Gibbs, R.V., and Rider, C.C., 1999, IL-12 is a heparin-binding cytokine, *J Immunol.* 162:1064–1070.

Hascall, V.C., Calabro, A., Midura, R.J., and Yanagishita, M., 1994, Isolation and characterization of proteoglycans, *Methods Enzymol.* 230:390–417.

He, Y.W. and Malek, T.R., 1998, The structure and function of γc-dependent cytokines and receptors: regulation of T lymphocyte development and homeostasis, *Crit Rev Immunol.* 18:503–524.

Hileman, R.E., Fromm, J.R., Weiler, J.M., and Linhardt, R.J., 1998, Glycosaminoglycan-protein interactions: definition of consensus sites in glycosaminoglycan binding proteins, *BioEssays.* 20:156–167.

Heney, C.S., Kuribayashi, K., Kern, D.E., and Gillis, S., 1981, Interleukin-2 augments natural killer cell activity, *Nature.* 291:335–338.

Hoogewerf, A.J., Kuschert, G.S.V., Proudfoot, A.E.I., Borlat, F., Clark-Lewis, I., Power, C.A., and Wells, T.N.C., 1997, Glycosaminoglycans mediate cell surface oligomerization of chemokines, *Biochemistry.* 36:13570–13578.

Kjellen, L. and Lindahl, U., 1991, Proteoglycans: structures and interactions, *Annu Rev Biochem.* 60:443–475.

Kolset, S.O. and Gallagher, J.F., 1990, Proteoglycans in haemopoietic cells, *Biochim Biophys Acta.* 1032:191–211.

Kroemer, R.T., Doughty, S.W., Robinson, A.J., and Richards, W.G., 1996, Prediction of the three-dimensional structure of human interleukin-7 by homology modeling, *Prot Engin.* 9:493–498.

Kuschert, G.S.V., Hoogewerf, A.J., Proudfoot, A.E.I., Chung, C.-W., Coooke, R.M., Hubbard, R.E., Wells, T.N.C., and Sanderson, P.N., 1998, Identification of a glycosaminoglycan binding surface on human interleukin-8, *Biochemistry.* 37:11193–11201.

Laurent, T.C. and Fraser, J.R., 1992, Hyaluronan, *FASEB J.* 6:2397–2404.

Scott, J.E., 1993, The nomenclature of glycosaminoglycans and proteoglycans, *Glycoconj J.* 10:419–421.

Lee, K.K. and Lander, A.D., 1991, Analysis of affinity and structural selectivity in the binding of proteins to glycosaminoglycans: development of a sensitive electrophoretic approach, *Proc Natl Acad Sci USA.* 88:2768–2772.

Lindahl, U., Thunberg, L.L., Backstrom, G., Riesenfeld, J., Nordling, K., and Bjork, I., 1984, Extension and structural variability of the antithrombin-binding sequence in heparin, *J Biol Chem.* 259:12368–12376.

Lipscombe, R.J., Nalchoul, A.M., Sanderson, C.J., and Coombe, D.R., 1998, Interleukin-5 binds to heparin/heparan sulfate. A model for an interaction with extracellular matrix, *J Leukoc Biol.* 63:342–350.

Lortat-Jacob, H., Baltzer, F., and Grimaud, J.-A., 1996a, Heparin decreases the blood clearance of interferon-γ and increases its activity by limiting the processing of its carboxyl-terminal sequence, *J Biol Chem.* 271:16139–16143.

Lortat-Jacob, H., Brisson, C., Guerret, S., and Morel, G., 1996b, Non-receptor-mediated tissue localization of human interferon-γ: role of heparan sulphate/heparin-like molecules, *Cytokine.* 8:557–566.

Lortat-Jacob, H., Garrone, P., Banchereau, J., and Grimaud, J.-A., 1997, Human interleukin 4 is a glycosaminoglycan-binding protein, *Cytokine.* 9:101–105.

Lortat-Jacob, H. and Grimaud, J.-A., 1991, Interferon-γ binds to heparan sulfate by a cluster of amino acids located in the C-terminal part of the molecule, *FEBS Letts.* 280:152–154.

Lortat-Jacob, H., Kleinman, H.K., and Grimaud, J.-A., 1991, High-affinity binding of interferon-γ to a basement membrane complex (matrigel), *J Clin Invest.* 87:878–883.

Lortat-Jacob, H., Turnbull, J.E., and Grimaud, J.-A., 1995, Molecular organisation of the interferon γ-binding domain in heparan sulphate, *Biochem J.* 310:497–505.

Lustig, F., Hoebeke, J., Olsson, U., and Fager, G.G., 1996, Alternative splicing determines the binding of platelet-derived growth factor to glycosaminoglycans, *Biochemistry.* 35:12077–12085.

Lynch, D.H. and Miller, R.E., 1992, Induction of murine lymphokine-activated killer cells by recombinant IL-7, *J Immunol.* 145:1983–1990.

Mach, H., Volkin, D.B., Burke, C.J., and Middaugh, R.C., 1993, Nature of the interaction of heparin with acidic fibroblast growth factor, *Biochemistry.* 32:5480–5489.

Maimone, M.M. and Tollefsen, D.M., 1990, Structure of a dermatan sulfate hexasaccharide that binds to heparin cofactor II with high affinity, *J Biol Chem.* 265:18263–18271.

Martin, L., Blanpain, C., Garnier, P., Wittamer, V., Parmentier, M., and Vita, C., 2001, Structural and functional analysis of the RANTES-glycosaminoglycans interactions, *Biochemistry.* 40:6303–6318.

McFadden, G. and Kelvin, D., 1997, New strategies for chemokine inhibition and modulation, *Biochem Pharmacol.* 54:1271–1280.

McKay, D.B., 1992, Unraveling the structure of IL-2 Response, *Science.* 257:412–413.

Meager, T., 1998, *The Molecular Biology of Cytokines.* John Wiley & Sons, West Sussex, UK.

Meuer, S.C., Hussey, R.E., Cantrell, D.A., Hodgon, J.C., Schlossman, S.F., Smith, K.A., and Reinherz, E.L., 1984, Triggering of the T3-Ti antigen-receptor complex results in clonal T-cell proliferation through an interleukin-2-dependent autocrine pathway, *Proc Natl Acad Sci.* 81:1509–1513.

Mummery, R.S. and Rider C.C., 2000, Characterization of the heparin-binding properties of IL-6, *J Immunol.* 165:5671–5679.

Najjam, S., Gibbs R.V., Gordon, M.Y., and Rider, C.C., 1997, Characterization of human recombinant interleukin 2 binding to heparin and heparan sulfate using an ELISA approach, *Cytokine.* 9:1013–1022.

Najjam, S., Mulloy, B., Theze, J., Gordon, M., Gibbs, R., and Rider, C.C., 1998, Further characterization of the binding of human recombinant interleukin 2 to heparin and identification of putative binding sites, *Glycobiology.* 8:509–516.

Naume, B. and Espevik, T., 1991, Effects of IL-7 and IL-2 on highly enriched $CD56^+$ natural killer cells, *J Immunol.* 147:2208–2214.

Nice, E.C. and Catimel, B., 1999, Instrumental biosensors: new perspectives for the analysis of biomolecular interactions, *BioEssays.* 21:339–352.

Otsu, K., Inoue, H., Tsuzuki, Y., Yonekura, H., Nakanishi, Y., and Suzuki, S., 1985, A distinct terminal structure in newly synthesised chondroitin sulphate chains, *Biochem J.* 227:37–48.

Pace, J.L., Russell, S.W., Torres, B.A., Johnson, H.M., and Gray, P.W., 1983, Recombinant mouse γ interferon induces the priming step in macrophage activation for tumour cell killing, *J Immunol.* 130:2011–2013.

Petroni, K.C., Shen, L., and Guyer, P.M., 1988, Modulation of human polymorphonuclear leucocyte IgG Fc receptors and Fc receptor-mediated functions by IFN-γ and glucocorticoids, *J Immunol.* 140:3467–3472.

Ramsden, L. and Rider, C.C., 1992, Selective and differential binding of interleukin-1α, interleukin-2 and interleukin-6 to glycosaminoglycans, *Eur J Immunol.* 22:3027–3031.

Rich, R.L. and Myszka, D.G., 2000, Advances in surface plasmon resonance biosensor analysis, *Curr Opin Biotech.* 11:54–61.

Rider, C.C., Coombe, D.R., Harrop, H.A., Hounsel, E.F., Bauer, C., Feeney, J., Mulloy, B., Mahmood, N., Hay, A., and Parish, C.R., 1994, Anti-HIV-1 activity of chemically modified heparins: correlation between binding to the V3 loop of gp120 and inhibition of cellular HIV-1 infection *in vitro, Biochemistry.* 33:6974–6980.

Rinderknecht, E., O'Connor, B.H., and Rodriguez, H., 1984, Natural human interferon-γ, *J Biol Chem.* 259:6790–6797.

Robb, R.J. and Smith, C., 1981, Heterogeneity of human T-cell growth factor(s) due to variable glycosylation, *Mol Immunol.* 18:1087–1094.

Roberts, R., Gallagher, J., Spooncer, E., Allen, T.D., Bloomfield, F., and Dexter, T.M., 1988, Heparan sulphate bound growth factors: a mechanism for stromal cell mediated haemopoiesis, *Nature.* 332:376–378.

Rot, A., 1992, Endothelial cell binding of NAP-1/IL-8: role in neutrophil emigration, *Immunol Today.* 13:291–294.

Rot, A., Hub, E., Middleton, J., Pons, F., Rabeck, C., Thierer, K., Wintle, J., Wolff, B., Zsak, M., and Dukor, P., 1996, Some aspects of IL-8 pathophysiology III: chemokine interaction with endothelial cells, *J Leukoc Biol.* 59:39–44.

Sadir, R., Forest, E., and Lortat-Jacob, H., 1998, The heparan sulfate binding sequence of interferon-γ increased the on rate of the interferon-γ-interferon-γ receptor complex formation, *J Biol Chem.* 273:10919–10925.

Salaman, Z., Macleod, A.H., and Tollin, G., 1997, Surface plasmon resonance spectroscopy as a tool for investigating the biochemical and biophysical properties of membrane protein systems. I: Theoretical principles, *Biochim Biophys Acta.* 1331:117–129.

Salek-Ardakani, S., Arrand, J.R., Shaw, D., and Mackett, M., 2000, Heparin and heparan sulphate bind IL-10 and modulate its activity, *Blood.* 96:1879–1888.

San Antonio, J.D., Slover, J., Lawler, J., Karnovsky, M.J., and Lander, A.D., 1993, Specificity in the interactions of extracellular matrix proteins with subpopulations of the glycosaminoglycan heparin, *Biochemistry.* 32:4746–4755.

Scott, J.E., 1993, The nomenclature of glycosaminoglycans and proteoglycans, *Glycoconj J.* 10:419–421.

Solid, L.M., Kvale, D., Brrandtzaeg, P., Markussen, G., and Thorsby, E., 1987, Interferon-γ enhances expression of secretory component, the epithelial receptor for polymeric immunoglobulins, *J Immunol.* 138:4303–4306.

Spillmann, D. and Lindahl, U., 1994, Glycosaminoglycan-protein interactions: a question of specificity, *Curr Opin Struc Biol.* 4:677–682.

Spillmann, D., Witt, D., and Lindahl, U., 1998, Defining the interleukin-8-binding domain of heparan sulfate, *J Biol Chem.* 273:15487–15493.

Stearns, N.A., Prigent-Richard, S., Letourneur, D., and Castellot, J.J., 1997, Synthesis and characterisation of highly sensitive heparin probes for detection of heparin-binding proteins, *Anal Biochem.* 247:348–356.

Tanaka, Y., Adams, D.H., and Shaw, S., 1993, Proteoglycans on endothelial cells present adhesion-inducing cytokines to leucocytes, *Immunol Today.* 14:111–115.

Toida, T., Yoshida, H., Toyoda, H., Koshiishi, I., Imanari, T., Hileman, R.E., Fromm, J.R., and Lindardt, R.J., 1997, Structural differences and the presence of unsubstituted amino groups in heparan sulphates from different tissues and species, *Biochem J.* 332:499–506.

Trinchieri, G., 1995, Interleukin-12: a proinflammatory cytokine with immunoregulatory functions that bridge innate resistance and antigen-specific adaptive immunity, *Annu Rev Immunol.* 13:251–276.

Trinchieri, G. and Perussia, B., 1985, Immune interferon: a pleiotropic lymphokine with multiple effects, *Immunol Today.* 6:131–136.

Turnbull, J.E. and Gallagher, J.T., 1990a, Distribution of iduronate 2-sulphate residues in heparan sulphate, *Biochem J.* 273:553–559.

Turnbull, J.E. and Gallagher, J.T., 1990b, Molecular organisation of heparan sulphate from human skin fibroblasts, *Biochem J.* 265:715–724.

Vives, R.R., Pye, D.A., Salmivirta, M., Hopwoods, J.J., Lindahl, U., and Gallagher, J.T., 1999, Sequence analysis of heparan sulphate and heparin oligosaccharides, *Biochem J.* 339:767–773.

Webb, L.M.C., Ehrengruber, M.U., Clark-Lewis, I., Baggiolini, M., and Rot, A., 1993, Binding to heparan sulfate or heparin enhances neutrophil responses to IL-8, *Proc Natl Acad Sci USA.* 90:7158–7162.

Witt, D.P. and Lander, A.D., 1994, Differential binding of chemokine to glycosaminoglycan subpopulations, *Curr Biol.* 4:394–400.

Wolf, S.F., Temple, P.A., Kobayashi, M., Young, D., Dicig, M., Lowe, L., Dzialo, R., Fitz, L., Ferenz, C., Hewick, R.M., Kelleher, K., Herrmann, S.H., Clark, S.C., Azzoni, L., Chan, S.H., Trinchieri, G., and Perussia, B., 1991, Cloning of cDNA for natural killer cell stimulatory factor, a heterodimeric cytokine with multiple effects on T and natural killer cells, *J Immunol.* 146:3074–3081.

Wrenshall, L.E. and Platt, J.L., 1999, Regulation of T cell homeostasis by heparan sulfate-bound IL-2, *J Immunol.* 163:3793–3800.

Yamada, S., Yoshida, K., Sugiura, M., and Sugahara, K., 1992, One- and two-dimensional H-NMR characterization of two series of sulfated oligosaccharides prepared from chondroitin sulfate and heparan sulfate/heparin by bacterial eliminase digestion, *J Biochem.* 112:440–447.

Yanagishita, M., 1994, A brief history of proteoglycans, *EXS.* 70:3–7.

Yanagishita, M. and Hascall, V.C., 1992, Cell surface heparan sulfate proteoglycans, *J Biol Chem.* 287:9451–9454.

Yayon, A., Klagsburn, M., Esko, J.D., Leder, P., and Ornitz, D.M., 1991, Cell surface heparin-like molecules are required for binding of basic fibroblast growth factor to its high affinity receptor, *Cell.* 64:841–848.

Yip, Y.K., Barrowclough, B.S., Urban, C., and Vilcek, J., 1982, Purification of two subspecies of human γ (immune) interferon, *Proc Natl Acad Sci USA.* 79:1820–1824.

GLYCOPATHOLOGY

10

THOMSEN-FRIEDENREICH ANTIGEN: THE "HIDDEN" TUMOR ANTIGEN

Goletz S., Cao Y., Danielczyk A., Ravn P., Schoeber U., and Karsten U.

NEMOD Immuntherapie AG and Max Delbrück Centre for Molecular Medicine, Berlin-Buch, Germany

1. INTRODUCTION: CARBOHYDRATE TUMOR ANTIGENS

Carbohydrate tumor antigens on glycoproteins and glycolipids are targets for active and passive cancer immunotherapy. These highly abundant antigens are *de novo* expressed or up-regulated due to changes in the complex glycosylation apparatus of tumor cells, involving sets of enzymes like glycosyltransferases, glycosidases, epimerases, and nucleotide sugar transporters. Various lipid or protein bound carbohydrate tumor antigens are described, for example, GM2, GD2, GD3, fucosylated GM1, Globo H, Le^Y, Le^a, Sialyl-Le^a and the mucin core structures Tn, Sialyl-Tn, and the Thomsen-Friedenreich Antigen (TF). Carbohydrate tumor antigens are far more abundant than protein tumor antigens rendering them suitable targets especially for antibodies, for example, highly expressed protein tumor markers as Her-2/neu express about 10^6 and TF about 10^7 copies per cell. More recent data show that certain carbohydrate structures are not only targets for humoral but also cellular immune responses.

Amongst carbohydrate and protein based tumor antigens, TF is an oncofetal antigen with an outstanding tumor specificity occurring almost exclusively in tumors. This chapter aims to give an overview of the knowledge about TF in view of its potential for cancer diagnostics and immunotherapy.

2. HISTORY OF THE THOMSEN-FRIEDENREICH ANTIGEN

TF is a "hidden" antigen mainly due to two aspects. On the one hand, TF is a disaccharide cryptically hidden as an antigen in longer carbohydrate chains of normal tissues. On the other hand, the discovery of TF was long ago and its use as a target for tumor diagnosis and immunotherapy was only investigated to a comparatively small extent mainly due to the

Glycobiology and Medicine, edited by John S. Axford
Kluwer Academic / Plenum Publishers, New York, 2003

lack of appropriate TF-specific reagents, the lack of knowledge about suitable TF-densities and TF-glycosylation sites for TF-based therapeutics as well as the until recently widespread opinion that carbohydrates are not able to elicit specific T cell responses.

TF was detected by serendipity by O. Thomsen in Copenhagen in 1926, when a contaminated blood sample became "polyagglutinable." This observation was examined in depth by O. Thomsen and V. Friedenreich (Friedenreich, 1930). The reason for the phenomenon was that neuraminidase producing bacteria unmasked TF on red blood cells, and that natural anti-TF antibodies present in all human sera led to ABO-independent agglutination. The disaccharidic chemical nature of the TF antigen as a blood group precursor was disclosed by E. Klenk and G. Uhlenbruck in 1960 (Klenk and Uhlenbruck, 1960), but it took another 15 years until G. Springer detected that TF was also a tumor antigen of epithelial cancers (Springer *et al.*, 1975). In the following years until his untimely death in 1998, TF (and its precursor antigen, Tn) was to a large extent G. Springer's personal domain, to which he contributed numerous studies (Springer, 1984, 1997). With a TF-based vaccine he obtained remarkable results with breast cancer patients (Springer, 1997). The generation of highly TF-specific antibodies (Clausen *et al.*, 1988; Karsten *et al.*, 1995; Steuden *et al.*, 1985) which showed the outstanding tumor specificity of TF, and new recombinant technologies renewed the interest in this old antigen for tumor diagnosis and therapy.

3. TF-STRUCTURE AND BIOCHEMICAL PATHWAY

TF is the disaccharide Galβ1-3GalNAc which is O-glycosidically linked in an α-anomeric configuration to the hydroxy amino acids serine or threonine of proteins (Galβ1-3GalNAcα-1-O-Ser/Thr). Galβ1-3GalNAc can also be bound in a β-anomeric configuration as terminal sugars on glycosphingolipids (TFβ) which are not expressed in epithelial cancer. In the following we concentrate on the carcinoma-specific protein-bound TFα which is further referred to as TF.

3.1. TF as a Hidden Structure in O-Glycan Chains of Normal Tissues

TF equals the core-1 structure of O-linked carbohydrate chains which were originally described as mucin-type glycans. In contrast to N-glycans, O-glycans do not have a consensus peptide sequence for attachment and upto about 20 monosaccharides are sequentially added in the golgi-apparatus. O-glycosylation is a complex phenomenon which is far from being clearly understood. Nearly each step in the glycosylation cascade of an O-glycan chain can be performed by several glycosyltransferases (for overview see, Brockhausen, 2000). The intial step is the attachment of the monosaccharide *N*-acetylgalactosamine (GalNAc) to serine or threonine which is performed throughout the Golgi but not in the ER (Rottger *et al.*, 1988) by UDP-GalNAc:polypeptide *N*-acetyl-galactosaminyltransferases. The members of the large family of homologous GalNAc-transferases (Schwientek *et al.*, 2002) show an array of different tissue-specific expression patterns, activities, kinetic properties, and substrate fine-specificities which can be influenced by critical amino acids in the vicinity and carbohydrate structures at different attachment sites (Goletz *et al.*, 1997; Hassan *et al.*, 2000; Muller *et al.*, 1997; Wandall *et al.*, 1997). For example, threonine of the DTR-motif in

MUC1 tandem repeat peptides can only be glycosylated by GalNAc-T4 and only if other potential glycosylation sites are already glycosylated by GalNAc which can be performed by various other GalNAc transferases with different activities. The complexity of this first step in O-glycosylation and the differential expression of these enzymes reflect the high variability of O-glycan chains in relation to their attachment sites and densities which is development-, differentiation- and tissue-specific. The innermost core-sugar of O-glycan chains is also called Tn when exposed in carcinomas. Tn is the substrate for further glycosyltransferases leading to di- or trisaccharides which represent the O-glycan core structures Core-1 to Core-6 (Schachter, 1986).

The Core-1 (TF) is generated by the activity of UDP-Gal:GalNAc-R-β-galactosyltransferase (β3-galactosyltransferase). Peptide sequences of glycopeptide substrates, existing glycosylation, the availability and density of Tn, activities of competing glycosyltransferases for the same substrate position, α3-*N*-acetylgalactosaminyl-(Core-5) or β3-*N*-acetylglucosaminyltransferases (Core-3), or other substrate positions, β6-*N*-acetylglucosaminyltransferases (Core-2) or the termination signal by α2,6-sialyltransferases (Sialyl-Tn) and α2,3-sialyltransferase, influence the synthesis of Core-1.

In normal tissues Core-1 is further elongated either to long O-glycan chains of up to about 20 sugar units or to short sialylated Core-1 structures. Long chains contain the Core-2 structure and often polylactosamines, i antigens, L-fucose, and/or sialic acids. The chains can be terminated by blood group or tissue antigens (ABO and Lewis Ags), sulfate or sialic acid (which is often further O-acetylated). Sialylation and sulfatation of Core-1 are also stop factors resulting in mono-, di-, and few tri-sialylated TF (Sialyl-TF) or SO_4-3-TF. While the transferases for the Core-structures are O-glycan specific, transferases for further elongation and modification can often act on O- and N-glycans as well as on glycolipids.

3.2. TF Antigen in Tumors

The Core-1 structure which is "hidden" in O-glycans of normal tissues becomes exposed during tumorgenesis as a *quasi de novo* tumor antigen.

It seems that various factors and pathway changes can lead to the demasking of TF in different tumor cells, however, the processes are not completely understood. In addition to changes in the balance of the complex transferase system, other critical factors influencing the O-glycosylation can be involved, for example, sugar nucleotide transporters and epimerases. Defects occur in various forms and might act synergistically, for example, as mutated enzymes or on the expression level. In one model system the biosythesis of Core-3 structures was shifted to Core-1 and Core-2 structures, because the competing Core-3 β3-GlcNAc transferase was downregulated (Brockhausen *et al.*, 1991); in other cases the Core-2 enzymatic activity (β6-*N*-acetylglucosaminyltransferases) was deleted resulting in accumulation of Core-1 structures (Brockhausen *et al.*, 1995). In addition, tumors often express more sialyltransferases (Schneider *et al.*, 2001). It is unclear why the latter case results in TF structures where it would be expected that the lack of Core-2 transferase activity and the increase in competing α2,6-sialyltransferases would result mainly in Sialyl-TF and not TF structures. In other cases, the UDP-galactose transporter seems to be up-regulated and transfection of cells with this transporter can enhance the expression of TF in these cells (Kumamoto *et al.*, 2001). New results indicate that in another case a defect in the UDP-*N*-acetylglucosamine 2-epimerase leads to an undersialylation and therefore exposure of TF (manuscript in preparation). It was also reported that TF carrying molecules emerge faster on the cell surface (Hull and Carraway, 1988). If this is a result of the above described enzymatic defects resulting in a faster transit, or if in certain cases an increased transit rate on its own causes TF exposure remains to be shown. Finally, we showed recently that overexpression of a nonenzymatic Golgi protein also results in TF expression at the cell membrane (Engelsberg *et al.*, 2003).

3.3. TF Carrier Molecules

TF is cryptically present in O-glycans of a variety of mucin-like membrane proteins, and becomes exposed during transition to malignancy as described above. It is intriguing that unmasking of TF seems to be a selective process which involves only one or a few of the possible candidate glycoproteins present at the cell membrane. The most prominent carrier molecule identified so far is the polymorphic epithelial mucin, MUC1, for example in breast cancer. Double staining data suggest that MUC3 may also be a carrier in some cases (Cao *et al.*, 1997). On colorectal carcinomas a variant of CD44 has also been identified as a carrier of TF (Singh *et al.*, 2001). In case of TF-positive leukemic cells, carrier molecules different from MUC1 are to be expected but have not yet been unequivocally identified (Cao *et al.*, 2002; Karsten, 2002).

The only known TF carrier molecule on non-tumor cells is glycophorin A on aging blood cells. TF becomes exposed on aging erythrocytes, which leads to their attachment to asialoglycoprotein receptors (ASGPR) in the liver and to their subsequent degradation.

4. TF-SPECIFIC MOLECULES

Lectins, polyclonal and monoclonal antibodies are available which recognize TF. Peanut agglutinin (PNA) has been for years the classical TF reagent. It is, however, not strictly TF-specific since it also binds to other glycans with terminal Galβ and shows a rather broad reactivity with normal tissues (Cao *et al.*, 1996). Monoclonal antibodies (mAbs) are known to be the most specific reagents. The generation of hybridomas producing mAbs against TF proved to be difficult resulting in rather few suitable mouse mAbs to TF (Clausen *et al.*, 1998; Karsten *et al.*, 1995; Stein *et al.*, 1997; Steuden *et al.*, 1995). All are IgM antibodies. The mAbs reveal subtle differences in their fine specificities and only very few are highly specific to TF. Most show cross-reactivities or specificities against synthetic TF-conjugates which render them unsuitable as TF-reagents for diagnosis or therapy.

Among our own anti-TF antibodies, which we believe are the most thoroughly examined, are A78-G/A7 (Karsten *et al.*, 1995), the standard mAb defining CD176 (Karsten, 2002), Nemod-TF1, Nemod-TF2, and A63-C/A9. A78-G/A7 has been described first in Karsten *et al.* (1995) and in many subsequent studies in immunohistochemistry (IHC) (Cao *et al.*, 1995, 1996, 1997a, b, 1999, 2000; Baldus *et al.*, 1998, 1999, 2000, 2001; Jeschke *et al.*, 2002) where it has emerged as the gold standard. This antibody binds strongly to TFα and has a slight cross-reactivity with TFβ, which is part of the glycan moiety of several glycosphingolipids (exposed in GA1 = asialo-GM1, GM1, GD1b, and masked in GM1b, GD1a, etc). Since TFβ is not present in normal or malignant epithelial cells in amounts detectable by IHC, this does not influence the detection of tumor TF. Nemod-TF1 and Nemod-TF2 are novel TF antibodies with particular high affinity and specificity. Their specificity was tested with a large panel of synthetic oligosaccharides bound to

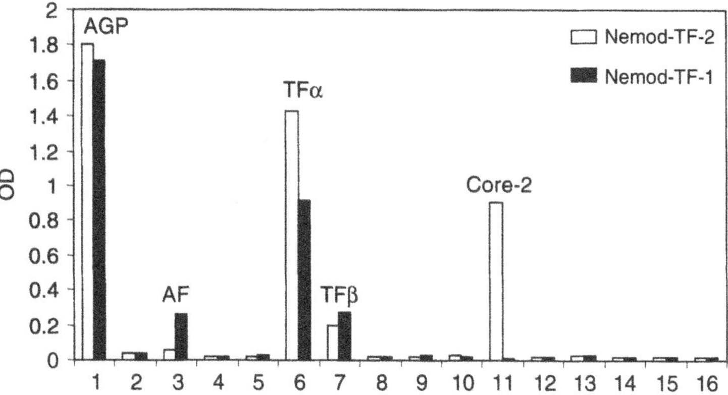

Figure 1. Specificity of antibodies Nemod-TF1 and Nemod-TF2. The binding of the mAb (3,3 μg/ml) was tested in ELISA experiments on immobilized glycoproteins (**1–4**) or synthetic polyacrylamide-carbohydrate conjugates (PAA-conjugates; **4–16**). *Glycoproteins*: Asialoglycophorin (AGP; **1**) and asialofetuin (AF, **3**) have high and low densities of TFα respectively; glycophorin (**2**), the sialylated form of AGP, and BSA (**4**) are controls. *PAA-conjugates*: Tn (GalNAcα1; **5**); TF-related disaccharides: TFα (Galβ1-3GalNAcα1; **6**), TFβ (Galβ1-3GalNAcβ1-; **7**); Galα1-3GalNAcα1- (**8**); Galα1-3-GalNAcβ1- (**9**); GalNAcα1-3Galβ1- (**10**); TF-related trisaccharides: Core-2 (Galβ1-3(GlcNAcβ1-6)GalNAcα1-; **11**); GlcNAcα1-3-TFα (GlcNAcα1-3Galβ1-3GalNAcα1-; **12**); GlcNAcβ1-3-TFα (GlcNAcα1-3Galβ1-3GalNAcα1-; **13**); 3′-Sialyl-TFα (Neu5Acα2-3Galb1-3GalNAcα1-; **14**) 6′-Sialyl-TFα (Galβ1-3(Neu5Acα2-6)GalNAcα1-; **15**); 3′-O-Sulfo-TFα (3′-O-Su-Galβ1-3GalNAcα1-; **16**).

polyacrylamide (PAA) (Figure 1). Nemod-TF2 has a very similar specificity compared to A78-G/A7 and is virtually identical in IHC. Nemod-TF2 is cross-rective to Core-2 and to some extent to TFβ, has a high avidity, and does not bind to asialofetuin (which we interpret as requiring more dense clusters of TF than Nemod-TF1). Cross-reactivity with Core-2 apparently does not interfere with IHC for TF. Core-2 has not been reported to be exposed anywhere in the body, whether in health or disease. Nemod-TF1 has a high affinity and specificity for the α-anomer of TF, and is able to detect lower densities of TF such as found in asialofetuin and therefore seems to be the most specific antibody for TF. Mab A63-C/A9 detects TF on glycophorin (in a mixed epitope) very specifically and with high avidity (unpublished data).

5. THOMSEN-FRIEDENREICH ANTIGEN AS A TUMOR-SPECIFIC ANTIGEN

With the above-mentioned mAbs, preferably A78-G/A7, we were able to reexamine and extend earlier data on the distribution of TF in normal and tumorous tissues done by Springer and other groups. We started with a comprehensive examination of normal human tissues for the expression of TF and related structures which resulted in a data matrix showing that TF and to a large extent also Tn, but not sialyl-Tn are virtually absent from normal adult tissues (Cao *et al.*, 1996). Table 1 summarizes the TF data. The few sites where some staining was found were rather weak and were located either on outer membranes or otherwise immunologically privileged sites. Intracellular TF staining of some macrophages can be interpreted differently including representing steps in biosynthesis or degradation of

Table 1. Distribution of TF as detected by A78-G/A7 among normal adult human tissues (modified from Cao *et al.*, 1996)

Tissue	TF staining	Tissue	TF staining	Tissue	TF staining	Tissue	TF staining
Epidermis		Pancreas (continued)		Brain		Muscle tissues	
Basal layer	–	Acini	–	Cerebral cortex	–	Smooth muscle	–
Spinous layer	–	Islets	–	Cerebral medulla	(+)	Skeletal muscle	–
Granular layer	–	Heart		Glial cells	–	Connective tissue	–
Non-epid. cells	–	Endocardium	–	Ependymal cells	–	Synovial tissue	–
Hair follicles	–	Myocardium	–	Meninges		Endothelium	–
Sebaceous gland	+	Mesothelium	–	Arachnoid membrane	–	Enthrocytes	–
Sweat gland		Trachea and lung		Pacchioni granules	–		
Ducts	–	Serous gland	–	Peripheral nerve			
Secretory cells	–	Mucous gland	–	Axons	–		
Submandibular gland		Ciliated epithelium	–	Schwann cells	–		
		Respiratory epithelium	–	Perineurium	–		
Serous acini	–	Kidney		Thymus gland			
Mucous acini	–	Glomerulus	–	Epithelial reticulum			
Ducts	–	Bowman's capsules	–	cells	–		
Myoepith. cells	–	Proximal tubules	–	Hassall's bodies	–		
Esophagus		Distal tubules	+/–a	Lymphocytes	–		
Squamous epith.	–	Collecting ducts	+/–a	Macrophages	+b		
Mucous acini	–	Bladder urothelium	–	Spleen			
Ducts	–	Prostate	–	Trabeculae	–		
Stomach		Uterus		Reticular cells	–		
Foveolar epith.	–	Corpus endometrium	–	Lymphocytes	–		
Fundus glands	–	Cervix endometrium	–	Macrophages	+b		
Corpus glands	–	Cervix glandular cells	–	Lymph nodes			
Jejunum	–	Mammary gland		Lymphocytes	–		
Ileum	–	Ducts	–	Macrophages	+b		
Descending colon	–	Acini	–	Reticular cells	–		
Gall bladder	–	Myoepithelial cells	–	Thyroid gland			
Liver		Testis		Epithelium	–		
Hepatocytes	–	Spermatids	+	Colloid	–		
Kupffer cells	–	Spermatocytes	+	Adrenal gland			
Bile ducts	–	Spermatogonia	–	Adrenal cortex	–		
Pancreas		Sertoli's cells	–	Adrenal medulla	–		
Ducts	+a	Interstitial cells	–				

a, Lumenal surface only; b, Not all cells, cytoplasma.

Table 2. Frequency of TF-positive carcinomas among different locations

Type of tumour	TF-positive cases (%)	Reference
Breast cancer	86	Cao *et al.* (unpublished)
Colon cancer	60–70	28, 33, 37
Gastric cancer	69	35
Prostate cancer	56	38[a]
Lung cancer	48	39
Liver cancer	38	31
Kidney cancer	15	34

[a]In this case staining was done with PNA which is not strictly TF-specific.

glycans. An additional TF-positive site was recently detected in the placenta (Jeschke *et al.*, 2002). In a number of consecutive studies we then examined the distribution of TF and TF-related structures in different types of carcinoma tissues (Baldus *et al.*, 1998, 1999, 2000, 2001; Cao *et al.*, 1995, 1997b, 1999, 2000). A brief summary of the results is given in Table 2, which also contains some data from other groups (Ghazizadeh *et al.*, 1984; Itzkowitz *et al.*, 1989; Springer, 1984; Takanami, 1999). We conclude, in accord with the results published by Springer, that TF is a highly specific, early marker of carcinogenesis virtually not occurring on normal tissues as demonstrated in more detail with the colorectal adenoma-carcinoma sequence (Cao *et al.*, 1997b), and made visual in a case of early breast cancer in Figure 2a. It is present on various carcinomas including the major indications. It is also obvious that the percentage of TF-positive cases varies among carcinomas of different organs, and that the positivity of cells within a given tumor is not homogeneous as seen with the vast majority of tumor markers (Figure 2b). The latter may reflect a fluctuation of TF expression over time, an observation not unusual with carbohydrate antigens. Sialylation, fucosylation, or sulfatation are possible reasons for masking TF. Interestingly, we found that pretreatment of tissue sections with sialidase in some but not all tissues led to increased TF staining. Although TF is mainly found on epithelial cancers, we have found it on a number of leukemias, especially acute myeloic leukemias including several leukemic cell lines like KG-1 (Figure 2c). This field has so far not been extensively examined.

TF expression in primary colorectal and gastric carcinomas has been found to present an independent prognostic marker indicating a less favorable prognosis (Baldus *et al.*, 2000; Singh *et al.*, 2001). This is in accordance with our earlier publication in which we were able to demonstrate for the first time in a clinical study that the expression of TF in primary colorectal cancer is correlated with a significantly higher risk for the development of liver metastases (Cao *et al.*, 1995). It was also observed in this study that the percentage of TF-positive cases was clearly higher in liver metastases (91%) than in primary colorectal tumors (60%).

6. TF INVOLVEMENT IN METASTASIS

6.1. Mechanism for Liver Metastasis

The above described histological investigations show that colon cancer patients with TF-positive primary tumors have a high risk of nearly 60% for developing liver metastasis

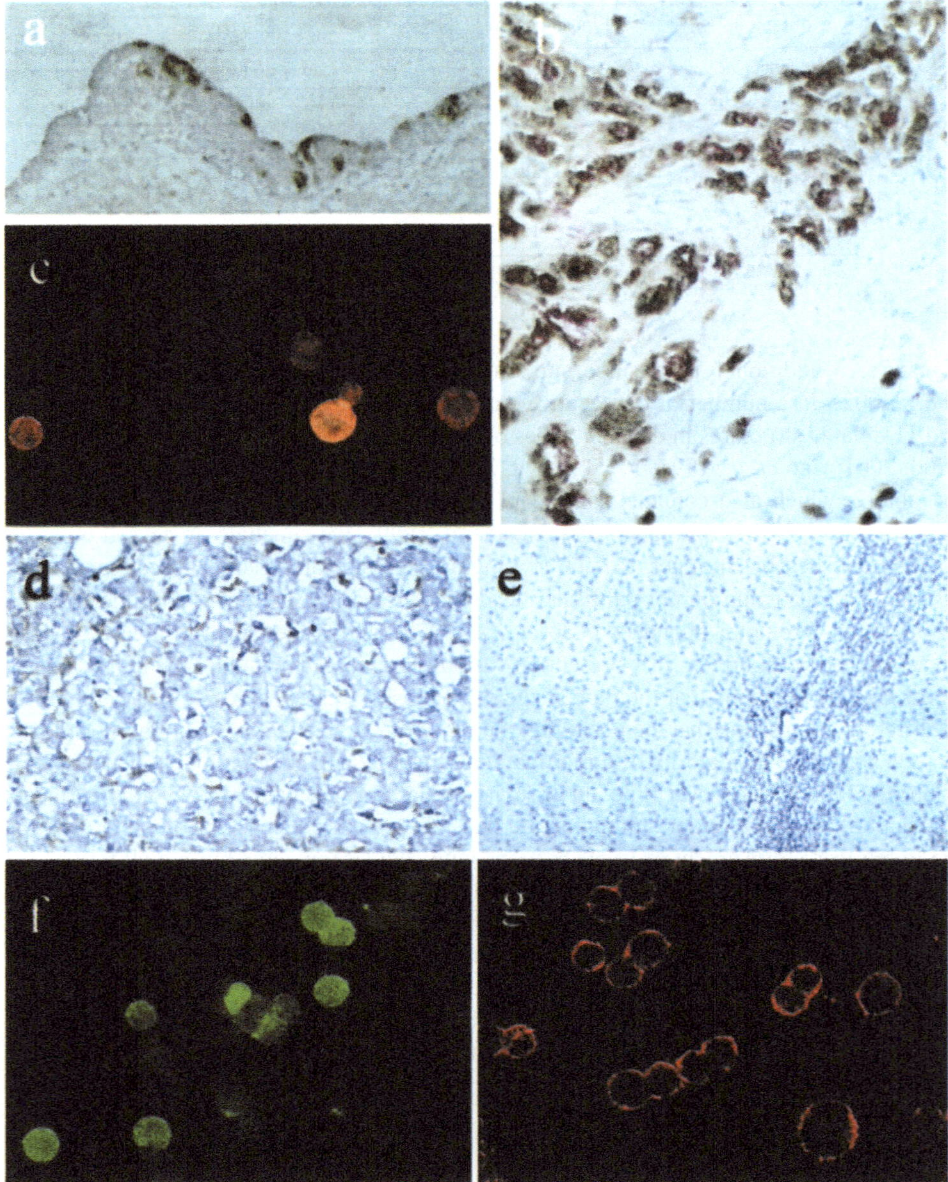

Figure 2. Immunohistochemistry and -cytochemistry of TF and of ASGPR. a: Breast duct with carcinoma *in situ* stained with Nemod-TF2. b: Breast carcinoma double-stained with Nemod-TF2 (red) and a MUC1-specific antibody (brown). c: AML cell line KG-1 double-stained with A78-G/A7 (red) and a CD45-specific antibody (green). d: Normal liver stained with biotinylated (TF-carrying) asialoglycophorin (brown). e: Cirrhotic liver stained as in d (no staining). f: Cell line KG-1 (as in c) stained with chimaeric anti-TF IgM antibody PankoMetin. g: Cell line KG-1 stained with the TF-specific multibody KaroMab.

whereas those with TF-negative tumors develop only in about 15% of the cases of liver metastasis (Cao *et al.*, 1995). Furthermore, new unpublished results show that the hepatocytes of patients with cirrhotic liver, who are known not to develop liver metastasis, do not bind the TF-carrying glycoprotein asialoglycophorin in immunohistological stainings with labelled asialoglycophorin. In contrast, hepatocytes from normal liver or from the liver of colon cancer patients are bound by asialoglycophorin (Figure 2d). Hepatocytes of cirrhotic livers lack the asialoglycoprotein receptors (ASGPR) on their surface which are important for the clearance of aging red blood cells and serum proteins. TF on glycophorin A, the major membrane glycoprotein of aging red blood cells, which can be exposed after desialylation of glycophorin A (asialoglycophorin), and TF on other glycoproteins is a ligand for ASGPR which binds to desialylated terminal galactose-containing carbohydrate structures (Schlepper-Schafer and Springer, 1989). Our hypothesis is that tumor cells circulating in the blood enter the liver via the portal vein where they come into direct contact with hepatocytes through "holes" in the endothelial layer of liver sinusoids. The contact through the gaps in the loose endothelial layer is supposed to be mediated by the interaction of TF groups on the tumor cell with ASGPR molecules on the hepatocytes leading to the adhesion of the tumor cells and the subsequent formation of metastases.

6.2. Mechanism for Metastasis via the Endothelium

It has been recently proposed that TF binds to galectin-3, a member of the β-galactoside-binding family of endogenous lectins, which was assumed to mediate the docking of tumor cells to the endothelium thereby mediating metastasis (Glinsky *et al.*, 2001).

7. NATURAL ANTI-TF ANTIBODIES

It was obvious from the beginning that anti-TF antibodies exist in sera of healthy individuals. Springer established that they originate from cross-reacting epitopes of gastrointestinal bacteria similar to blood group antibodies (Springer and Tegtmeyer, 1981). Very recently we have employed novel neoglycoconjugates (polyacrylamide-bound synthetic oligosaccharides) for the purpose of purification, quantification, and in-depth specificity analysis of anti-TF antibodies present in pools of normal human sera as well as in individual sera (Butschak and Karsten, 2002). Mean values of 8.7 μg/ml and 0.77 μg/ml of TF-specific IgM and IgG, respectively, were found in normal sera with a high degree of specificity, and almost no cross-reactivity towards galactosides, which makes them more specific than the classical TF reagent, peanut agglutinin. Although the presence of these natural anti-TF antibodies leads to difficulties in the development of a serum tumor marker test based on TF, they may be beneficial for the patient. We propose that they provide a barrier against TF-carrying tumor cells entering the circulation. We were supported in this hypothesis by the results of an experimental study in which we were able to suppress liver colonization by TF-positive syngeneic tumor cells by pretreatment with an anti-TF antibody (Shigeoka *et al.*, 1999).

8. TF FOR TUMOR THERAPY AND DIAGNOSIS

The outstanding specificity of TF for various carcinomas renders it a suitable target for tumor diagnosis and therapy.

8.1. Diagnosis

The presented immunohistochemical data demonstrate that TF based on the recognition by highly specific TF mAbs is a valuable marker for the detection and prognostic evaluation of various carcinomas (Table 2). Useful diagnostic approaches include the immunohistochemical (IHC) examination of TF on gastrointestinal tumors as an indicator of prognosis, while its prognostic suitability for other carcinomas has to be further investigated. In addition, IHC examinations could also be the basis for metastasis prevention (see below). Based on lectin studies correlations between TF expression and grading have also been found in other carcinomas which would be worth reexamining with specific mAbs. The promising data on TF renders it a valuable tool for further development for clinical routine diagnostic and prognostic tests.

An interesting non-IHC test for colorectal carcinoma based on TF on fecal mucin has been developed by Shamsuddin (Shamsuddin and Elsayed, 1988).

An assay based on measuring the level of anti-TF antibodies has been proposed as a serum tumor marker (Desai *et al.*, 1995). However, TF is a rather complicated system since TF shed into the circulation is caught by natural anti-TF antibodies present in the serum of healthy individuals (Butschak and Karsten, 2002; Springer and Tegtmeyer, 1981). The level of anti-TF antibodies may vary according to the rather unpredictable number and kind of cross-reacting gastrointestinal bacteria, and in cancer patients in addition depend on the amount of shed TF and its own immunogenicity. It appears that the presently available serum test for TF may be useful only in follow-up settings.

8.2. Cancer immunotherapy Using the TF Antigen

8.2.1. Antibody Therapeutics. The pan-carcinomic specificity of TF as well as its involvement in liver metastasis, and perhaps metastasis via the endothelium, also renders it a promising target for antibody therapeutics. Due to the difficulties in generating highly specific anti-TF antibodies and because of their nature as IgM isotypes with comparably lower intrinsic affinities of single binding domains, TF-specific antibodies were not further developed so far. We recently applied and developed novel recombinant antibody technologies enabling us to construct chimaeric and humanized recombinant TF-specific antibodies and antibody fragments including IgG, IgM, and multibodies.

8.2.1.1. Metastasis Prevention. TF appears to be a key element in the liver metastasis of tumor cells from colon carcinomas by mediating the attachment of the tumor cells to the ASGPR of the hepatocytes. Blocking the interaction of TF with ASGPR is one strategy to prevent liver metastasis. Theoretically this could be done in two ways. The approach to block the receptor was used by Uhlenbruck and co-workers applying large concentrations of the soluble disaccharide lactose intravenously (Beuth *et al.*, 1988). Stahn and Zeisig used lactosylated glycoliposomes for the inhibition of the attachment of TF-tumor cells to the ASGPR of hepatocytes *in vitro* (Stahn and Zeisig, 2000). One of the disadvantages of this approach is the blockade of the normal function of the ASGPR restricting corresponding therapeutics to a small time window. In addition, the tumor cells are only prevented from attachment but no additional anti-tumor immune response which destroys the circulating cells is activated. Therefore the second approach aims to block the TF-ASGPR interaction with an anti-TF antibody without blocking the receptor for most of its normal function and recruiting the immune system via its effector domains. In a mouse model murine TF-positive tumor cells were generated which metastasize to the liver. Application of TF-specific murine IgM reduced the liver metastasis rate to the lower rate seen with TF negative cells. Control antibodies of the same isotype did not reduce liver metastasis (Shigeoka, 1999; and unpublished data). These preclinical tests show the suitability of this approach for the prevention of liver metastasis. For development of an antibody therapeutic for clinical use in humans we developed PankoMetin, a TF-specific chimaeric multivalent IgM (cIgM) by using a novel vector system. All constant parts of the cIgM, including the J-chain, are human sequences and the variable domains VL and VH, which are responsible for the binding specificity and affinity, are murine. PankoMetin which is recombinantly expressed in CHO-cells combines the necessary high specificity against TF-structures and TF-carrying tumor cells with a high affinity constant (1×10^{10} M^{-1}) (unpublished data). The human constant regions of the cIgM are important in order to effectively recruit immune effector functions, to increase the serum half life and to be repetitively applicable to humans by reduction of HAMA responses. Importantly, PankoMetin did not show a stimulation of the proliferation of colon carcinoma cells *in vitro* as described for certain TF-binding lectins as for example, PNA and ACA (Irazoqui *et al.*, 2001; Yu *et al.*, 2001).

Anti-TF antibody-based prevention of metastasis might not be restricted to the prevention of liver metastasis. The recently proposed role of TF in mediating metastasis via the endothelium by interaction with Galectin-3 (Glinsky *et al.*, 2001) would also likely be inhibited by anti-TF antibodies as PankoMetin.

The frequent occurrence of TF-positive primary tumors from various carcinoma types implies the development of anti-TF antibodies with immune effector domains, as PankoMetin or other formats like IgG, as more general passive immunotherapeutics for the prevention of metastasis. An important feature will be the application in a clinical setting where metastasis is most likely to occur. One critical stage is during and after surgery where tumor cells circulate in the blood stream.

8.2.1.2. Radiolabelled TF-Specific Antibodies. The high specificity of TF with its virtual absence on normal tissues and its occurrence as a pan-carcinomic marker on carcinomas of the major indications renders it a promising target for radioimmunotherapy and -diagnosis. The TF-specific antibodies are murine IgM which are associated with unfavorable pharmacokinetic properties for a treatment of solid cancers or residual disease. In order to achieve a small antibody with a high affinity and advantageous pharmacokinetic properties, KaroMab was developed as a multibody. KaroMab contains the variable domains of the light and heavy chains of the antibody which are fused without a linker. In contrast, single chain antibody fragments (scFv) have an additional linker between VH and VL allowing enough flexibility between the two domains to fold as an active monomer. Without a linker this flexibility is lost resulting in multibodies where VL and VH domains of different polypeptides pair to form binding sites. The resulting multimer, dia-, or triabodies, has two or three binding sites and, especially in the case of the triabody, a highly increased avidity (Figure 3). The high avidity and medium size assume the triabody to be the preferred format for radioimmunotherapy and -diagnosis with improved pharmacokinetics compared to scFv and the whole antibody combining good tumor penetration, tumor retention, and blood clearance (Figure 4). This has to be confirmed in pre-clinical studies in a xenograft model comparing various radiolabelled

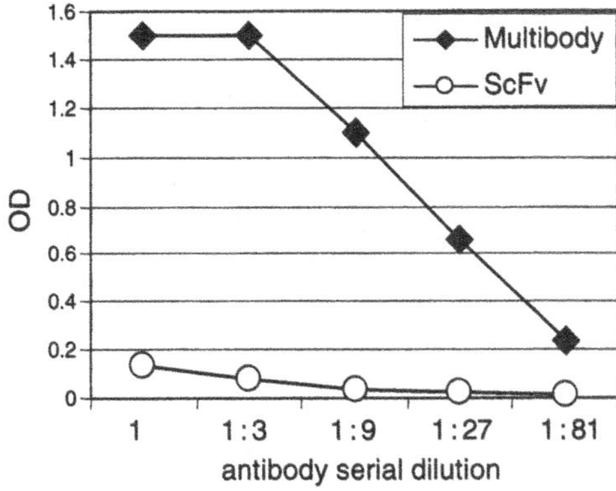

Figure 3. High-affinity binding of TF multibody. Binding of the TF-specific multibody (no linker) and scFv (15 amino acid linker) to immobilized asialo-glycophorin (15 TFα groups) in ELISA experiments reflect the high affinity of the multimer.

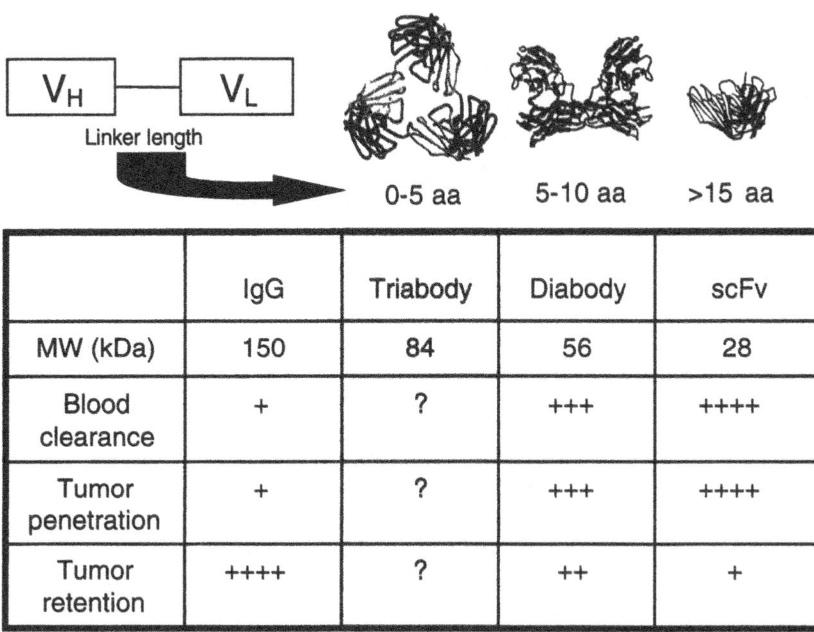

	IgG	Triabody	Diabody	scFv
MW (kDa)	150	84	56	28
Blood clearance	+	?	+++	++++
Tumor penetration	+	?	+++	++++
Tumor retention	++++	?	++	+

Figure 4. Schematic demonstration of the relationship between linker length and the formation of multibodies (upper panel), and the pharmacokinetic properties of the different antibody formats (lower panel).

antibody formats including multibodies and the bivalent chimaeric IgG1 (cIgG1) and humanized IgG1 (hIgG1) which are underway. As a prerequisite for these studies and for the use in humans, the cIgG1 as well as the multibodies are TF-specific and show in immunocytochemical and -histological stainings the characteristic pattern of TF-specific IgM antibodies (Figures 2f, g).

These recombinant antibody constructs are in principle not only suitable for radio-immunotherapy and -imaging but also for toxin delivery or coupling to other effector molecules or cells. However, the radioactive approach has at least two advantages: an irradiation bystander effect is assumed to be beneficial due to the fact that TF is, as nearly all other tumor markers, not always homogeneously expressed in a given tumor; and, radioactivity has not to be taken up by the cell in order to act.

8.2.2. TF Vaccines. Early during the development of tumor vaccines the pioneer in the TF field, G. F. Springer, conducted non-randomized clinical studies with advanced breast cancer patients (stages II–IV) which were highly successful compared to historical controls. The vaccine was a crude preparation of enzymatically desialylated glycophorin from O RBC carrying high densities of TF in a formulation with *Salmonella typhi* vaccine as an adjuvant administered intradermally every 6–20 weeks after surgery and radio- or chemotherapy. In a first study of 32 patients all survived the first 5 years and 14 more than 10 years; in a second study of 20 patients 15 had no evidence of disease after 3 to 5 years and 5 showed distant metastases (Springer, 1997). During his studies beside the activation of TF-specific antibody responses he showed a TF-specific DTH response which reflects the activation of specific T cells. This was one of the first hints that the cellular immune system, in contrast to a wide spread opinion, can evoke specific anti-carbohydrate responses. More recent studies show that certain carbohydrate epitopes on glycopeptides can be recognized by T cell receptors on CD8 and CD4 positive T cells (Abdel-Motal *et al.*, 1996; Galli-Stampino *et al.*, 1997; Haurum *et al.*, 1994; Rudd *et al.*, 2001; Speir *et al.*, 1999). Some of them mediate specific T helper or CTL responses in a MHC restricted or non-restricted manner.

Beside Springer's experiments most carbohydrate-based tumor vaccines were focusing on the induction of antibody responses (Livingston and Lloyd, 2000). However, they did not by far reach the success obtained with Springer's vaccine. One reason is that the vaccines consisted of non-natural TF-conjugates, for example, synthetic TF coupled randomly to KLH which did raise a humoral immune response against synthetic TF but not against TF on natural ligands (Adluri *et al.*, 1995).

These studies encourage the development of new tumor vaccines addressing TF. Important aspects are certainly to use suitable densities of TF with natural glycosylation sites and the choice of a potent adjuvant in order to activate specific humoral as well as cellular anti-tumor responses.

9. CONCLUSIONS

TF is a tumor marker with exceptional specificity and promise for passive and active immunotherapies and a valuable prognostic marker for diagnosis. TF-specific antibodies are reliable tools in immunohistochemistry; they can detect early stages of cancer. The development of a dependable serum tumor test based on TF has not yet been achieved. We are developing recombinant antibodies in different formats for metastasis prevention and radioimmunotherapy and diagnosis, and polyvalent vaccines with TF epitopes as one of its major ingredients.

10. REFERENCES

Abdel-Motal, U.M., Berg, L., Rosen, A., Bengtsson, M., Thorpe, C.J., Kihlberg, J., Dahmen, J., Magnusson, G., Karlsson, K.A., and Jondal, M., 1996, Immunization with glycosylated Kb-binding peptides generates carbohydrate-specific, unrestricted cytotoxic T cells, *Eur. J. Immunol.* 26(3):544–551.

Adluri, S., Helling, F., Ogata, S., Zhang, S., Itzkowitz, S.H., Lloyd, K.O., and Livingston, P.O., 1995, Immunogenicity of synthetic TF-KLH (keyhole limpet hemocyanin) and sTn-KLH conjugates in colorectal carcinoma patients, *Cancer Immunol Immunother.* 41(3):185–192.

Baldus, S.E., Hanisch, F.-G., Kotlarek, G.M., Zirbes, T.K., Thiele, J., Isenberg, J., Karsten, U., Devine, P.L., and Dienes, H.P., 1998, Coexpression of MUC1 mucin peptide core and the Thomsen-Friedenreich antigen in colorectal neoplasms, *Cancer.* 82:1019–1027.

Baldus, S.E., Hanisch, F.G., Monaca, E., Karsten, U., Zirbes, T.K., Thiele, J., and Dienes, H.P., 1999, Immunoreactivity of Thomsen-Friedenreich (TF) antigen in human neoplasms: the importance of carrier-specific glycotope expression on MUC1, *Histol. Histopathol.* 14:1153–1158.

Baldus, S.E., Zirbes, T.K., Hanisch, F.-G., Kunze, D., Shafizadeh, S.T., Nolden, S., Mönig, S.P., Schneider, P.M., Karsten, U.R., Thiele, J., Hölscher, A.H., and Dienes, H.P., 2000, Thomsen-Friedenreich (TF) antigen presents as a prognostic factor in colorectal carcinoma: a clinico-pathological study including 264 patients, *Cancer.* 88:1536–1543.

Baldus, S.E., Zirbes, T.K., Glossmann, J., Fromm, S., Hanisch, F.-G., Mönig, S.P., Schröder, W., Schneider, P.M., Flucke, U., Karsten, U., Thiele, J., Hölscher, A.H., and Dienes, H.P., 2001, Immunoreactivity of monoclonal antibody BW835 represents a marker of progression and prognosis in early gastric cancer, *Oncology.* 61:147–155.

Beuth, J., Ko, H.L., Schirrmacher, V., Uhlenbruck, G., and Pulverer, G., 1988, Inhibition of liver tumor cell colonization in two animal tumor models by lectin blocking with D-galactose or arabinogalactan, *Clin Expl Metastasis.* 6:115–120.

Brockhausen, I., 2000, O-linked chain glycosyltransferases, *Methods Mol Biol.* 125:273–293.

Brockhausen, I., Romero, P.A., and Herscovics, A., 1991, Glycosyltransferase changes upon differentiation of CaCo-2 human colonic adenocarcinoma cells, *Cancer Res.* 51(12):3136–3142.

Brockhausen, I., Yang, J.M., Burchell, J., Whitehouse, C., and Taylor-Papadimitriou, J., 1995, Mechanisms underlying aberrant glycosylation of MUC1 mucin in breast cancer cells, *Eur J Biochem.* 233(2):607–617.

Butschak, G. and Karsten, U., 2002, Isolation and characterization of Thomsen-Friedenreich-specific antibodies from human serum, *Tumor Biol.* 23:113–122.

Cao, Y., Blohm, D., Ghadimi, B.M., Stosiek, P., Xing, P.X., and Karsten, U., 1997a, Mucins (MUC1 and MUC3) of gastrointestinal and breast epithelia reveal different and heterogeneous tumor-associated aberrations in glycosylation, *J Histochem Cytochem.* 45(11):1547–1557.

Cao, Y., Karsten, U., Liebrich, W., Haensch, R., Springer, G.F., and Schlag, P., 1995, Expression of Thomsen-Friedenreich-related antigens in primary and metastatic colorectal carcinomas: a reevaluation, *Cancer.* 76:1700–1708.

Cao, Y, Karsten, U., Otto, G., and Bannasch, P., 1999, Expression of MUC1, Thomsen-Friedenreich antigen, Tn, sialosyl-Tn, and α2,6-linked sialic acid in hepatocellular carcinomas and preneoplastic hepatocellular lesions, *Virchows Arch.* 434:503–509.

Cao, Y., Karsten, U., Zerban, H., and Bannasch, P., 2000, Expression of MUC1, Thomsen-Friedenreich-related antigens, and cytokeratin 19 in human renal cell carcinomas and tubular clear cell lesions, *Virchows Arch.* 436:119–126.

Cao, Y., Karsten, U., and Schwartz-Albiez, R., 2002, Expression of Thomsen-Friedenreich-related carbohydrate antigens on human leukemia cells, in: *Leucocyte Typing VII*, D. Mason *et al.*, Eds., Oxford University Press, Oxford, pp. 204–205.

Cao, Y., Schlag, P.M., and Karsten, U., 1997b, Immunodetection of epithelial mucin (MUC1, MUC3) and mucin-associated glycotopes (TF, Tn, and sialosyl-Tn) in benign and malignant lesions of colonic epithelium: apolar localization corresponds to malignant transformation, *Virchows Arch.* 431:159–166.

Cao, Y., Stosiek, P., Springer, G.F., and Karsten, U., 1996, Thomsen-Friedenreich-related carbohydrate antigens in normal adult human tissues: a systematic and comparative study, *Histochem Cell Biol.* 106:197–207.

Clausen, H., Stroud, M., Parker, J., Springer, G., and Hakomori, S., 1988, Monoclonal antibodies directed to the blood group A associated structure, galactosyl-A: specificity and relation to the Thomsen-Friedenreich antigen, *Mol Immunol.* 25(2):199–204.

Desai, P.R., Ujjainwala, L.H., Carlstedt, S.C., and Springer, G.F., 1995, Anti-Thomsen-Friedenreich (T) antibody-based ELISA and its application to human breast carcinoma detection, *J Immunol Methods.* 188: 175–185.

Engelsberg, A., Hermosilla, R., Karsten, U., Shulein, R., Dorken, B., and Rehm, A., 2003, The Golgi protein RCAS1 controls cell surface expression of tumor-associated O-linked glycan antigens. *J Biol Chem.* 278(25):22998–23007. Epub 2003 Apr 02.

Friedenreich, V., 1930, *The Thomsen Haemagglutiantion Phenomenon*, Levin & Munksgaard, Copenhagen.

Galli-Stampino, L., Meinjohanns, E., Frische, K., Meldal, M., Jensen, T., Werdelin, O., and Mouritsen, S., 1997, T-cell recognition of tumor-associated carbohydrates: the nature of the glycan moiety plays a decisive role in determining glycopeptide immunogenicity, *Cancer Res.* 57(15):3214–3222.

Ghazizadeh, M., Kagawa, S., Izumi, K., and Kurokawa, K., 1984, Immunohistochemical localization of antigen-like substance in benign hyperplasia and adenocarcinoma of the prostate, *J Urol.* 132:1127–1130.

Glinsky, V.V., Glinsky, G.V., Rittenhouse-Olson, K., Huflejt, M.E., Ginskii, O.V., Deutscher, S.L., and Quinn, T.P., 2001, The role of Thomsen-Friedenreich antigen in adhesion of human breast and prostate cancer cells to the endothelium, *Cancer Res.* 61(12):4851–4857.

Goletz, S., Thiede, B., Hanisch, F.G., Schultz, M., Peter-Katalinic, J., Muller, S., Seitz, O., and Karsten, U., 1997, A sequencing strategy for the localization of O-glycosylation sites of MUC1 tandem repeats by PSD-MALDI mass spectrometry. *Glycobiology.* 7(7):881–896.

Hassan, H., Reis, C.A., Bennett, E.P., Mirgorodskaya, E., Roepstorff, P., Hollingsworth, M.A., Burchell, J., Taylor-Papadimitriou, J., and Clausen, H., 2000, The lectin domain of UDP-N-acetyl-D-galactosamine: polypeptide N-acetylgalactosaminyltransferase-T4 directs its glycopeptide specificities, *J Biol Chem.* 275(49):38197–38205.

Haurum, J.S., Arsequell, G., Lellouch, A.C., Wong, S.Y., Dwek, R.A., McMichael, A.J., and Elliott, T., 1994, Recognition of carbohydrate by major histocompatibility complex class I-restricted, glycopeptide-specific cytotoxic T lymphocytes, *J Exp Med.* 180(2):739–744.

Hull, S.R., and Carraway, K.L., 1988, Mechanism of expression of Thomsen-Friedenreich (T) antigen at the cell surface of a mammary adenocarcinoma, *FASEB J.* 2(8):2380–2384.

Irazoqui, F.J., Jansson, B., Lopez, P.H., and Nores, G.A., 2001, Correlative fine specificity of several Thomsen-Friedenreich disaccharide-binding proteins with an effect on tumor cell proliferation, *J Biochem* (Tokyo). 130(1):33–37.

Itzkowitz, S.H., Yuan, M., Montgomery, C.K., Kjeldsen, T., Takahashi, H.K., Bigbee, W.L., and Kim, Y.S., 1989, Expression of Tn, sialosyl-Tn, and T antigens in human colon cancer, *Cancer Res.* 49(1):197–204.

Jeschke, U., Richter, D.U., Hammer, A., Briese, V., Friese, K., and Karsten, U., 2002, Expression of the Thomsen-Friedenreich antigen and of its putative carrier protein mucin 1 in the human placenta and in trophoblast cells *in vitro*, *Histochem Cell Biol.* 117:219–226.

Karsten, U., Butschak, G., Cao, Y., Goletz, S., and Hanisch, F.-G., 1995, A new monoclonal antibody (A78-G/A7) to the Thomsen-Friedenreich pan-tumor antigen, *Hybridoma.* 14:37–44.

Karsten, U., 2002, CD176 Workshop Panel report, in: *Leucocyte Typing VII.* D. Mason *et al.*, Eds., Oxford University Press, Oxford, pp. 202–203.

Klenk, E. and Uhlenbruck, G., 1960, Über neuraminsäurehaltige Mucoide aus Menschenerythrocytenstroma, ein Beitrag zur Chemie der Agglutinogene, *Z Physiol Chem.* 319:151–160.

Kumamoto, K., Goto, Y., Sekikawa, K., Takenoshita, S., Ishida, N., Kawakita, M., and Kannagi, R., 2001, Increased expression of UDP-galactose transporter messenger RNA in human colon cancer tissues and its implication in synthesis of Thomsen-Friedenreich antigen and sialyl Lewis A/X determinants, *Cancer Res.* 61(11):4620–4627.

Livingston, P.O. and Lloyd, K.O., 2000, Carbohydrate antigens on glycolipids and glycoproteins, Principles and practice of the biologic therapy of cancer, Rosenberg, S.A., 3rd edn, Lippincott Williams & Wilkins.

Muller, S., Goletz, S., Packer, N., Gooley, A., Lawson, A.M., and Hanisch, F.G., 1997, Localization of O-glycosylation sites on glycopeptide fragments from lactation-associated MUC1. All putative sites within the tandem repeat are glycosylation targets *in vivo*, *J Biol Chem.* 272(40):24780–24793.

Rottger, S., White, J., Wandall, H.H., Olivo, J.C., Stark, A., Bennett, E.P., Whitehouse, C., Berger, E.G., Clausen, H., and Nilsson, T., 1998, Localization of three human polypeptide GalNAc-transferases in HeLa cells suggests initiation of O-linked glycosylation throughout the Golgi apparatus, *J Cell Sci.* 111 (Pt 1):45–60.

Rudd, P.M., Elliott, T., Cresswell, P., Wilson, I.A., and Dwek, R.A., 2001, Glycosylation and the immune system, *Science.* 291(5512):2370–2376.

Schachter, H., 1986, Biosynthetic controls that determine the branching and microheterogeneity of protein-bound oligosaccharides, *Biochem Cell Biol.* 64(3):163–181.

Schlepper-Schafer, J. and Springer, G.F., 1989, Carcinoma autoantigens T and Tn and their cleavage products interact with Gal/GalNAc-specific receptors on rat Kupffer cells and hepatocytes, *Biochim Biophys Acta.* 1013(3):266–272.

Schneider, F., Kemmner, W., Haensch, W., Franke, G., Gretschel, S., Karsten, U., and Schlag, P.M., 2001, Overexpression of sialyltransferase CMP-sialic acid: Galbeta1,3GalNAc-R alpha6-Sialyltransferase is related to poor patient survival in human colorectal carcinomas, *Cancer Res.* 61(11):4605–4611.

Schwientek, T., Bennett, E.P., Flores, C., Thacker, J., Hollmann, M., Reis, C.A., Behrens, J., Mandel, U., Keck, B., Schafer, M.A., Haselmann, K., Zubarev, R., Roepstorff, P., Burchell, J.M., Taylor-Papadimitriou, J.,

Hollingsworth, M.A., and Clausen, H., 2002, Functional conservation of subfamilies of putative UDP-N-acetylgalactosamine: polypeptide N-acetylgalactosaminyltransferases in Drosophila, *Caenorhabditis elegans*, and mammals. One subfamily composed of l(2)35Aa is essential in Drosophila *J Biol Chem.* 277(25):22623–22638.

Shamsuddin, A.M. and Elsayed, A.M., 1988, A test for detection of colorectal cancer, *Human Pathol.* 19:7–10.

Shigeoka, H., Karsten, U., Okuno, K., and Yasutomi, M., 1999, Inhibition of liver metastases from neuraminidase-treated Colon 26 cells by an anti-Thomsen-Friedenreich-specific monoclonal antibody, *Tumor Biol.* 20:139–146.

Singh, R., Campbell, B.J., Yu, L.G., Fernig, D.G., Milton, J.D., Goodlad, R.A., FitzGerald, A.J., and Rhodes, J.M., 2001, Cell surface-expressed Thomsen-Friedenreich antigen in colon cancer is predominantly carried on high molecular weight splice variants of CD44, *Glycobiology.* 11(7):587–592.

Speir, J.A., Abdel-Motal, U.M., Jondal, M., and Wilson, I.A., 1999, Crystal structure of an MHC class I presented glycopeptide that generates carbohydrate-specific CTL, *Immunity.* 10(1):51–61.

Springer, G.F., 1984, T and Tn, general carcinoma autoantigens, *Science.* 224:1198–1206.

Springer, G.F., 1997, Immunoreactive T and Tn epitopes in cancer diagnosis, prognosis, and immunotherapy, *J Mol Med.* 75:594–602.

Springer, G.F., Desai, P.R., and Banatwala, I., 1975, Blood group MN antigens and precursors in normal and malignant human breast glandular tissue, *J Natl Cancer Inst.* 54:335–339.

Springer, G.F. and Tegtmeyer, H., 1981, Origin of anti-Thomsen-Friedenreich (T) and Tn agglutinins in man and White Leghorn chicks, *Br J Haematol.* 47:453–460.

Stahn, R. and Zeisig, R., 2000, Cell adhesion inhibition by glycoliposomes: effects of vesicle diameter and ligand density, *Tumour Biol.* 21(3):176–186.

Stein, R., Chen, S., Grossman, W., and Goldenberg, D.M., 1989, Human lung carcinoma monoclonal antibody specific for the Thomsen-Friedenreich antigen, *Cancer Res.* 49(1):32–37.

Steuden, I., Duk, M., Czerwinski, M., Radzikowski, C., and Lisowska, E., 1985, The monoclonal antibody anti-asialoglycophorin from human erythrocytes specific for beta-D-Gal-1-3-alpha-D-GalNac-chains (Thomsen-Friedenreich receptors), *Glycoconjugate J.* 2:303–314.

Takanami, I., 1999, Expression of Thomsen-Friedenreich antigen as a marker of poor prognosis in pulmonary adenocarcinoma, *Oncol. Rep.* 6(2):341–344.

Yu, L.G., Milton, J.D., Fernig, D.G., and Rhodes, J.M., 2001, Opposite effects on human colon cancer cell proliferation of two dietary Thomsen-Friedenreich antigen-binding lectins, *J Cell Physiol.* 186(2):282–287.

Wandall, H.H., Hassan, H., Mirgorodskaya, E., Kristensen, A.K., Roepstorff, P., Bennett, E.P., Nielsen, P.A., Hollingsworth, M.A., Burchell, J., Taylor-Papadimitriou, J., and Clausen, H., 1997, Substrate specificities of three members of the human UDP-N-acetyl-alpha-D-galactosamine: Polypeptide N-acetylgalactosaminyltransferase family, GalNAc-T1, -T2, and -T3, *J Biol Chem.* 272(38):23503–23514.

11

GLYCODYNAMICS OF MUCIN BIOSYNTHESIS IN GASTROINTESTINAL TUMOR CELLS

Inka Brockhausen

Department of Medicine and Department of Biochemistry,
and Human Mobility Research Centre,
Queen's University, Kingston, Ontario, K7L 3N6 Canada

1. ABSTRACT

Glycoproteins found in the secretions and on the surfaces of cancer cells include mucins and mucin-like glycoproteins. These molecules have been shown to carry antigens that are characteristically expressed on cancer cells, including Tn and T antigens and Lewis epitopes. The structures of O-glycans are often abnormal in gastrointestinal tumors, or else are present in abnormal amounts, and these structures greatly contribute to the phenotype and biology of cancer cells. It has been shown that glycans of cancer cells have functional importance in cell adhesion, invasion and metastasis. The possible mechanisms leading to these cancer-specific changes in carbohydrate structures (termed glycodynamics) involve altered mRNA expression and catalytic activities of glycosyltransferases and sulfotransferases found in tissues and cells of gastrointestinal tumors. In a number of cases it has been possible to correlate enzyme changes with oligosaccharide structures. Different mechanisms have been suggested leading to the synthesis of cancer-specific Lewis, T and Tn antigens, but the regulation of cancer mucin antigens generally appears to be very complex and is poorly understood. The expression levels of specific mucin antigens and enzymes in gastro-intestinal tumors have diagnostic as well as prognostic value. These antigens also have potential for cancer immunotherapy. However, we first need to unravel the complexity of the control of glycosylation in cancer cells. Most importantly, studies of the functional implications of the glycodynamics in cancer cells, as related to cell adhesion and impact on the immune system will provide promising directions for future research.

Glycobiology and Medicine, edited by John S. Axford
Kluwer Academic / Plenum Publishers, New York, 2003

2. INTRODUCTION

Colon and gastric cancers are major causes of cancer death. Approximately 5–10% of colon cancers have a clear genetic cause. Cancers with a hereditary component include familial adenomatous polyposis and hereditary non-polyposis colorectal cancer. There is also the inflammatory bowel disease-associated colon cancer, suggesting that inflammation may be a triggering factor in the development to cancer.

Approaches for disease prevention focus on genetic testing, diet, fiber, antioxidants, probiotics, physical activity, good health care, and education. Treatment of these diseases often consists of a combination of surgery, chemotherapy and radiation. Immunotherapy using carbohydrate based cancer antigens as targets is a novel approach to treatment, while inhibitors of glycosylation have not been highly successful in clinical trials. Recently, cyclooxygenase-2 (COX-2) inhibitors have shown success in the treatment of colon cancer with resulting decreased metastasis and invasiveness. COX-2 is a glycoprotein involved in prostaglandin metabolism, and is induced in inflammation, colon cancer and adenomas. COX-2 inhibits Fas-mediated apoptosis in human gastric cancer and other cell types (Hsueh *et al.*, 2000) and confers resistance to Fas-mediated apoptosis in HCT-15 human colon cancer cells (Tang *et al.*, 2002). Inhibitors of COX-2 induce apoptosis in colonic and other cancers, increase mitogen-induced apoptosis in gastric cancer cells and decrease cancer growth and liver metastases. COX-2 inhibitors therefore may protect against colon cancer and its metastases (Dang *et al.*, 2002; Kakiuchi *et al.*, 2000). In addition, the COX-2 inhibitor celecoxib decreased sialyl-Lewis^a expression (Table 1) in human colon cancer HT29 cells and adherence of these cells to endothelial cells (Kakiuchi *et al.*, 2000). It also inhibited the expression of several glycosyltransferases (β3-Gal-transferase 5 and α3-sialyltransferase III and IV, see below) while prostaglandin E2 enhanced the

Table 1. Carbohydrate structures of mucins found in gastrointestinal tissues.

Tn antigen	GalNAc α-Ser/Thr
Sialyl-Tn antigen	Sialyl α2-6GalNAc α-
Core 1, T antigen	Gal β1-3GalNAc α-
Core 2	GlcNAc β1-6(Gal β1-3)GalNAc α-
Core 3	GlcNAc β1-3GalNAc α-
Core 4	GlcNAc β1-6(GlcNAc β1-3)GalNAc α-
Core 5	GalNAc α1-3GalNAc α-
Core 6	GlcNAc β1-6GalNAc α-
Blood group O (H antigen)	Fuc α1-2Gal β-
Blood group A	GalNAc α1-3(Fuc α1-2)Gal β-
Blood group B	Gal α1-3(Fuc α1-2)Gal β-
i Antigen, type 2 chain	[GlcNAc β1-3 Gal β1-4]$_n$
Type 1 chain	[GlcNAc β1-3 Gal β1-3]$_n$
I antigen	Gal β1-4GlcNAc β1-3[Gal β1-4GlcNAc β1-6]Gal β-
Lewis^a	Gal β1-3(Fuc α1-4)GlcNAcβ1-3Gal-
Sialyl-Lewis^a	Sialyl α2-3Gal β1-3(Fuc α1-4)GlcNAcβ1-3Gal-
Lewis^b	Fuc α1-2Gal β1-3(Fuc α1-4)GlcNAc β1-3Gal-
Lewis^x	Gal β1-4(Fuc α1-3)GlcNAcβ1-3Gal-
Sialyl-Lewis^x	Sialyl α2-3Gal β1-4(Fuc α1-3)GlcNAcβ1-3Gal-
Sialyl-dimeric Lewis^x	Sialyl α2-3 Gal β1-4(Fuc α1-3)GlcNAcβ1-3 Gal β1-4(Fuc α1-3)GlcNAcβ1-3Gal-

expression of these enzymes. Liver metastases of HT29 cells were inhibited by the COX-2 inhibitor as well as by anti-sialyl-Lewis antibodies.

Cancer cells are rich in membrane-bound glycoproteins that carry GalNAc-Ser/Thr O-linked oligosaccharides (O-glycans), as well as Asn-linked N-glycans. Mucins, the main class of O-glycosylated glycoproteins of cancer cells, are of very large molecular weight with more than 50% carbohydrate by weight, and have Ser/Thr/Pro-rich variable tandem repeat (TR) regions that are heavily O-glycosylated. Mucins are found in the secretions as well as on the surfaces of cancer cells. Mucins and mucin-like glycoproteins are involved in the regulation of cell adhesion and the immune system. Eighteen different genes encoding the protein backbone of mucins (MUC1 to MUC17) have been cloned to date, and at least ten of these are expressed in epithelial cells of the gastrointestinal tract. The gene expression of the protein backbone of glycoproteins is often altered in gastrointestinal cancers (Sylvester *et al.*, 2001; Kim, 1998). For example, expression of the MUC1 gene is upregulated and that of MUC2 is down regulated in colon cancer (Jang *et al.*, 2002). It also appears that the co-expression of MUC1 with MUC2 reflects regional lymph node metastases. In the gastric mucus epithelium, MUC5AC and MUC6 mucins are prevalent, while in gastric tumors, MUC2 and MUC4 are expressed and the glycosylation pattern has changed (de Bolos *et al.*, 2001). These alterations of the peptide backbone are accompanied by qualitative and quantitative structural changes of the O-glycan chains of glycoproteins, as well as tissue and blood group antigens. O-glycans are often truncated, less sulfated and highly sialylated in cancer cells (Brockhausen and Kuhns, 1997a; Brockhausen *et al.*, 1998a; Kim *et al.*, 1996). For example, cancer cells and tissues characteristically express Tn- and T antigens (Table 1) that are only found in a selected number of normal tissues. Most cancer glycan antigens, however, occur also on glycoproteins of healthy tissues but are quantitatively altered in cancer. Unusual or novel mucin epitopes may be useful for diagnosis and assessment of prognosis. In addition, mucin-like antigens with cancer-specific epitopes have been developed as cancer vaccines (Fung *et al.*, 1990; Singhal *et al.*, 1991; Springer *et al.*, 1995).

Mucin gene expression and glycosylation is tissue specific and differs along sections of the intestines and along colonic crypts. Transitional mucosa between normal and cancerous tissues may exhibit intermediate, mixed normal and cancerous antigen expression (Orntoft *et al.*, 1990). Thus cell growth and differentiation affects mucins and the structures of their glycan chains. It is therefore not surprising that cancer tissues exhibit different glycosylation patterns since these cells tend to be in an abnormal state of growth, differentiation and apoptosis.

The mechanisms underlying the changing N- and O-glycosylation was termed ‘glycodynamics’. This review focuses on the glycodynamics leading to abnormalities of N- and O-glycan structures in gastrointestinal tumor cells.

3. FUNCTIONS OF INTESTINAL MUCIN GLYCANS

In the intestinal tract, gel forming mucins MUC2, 5AC, 5B and 6, and cell surface mucins MUC1, MUC3, MUC4, MUC11 to MUC 13, MUC15, and MUC17 are expressed. Mucins change their expression along the intestinal tract which implies functional differences between various mucins (Corfield *et al.*, 2001; Gum *et al.*, 2002; Kim *et al.*, 1996; Moniaux *et al.*, 2001; Van Klinken *et al.*, 1997). The main function of these tissue- and cell-specific intestinal mucins in the mucus gel is to lubricate and protect the underlying epithelium. Specific mucin structures have been identified as adhesion sites for bacteria and viruses. For example, Lewis antigens function as cell adhesion molecules, while other glycans have anti-adhesive properties. Mucins and mucin-like molecules at cell surfaces protect underlying proteins, maintain protein conformations, control active epitopes and antigenicity (Dalziel *et al.*, 2001). They also participate in the control of the immune system and they bind to microbes. O-glycans have been shown to determine the cell surface expression and functions of cell surface receptors (Brockhausen and Kuhns, 1997a). N-glycans which can also be attached to mucins have similar functions, ranging from cell adhesion to anti-adhesion roles. For example, sialic acid-containing structures regulate adhesion of cancer cells to endothelium or extracellular matrix (Dimitroff *et al.*, 1999; Lin *et al.*, 2002). This may be important for the establishment of the metastatic potential of a cancer cell.

Mucin functions are mediated by hundreds of glycan chains with great structural heterogeneity. Why are mucin glycans so diverse in their structures? It is likely that the whole range of glycans is biologically important and forms a functional unit, similar to the sound of many individual notes that form the music of a symphony.

The abnormalities in intestinal cancer may be functionally important although it is still unknown how this is linked to the abnormal cell growth and cell death of cancer cells. When O-glycan extension of human colon cancer cells is inhibited with GalNAc α-benzyl, mucin antigens are altered and the adhesive properties of cells to E-selectin and endothelial cells are decreased (Huang *et al.*, 1992; Kojima *et al.*, 1992). O-glycans therefore contain important ligands for E-selectin-mediated cell adhesion. This indicates that O-glycans play important roles in the attachment of cancer cells to the endothelium.

Metastatic cells have been shown to be more highly sialylated and sialic acid has been implicated in the metastatic process (Bresalier *et al.*, 1996; Passaniti *et al.*, 1988; Takano *et al.*, 1994). Inhibition of sialylation (Kijima-Suda, 1986), and O-glycan extension (Bresalier *et al.*, 1991; Kuan *et al.*, 1987,1989) reduces the metastatic potential of cancer cells. It is possible that chains carrying sialic acid may regulate the interaction of cancer cells with other cells and with the cell matrix. The occurrence of the simple O-glycan, sialyl α2-6GalNAc α-linked to Thr/Ser of the mucin backbone (sialyl-Tn antigen), is associated with an unfavorable prognosis and formation of metastatic cancer. However, it has not been shown that sialyl-Tn is directly involved in the adhesion of metastatic cells to the endothelium, although it may interact with leukocytes expressing sialyl α2-6GalNAc-binding sialoadhesions. Sialyl-Tn may possibly function in the protection of cancer cells in the blood stream, or regulate the adhesion properties of associated molecules. Sialic acid can be both adhesive and anti-adhesive, and thus may promote the survival of cancer cells by several different mechanisms. For example, mucins carrying the sialyl-Tn antigen block NK cell lysis (Blottiere *et al.*, 1992; Ogata *et al.*, 1992). Sialic acid appears to block the induction of apoptosis (Keppler *et al.*, 1999), thus sialyl-Tn expressing cells may be less susceptible to cell death. Sialic acid may also be involved in growth regulation (Carraway *et al.*, 1992).

A rat colon cancer model was studied to determine the effects of increased amounts of blood group H (Fuc α1-2Gal) determinants. When rat colon carcinoma cells REG lacking α2-Fuc-transferase activity were transfected with blood group H α2-Fuc-transferase cDNA, their resistance to LAK, but not NK cell lysis, was increased (Marionneau *et al.*, 2000). These cells also grew more aggressively and were more tumorigenic (Goupille *et al.*, 1997). In addition, α2-Fuc-transferase transfected cells exhibited increased resistance to apoptosis (Goupille *et al.*, 2000). In human colon cancer cells

Colo 205, transfection with antisense DNA of α3-Fuc-transferases FUT3 and FUT6 suppressed the synthesis of sialyl-Lewis[a] and Lewis[x] determinants (Table 1) (Hiller *et al.*, 2000). In addition, cell proliferation was inhibited. This suggests that changes in fucosylation are relevant to tumor survival and progression.

The ability of cells to metastasize may depend on the presence of secreted mucins. Colon cancer cells that were producing large amounts of mucins showed greater adherence to basement membrane proteins and invasive properties, compared to cells secreting less mucin (Schwartz *et al.*, 1992). Inhibition of mucin O-glycan extension by GalNAc-benzyl affected these properties indicating that the glycan chains mediated these adhesive properties of cancer cells.

Altered glycosylation may affect the metastatic potential of cancer cells. Compared to primary tumors, the expression of Tn and T antigens (Table 1, Figure 1) is decreased in

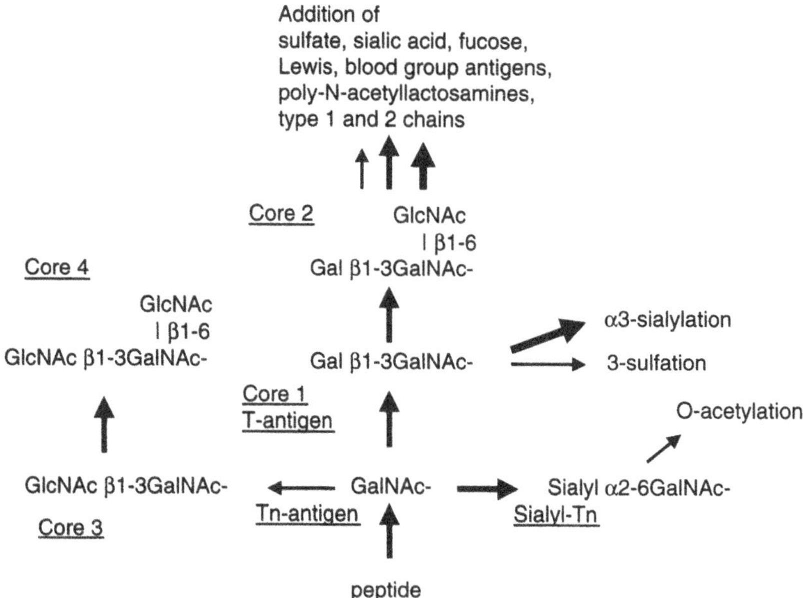

Figure 1. Glycodynamics of O-glycosylation in human gastrointestinal cancer tissues. Established reactions and pathways in the synthesis and processing of mucin O-glycan core structures are shown. Thick arrows indicate reactions that have been shown to be active in gastrointestinal cancer tissues, and thin arrows indicate reactions that are reduced in cancer. This scheme is mainly based on work with colon cancer tissue. Genes of at least some of the members of the glycosyltransferase and sulfotransferase families catalyzing these reactions have been cloned. GalNAc is first added to Ser/Thr of the peptide backbone by polypeptide GalNAc-transferase. GalNAc (Tn antigen) can then be converted to core 1 (T antigen) by core 1 β3-Gal-transferase, and core 1 is further converted to core 2 by core 2 β6-GlcNAc-transferase L (C2GnT1) and M (C2GnT2). GalNAc can also be acted on by α6-sialyltransferase synthesizing the sialyl-Tn antigen, which forms a stop signal and can only be modified further by O-acetylation. Core 3 synthesis from GalNAc by core 3 β3-GlcNAc-transferase is reduced in colon cancer. Core 3 can be converted to core 4 by C2GnT2. The core structures are either terminated by sulfation, which is reduced in colon cancer, or by sialylation which is increased in cancer, and by other reactions. Alternatively, core structures are processed to complex chains which are terminated by blood group and Lewis antigens. The relative activities of the enzymes involved in normal mucin synthesis are altered in cancer which causes shifts in the relative amounts of individual glycan structures in the heterogeneous mixture of chains on secreted mucins and cell surface glycoproteins. The glycodynamics of gastrointestinal cancer thus leads to antigenic and functional changes of mucins.

metastatic colon cancer cells, with a corresponding increase of sialyl-Tn, sialyl-T, sialyl-Lewis^a, and sialyl-Lewis^x. Sialyl-dimeric Lewis^x is also frequently increased in metastatic colon cancer (Hoff *et al.*, 1989). Antisense DNA of human α3/4-Fuc-transferase FUT3 which synthesized the Lewis^x and Lewis^a determinants inhibited adhesion of adenocarcinoma cells HT-29LMM to E-selectin (Weston *et al.*, 1999) and produced a less metastatic phenotype. This suggests that specifically Lewis structures play a role in cell adhesion and metastasis.

4. CONTROL OF MUCIN BIOSYNTHESIS

In the mucin O-glycosylation biosynthetic pathways, the first sugar added to the peptide backbone is αGalNAc. There is a large family of polypeptide GalNAc-transferases catalyzing this reaction. These enzymes are expressed in a tissue-specific fashion. In many tumors, GalNAc-transferase is highly active and synthesizes the Tn antigen (Figure 1, Table 1). The Tn antigen may be sialylated by α6-sialyltransferase to form the sialyl-Tn (STn) antigen, sialylα2-6GalNAc-. Alternatively, mucin O-glycan core structures may be synthesized, extended and terminated by sialic acid, Fuc, Gal, GlcNAc and GalNAc residues. These terminal structures include blood group and tissue antigens (Table 1). Sulfate esters, linked to the 6-position of GlcNAc, or the 3- or 6-position of Gal, are synthesized by sulfotransferases that transfer sulfate to glycans of glycoproteins in specific linkages. Glycosyltransferases and sulfotransferases occur as families of related proteins, with high sequence similarity within the family but sometimes different properties and specificities, and tissue-specific distribution (Brockhausen 1995a, 2003; Brockhausen and Kuhns, 1997a, b; Brockhausen and Schachter, 1997; Brockhausen *et al.*, 1998a; Schachter and Brockhausen, 1992).

The human and rat colon is rich in enzymes that synthesize the four common core structures one to four (Figure 1). The extensive heterogeneity of N- and O-glycan structures results from the competition of a number of different enzymes for the same glycoprotein substrate structure. While the types of enzymes define the final oligosaccharide structures by their distinct substrate specificities, the relative activities of these competing enzymes determine the amounts of structures occurring on a glycoprotein. For example, core 1, Gal β1-3GalNAc-, can be converted to the branched core 2 structure GlcNAc β1-6(Gal β1-3)GalNAc-. Alternatively, core 1 can be sialylated which would terminate chain growth (Figure 1). Transfection experiments have demonstrated that the relative activities of α3-sialyltransferase and core 2 β6-GlcNAc-transferase determine the complexity and sialylation status as well as antigenicity of cell surface mucins (Dalziel *et al.*, 2001).

An additional controlling factor is the arrangement of biosynthetic enzymes within the assembly line in the Golgi apparatus. The topology and localization of Golgi proteins is thought to be essential for the successful completion of glycoprotein biosynthesis. When the localization of specific reactions has changed in cancer cells (Egea *et al.*, 1993), the normal assembly line of enzymes may be disturbed leading to the premature synthesis of terminal structures, truncation of glycans and aberrant glycoprotein epitopes in cancer. It has been demonstrated that in HeLa cells, polypeptide GalNAc-transferases T1, T2, and T3 are present throughout early, medial and late Golgi compartments (Roettger *et al.*, 1998). However, in porcine and bovine submaxillary gland, polypeptide α-GalNAc-transferases have been demonstrated in the *cis*-Golgi compartment (Roth *et al.*, 1994) while T2 and T3 enzymes were preferentially localized to the medial and *trans*-Golgi compartments. It appears that the localization of individual enzymes may vary depending on the cell type and possibly also the differentiation status of the cells. If GalNAc is added in late Golgi compartments, these GalNAc substrates would not be available for the processing by glycosyltransferases localized to early compartments, and incomplete structures such as the Tn antigen (GalNAc-Ser/Thr) will result. Other glycans may be completely assembled, depending on the co-presence of enzymes and their respective substrates. A broad distribution of enzymes in the Golgi will thus give rise to structural heterogeneity.

Some glycosyltransferase activities are very low in *in vitro* assays. This is an apparent discrepancy to the finding that the enzyme products are often abundantly synthesized *in vivo*. Examples are low *in vitro* activities of β3-GlcNAc-transferases synthesizing mucin core 3, and the i antigen (poly-N-acetyllactosamine chains) in colonic tissues. It is interesting that in spite of these low activities, colonic mucins are rich in core 3 and i antigen. This suggests that the intracellular organization of enzymes and other membrane components is important for efficient synthesis, and that this assembly is disrupted upon homogenization and solubilization of membrane components *in vitro*.

In cancer, aberrant expression of mucin genes is often found and thus different peptide backbones are available for glycosylation. The enzymes catalyzing the first O-glycosylation reaction, polypeptide GalNAc-transferase and the enzymes synthesizing O-glycan core structures 1, 2, and 3 have been shown to be differently influenced by the peptide moieties of substrates (Brockhausen *et al.*, 1990, 1996; Granovsky *et al.*, 1994; Brockhausen *et al.*, unpublished). This site directed processing therefore may become a very important control mechanism in the synthesis of carbohydrate chains in cancer cells when the polypeptide backbones of glycoproteins have changed.

Cellular transformation, differentiation, growth, and apoptosis are often associated with changes in glycosylation. Thus the activation of specific signaling pathways and transcription factors could affect glycosyltransferase and sulfotransferase gene expression, and as a consequence, affect the synthesis of different glycoproteins.

Other factors may also directly control the activities of transferases. These include metal ion concentrations, posttranslational modifications of enzyme proteins and factors determining protein folding. The interactions of enzyme proteins with other protein or lipid components of Golgi membranes may be important factors in regulating enzyme activities as well as targeting of transferases to specific Golgi compartments.

5. N-GLYCOSYLATION IN GASTROINTESTINAL CANCER

In the N-glycosylation pathways in mammalian tissues, GlcNAc-transferases I to V can build up the antennary structures of complex N-glycans (Brockhausen and Schachter, 1997; Brockhausen *et al.*, 1998a). With higher branching patterns, chains become increasingly complex. When a biantennary substrate is acted upon by GlcNAc-transferase III, which introduces the bisecting GlcNAc, no further branching can occur. While most cells express GlcNAc-transferases I and II synthesizing biantennary N-glycan chains, the additional branching reactions and the synthesis of the bisecting GlcNAc appear to be regulated in a tissue-specific and differentiation-specific fashion. *In vitro* differentiation of human adenocarcinoma cells Caco-2 leads to increased activities of GlcNAc-transferases II to V, synthesizing complex types of N-glycans (Brockhausen *et al.*, 1991).

GlcNAc-transferase III changes its expression during the progression of the cell cycle. The activity, as well as enzyme protein levels, in synchronized human colon cancer cells Colo201 have been shown to increase in the M phase and to be five times higher than basal level in the G1 phase. Staining with PHA-E lectin which binds to bisected N-glycan structures confirmed the production of bisected N-glycans in glycoproteins produced in these phases (Kang *et al.*, 2000).

The expression of GlcNAc-transferase V was found to be increased in colorectal carcinoma and in their liver metastasis, as well as in gastric carcinoma (Petretti *et al.*, 1999, 2000). The metastatic potential correlated with GlcNAc-transferase V activity in several cancer models. Monoclonal antibody against GlcNAc-transferase V stained this enzyme specifically in colorectal cancer tissues. The expression of the enzyme was significantly correlated with distant metastasis and lower survival (Murata *et al.*, 2000). It is thought that the structures of the additional antennae synthesized by GlcNAc-transferase V contribute to the metastatic potential of cancer cells.

Many of the glycan structures elongating and terminating N-glycans are similar to those of O-glycans. The exceptions include sialyl α2-6Gal structures which are characteristic of N-glycans. The α6-sialyltransferase (ST6Gal) acting on the Gal-termini of N-glycans is increased in activity and mRNA expression in human and rat colonic cancer (Dall'Olio *et al.*, 2000, 2001; Petretti *et al.*, 2000). The level of ST6Gal expression in moderate and well-differerentiated adenocarcinomas correlates with poor survival and a poor prognosis (Lise *et al.*, 2000). Different transcript sizes were observed in colon cancer tissues which appeared to be characteristic of cancer (Dall'Olio *et al.*, 2000). Sambucus Nigra agglutinin (SNA) recognizes the sialyl α2-6Gal moieties and preferably stains colon cancer tissue. However, its reactivity in cancer samples did not always correlate with the levels of transcripts, which suggests that there is a complex regulation of the SNA epitope. Transfection of human colon cancer cells HT29 with antisense DNA for ST6Gal resulted in cells with a drastically lower ability to from colonies in soft agar and to invade Matrigel extracellular matrix (Zhu *et al.*, 2001). The sialyl α2-6Gal structure therefore appears to be involved in invasiveness.

6. SYNTHESIS OF O-GLYCANS IN GI CANCER

The first enzyme of the O-glycosylation pathways, polypeptide α-GalNAc-transferase, is encoded by several different genes (Clausen *et al.*, 1996). Terminal sugars and sulfate groups have important roles in the overall properties of O-glycans and glycoproteins, and control further glycan processing. In addition, specific terminal structures can serve as antigenic determinants or ligands for cell adhesion molecules such as selectins. Poly N-acetyllactosaminoglycans and the Lewis antigens (Table 1) attached to these chains have been implicated in adhesive interactions between tumor cells and activated endothelium.

7. TN AND SIALYL-TN ANTIGENS

Probably all mammalian cells are capable of synthesizing GalNAc α-Ser/Thr linkages (Tn antigen) which is the first step in the synthesis of all O-glycans. If no other glycosyltransferases act on GalNAc, a glycoprotein will be produced containing the Tn antigen, as seen in many advanced cancers. There is variability in the expression of GalNAc-transferases between individual normal and cancer patients. Immunohistochemistry, using antibodies against polypeptide GalNAc-transferase T1 and T2 showed that amounts of enzyme proteins are increased in tissues from colorectal cancer patients. Supranuclear staining suggested Golgi localization of the enzymes (Kohsaki *et al.*, 2000).

Polypeptide GalNAc-transferase T3 can completely glycosylate Thr residues of the MUC2 tandem repeat peptide *in vitro*. The enzyme was also shown to be increased in colonic adenocarcinoma (Inoue *et al.*, 2001). The presence of the polypeptide GalNAc-transferase T3 was correlated with histological differentiation and depth of invasion, as well as with a likelihood of five year survival. Thus, polypeptide GalNAc-transferase T3 is a useful indicator of tumor differentiation, disease aggressiveness, and its presence indicates a good prognosis in colorectal cancer patients (Shibao *et al.*, 2002).

In gastric cancer cells JRST, transfection with antisense cDNA of polypeptide GalNAc-transferase T1 decreased its mRNA and protein levels, and significantly increased the susceptibility of cells to NK and lymphokine-activated killer cells. The *in vivo* growth rate of transfected cells was lower than the mock-transfected control cells (Adachi *et al.*, 1997). This suggests that O-glycans may protect cancer cells. Antisense transfection to prevent O-glycosylation may thus be a new therapy to reduce the growth of cancer cells.

Sialyl-Tn antigen, sialylα2-6GalNAc-Thr/Ser (Table 1), is another cancer-associated structure found in glycoproteins from many tumors but not usually from normal mucosa (Cao *et al.*, 1996). Sialyl-Tn often occurs as a clustered epitope in colon cancer (Ogata *et al.*, 1998). Most colonic tumors have some cells within the heterogeneous cell population that express sialyl-Tn. Tn mucin antigens are virtually absent from normal colon but often occur in advanced stage tumors and are associated with a decreased survival in colon cancer patients (Itzkowitz *et al.*, 1989, 1990; Takahashi *et al.*, 1993). Sialyl-Tn correlates with progression to malignancy, and is found in polyps, chronic ulcerative colitis and intestinal metaplasia (Itzkowitz *et al.*, 1992). In intestinal and gastric tumors, sialyl-Tn and Tn expression appears to be a marker for poorly differentiated adenocarcinomas and mucinous carcinomas, and is associated with invasive and high proliferative properties of the tumors, metastasis and a poor clinical outcome (David *et al.*, 1992; Itzkowitz *et al.*, 1990; Kakeji *et al.*, 1995).

An increased expression of sialyl-Tn is seen in colon cancer metastasis. This is in apparent contrast to *in vitro* models of colon cancer where sialyl-Tn expressing cells are less metastatic than those expressing more complex O-glycans containing selectin ligands (Bresalier *et al.*, 1984, 1996; Brockhausen, 1999; Brockhausen *et al.*, 1998b). However, the sialic acid moiety of the sialyl-Tn antigen may serve to protect cells from NK cell lysis (Blottiere *et al.*, 1992). In inflammatory bowel disease and premalignant GI, the presence of sialyl-Tn was correlated with a higher risk for cancer. These cancer-associated antigens are useful for a prognosis and to evaluate the choice of treatment (Rhodes, 1996).

Several possible mechanisms can lead to sialyl-Tn antigen expression in cancer (Figure 1). Higher expression or activities of polypeptide α-GalNAc-transferase and the α6-sialyltransferase synthesizing sialyl-Tn may be responsible. If the enzymes would be co-localized, particularly in late Golgi compartments, α6-sialyl-GalNAc structures will be synthesized preventing any other O-glycan core structure from being formed. Another mechanism would be the block in further processing of GalNAc to complex O-glycans. Human colon cancer cells LSC express Tn and sialyl-Tn antigens because the synthesis of complex structures is blocked. These cells lack the enzymes synthesizing core 1 and

3 structures which are precursors for core structures 2 to 4, respectively. LSB cells derived from the same cell lineage have the enzyme synthesizing core 1 (but not core 3) and thus make glycoproteins containing complex O-glycans with Lewis antigens, which are presumably based on core 1 and 2 structures (Brockhausen *et al.*, 1998b). In rat colon cancer cells LMCR, a different mechanism for sialyl-Tn expression was apparent (Brockhausen *et al.*, 2001). While core 1 β3-Gal-transferase was highly active in these cells, it appeared that sialyl-Tn antigen expression was regulated by the relative activities of the α6-sialyltransferase synthesizing sialyl-Tn and core 2 β6-GlcNAc-transferase further processing core 1 to core 2.

In the normal colon, sialic acids of mucins are largely O-acetylated. O-acetylation of sialic acids in the gut is thought to protect carbohydrate chains from degradation by intestinal bacteria. This modification masks the recognition of sialylated epitopes such as sialyl-Tn or sialyl-Lewis structures (Muchmore *et al.*, 1987). In colon cancer, however, the degree of O-acetylation decreases, leading to the exposure of free sialic acid and recognition of the sialylated antigens (Jass and Walsh, 2001). A decrease in the activity of O-acetyltransferase in cancer is therefore one of the possible mechanisms responsible for the expression of sialylated antigens.

8. T ANTIGEN

O-glycan core 1, Galβ1-3GalNAcα-Thr/Ser, the Thomsen-Friedenreich antigen, TF antigen, is normally modified but is prevalent in cancer cells as the unmodified T antigen. The antigen is considered to be an oncofetal antigen since it occurs in meconium as well as in cancer cells and colonic adenocarcinoma (Campbell *et al.*, 1995). In a recent study, the T antigen was found in a small number of normal colonic tissues but in 57% of carcinoma samples (Schneider *et al.*, 2001). The T antigen is expressed in many tumors and is an early/intermediate cancer marker, useful for diagnosis, prognosis and monitoring of the disease stage. It is highly expressed in colon cancer metastases. Both the Tn and T antigens appear to confer increased sensitivity to NK cell lysis and can trigger immune responses. These antigens therefore have potential for use in immunotherapy (Blottiere *et al.*, 1992; Fung *et al.*, 1990; Singhal *et al.*, 1991).

Benign polyps may exhibit changed carbohydrate and altered glycosyltransferase activities that are reminiscent of cancer (Slomski *et al.*, 1986), and a great number of polyps express the truncated O-glycans, T and Tn antigens (Itzkowitz *et al.*, 1992). These human colorectal polyps tend to develop into carcinomas. The presence of the T antigen is also associated with risk for colon cancer in susceptible monkeys with colitis (Boland and Clapp, 1987).

The enzyme synthesizing core 1, core 1 β3-Gal-transferase, is a ubiquitous enzyme that occurs in most normal and cancer cells (Brockhausen *et al.*, 1995a; Schachter and Brockhausen, 1992). An unusual lack of the activity has been demonstrated in human colon cancer LSC cells leading to a lack of complex O-glycans and Tn antigen expression (Brockhausen *et al.*, 1998b). The defect in LSC cells cannot be reversed by differentiation or by *in vivo* tumor growth. A similar effect of suppressing core synthesis can be achieved with the O-glycosylation inhibitor GalNAcα-benzyl (Kuan *et al.*, 1987). In human colonic adenocarcinoma Caco-2 cells (Brockhausen *et al.*, 1991), core 1 β3-Gal-transferase activity decreases during enterocytic differentiation.

Possible mechanisms of T antigen expression include increased activities of polypeptide GalNAc-transferase and core 1 β3-Gal-transferase. Previous specificity studies of these enzymes using glycopeptide substrates (Brockhausen *et al.*, 1990; Granovsky *et al.*, 1994) showed that both the peptide sequence and its pre-existing glycosylation play a role in directing the synthesis of core 1. The expression of a different type of polypeptide GalNAc-transferase with a different specificity may also cause altered attachment sites of O-glycans and the exposure of T antigen clusters.

Alternatively, a loss of enzymes modifying core 1 may result in more core 1 being present in the unmodified form, as seen in cancer. Transfection experiments in human colon cancer cells SW480 show that the expression of T antigen is reduced by transfection with the enzyme synthesizing core 2, C2GnT1 while α6-sialyl-transferase (ST6GalNAc-II) transfection had the opposite effect and increased T antigen expression. Thus, these two enzymes apparently compete for the core 1 substrate and control T antigen expression in SW480 cells (Schneider *et al.*, 2001). It remains to be shown if these two enzymes reside in the same Golgi compartment in these cells.

9. SYNTHESIS OF MUCIN CORE STRUCTURES

Colonic mucins from human (Podolsky, 1985a,b) and rat (Slomiany *et al.*, 1980) have been reported to contain a variety of simple and complex elongated O-glycans with core 3 structures (Figure 1), and in another study of human colonic mucins, core structures 1 to 4 and mainly core 3 (Capon *et al.*, 2001). Gastric mucins from human, pig, and sheep have also been shown to contain complex elongated core structures 1 and 2, and sheep, in addition, has core 4 (Oates *et al.*, 1974; Slomiany *et al.*, 1994; van Halbeek *et al.*, 1982; Wood *et al.*, 1981).

From this structural information, we expect that there would be a high expression of core 3 β3-GlcNAc-transferase in the colon and a low expression in the stomach. *In vitro* assays of the enzyme in tissue homogenates from several species confirmed that the colon is the tissue with the highest (although modest) activity (Brockhausen *et al.*, 1985, 2001; Yang *et al.*, 1994). However, the mRNA levels of core 3 β3-GlcNAc-transferase measured by RT-PCR show that, in humans, the stomach has the highest level of expression, compared to colon and small intestine (Iwai *et al.*, 2002). Another discrepancy is the finding that colonic tissues, in addition to core 3 β3-GlcNAc-transferase, have high activities of the enzymes synthesizing mucin core structures 1, 2 to 4 (Brockhausen *et al.*, 1985, 2001; Yang *et al.*, 1994), although these structures apparently are not made *in vivo*. Factors other than mRNA and *in vitro* measured activities may regulate the *in vivo* activities.

The enzyme synthesizing core 3 is reduced in colon cancer tissues (King *et al.*, 1994; Yang *et al.*, 1994). Since cancer tissues consist of a heterogeneous population of cells, it is possible that advanced tumor cells have lost the enzyme expression. *In vitro*, human colon and intestinal carcinoma cells (Brockhausen *et al.*, 1991, 1998b; Vavasseur *et al.*, 1995), rat colon cancer cells (Brockhausen *et al.*, 2001) as well as tumorigenic polyposis cells (Vavasseur *et al.*, 1994) do not have any detectable activity of the enzyme. Since core 1 β3-Gal-transferase and core 3 β3-GlcNAc-transferase may compete for GalNAc-terminating substrates, the absence of core 3 synthesis in tumors would not only prevent the synthesis of core 4, but also shift the synthesis of O-glycans towards core 1 and 2 structures (Figure 1). This may partly contribute to the prevalence of the T antigen in colon cancer (Springer *et al.*, 1990).

Core 2 can be synthesized from core 1, and core 4 is synthesized from core 3 (Figure 1). The synthesis of core 2 is catalyzed by a family of core 2 β6-GlcNAc-transferases, C2GnT1, C2GnT2, and C1GnT3. C2GnT1 (L type enzyme) occurs in many tissues and mucin-producing as well as non-mucin producing cell types including leukocytes. A number of cultured colon cancer cells contain high activity of the L type core 2 β6-GlcNAc-transferase activity, for example SW403, SW48 cells which synthesize core 2 but not core 4 (Vavasseur *et al.*, 1995).

An M type enzyme (C2GnT2) is highly active in colonic tissues (Brockhausen *et al.*, 1985, 2001; Yang *et al.*, 1994), and is probably responsible for the high activity of a similar branching enzyme in gastric mucosa (Brockhausen *et al.*, 1986; Piller *et al.*, 1984). The activity synthesizes the branches of core 2, core 4 and the I antigen (Figure 1, Table 1) (Kuhns *et al.*, 1993; Yeh *et al.*, 1999). The enzyme is also found in human colon cancer cells LS180, HT29, SW1116, LSC, LSB, ileo-caecal cancer cells NCI498 and polyposis cells (Vavasseur *et al.*, 1995). The activity is lower in colon cancer tissues (Yang *et al.*, 1994), and diminishes in polyposis cells during progression to tumorigenic cells (Vavasseur *et al.*, 1994). The mRNA levels for C2GnT2 are high in colon and and low in small intestines, and in HT-29 cells, but are not found in human colon COLO357 cells (Schwientek *et al.*, 1999).

The L type activity is often altered in intestinal cancer cells (Vavasseur *et al.*, 1994, 1995; Yang *et al.*, 1994). The mRNA for L type core 2 β6-GlcNAc-transferase, GlcNAcT1, has been detected in most patients with colon cancer but not in normal mucosa and has been correlated with vessel invasion (Shimodaira *et al.*, 1997). The L enzyme may thus be induced, while the M enzyme may be suppressed in cancer. This could explain why the ratio of core 2 to core 4 synthesis increases in cancer cells while the total core 2 β6-GlcNAc-transferase activity decreases (Vavasseur *et al.*, 1994; Yang *et al.*, 1994). The functional implications of different core 2/4 ratios are not yet known.

Mucin isolated from human colon cancer cells LS174T have mainly core 1 and 2 structures (but not cores 3 and 4) which are partly extended, and terminated by sulfate and sialic acid (Capon *et al.*, 1997). The glycosyltransferase and sulfotransferase activities determined in LSB cells which are derived from LS174T cells, exactly correspond to these structures (Brockhausen *et al.*, 1998b) and support the structural data.

The sialylated O-glycan core 5 structure, sialyl α2-6 (GalNAc α1-3) GalNAc-, has been described from human meconium and mucin from colonic adenocarcinoma (Hounsell *et al.*, 1985; Kurosaka *et al.*, 1983). The enzyme activity synthesizing core 5, core 5 α3-GalNAc-transferase, has been demonstrated in the colonic mucosa of a patient with adenocarcinoma (Kurosaka *et al.*, 1985) but remains to be further characterized.

Glycoproteins isolated from human meconium have core structures 1 to 5, as well as core 6, GlcNAc β1-6GalNAc- (Capon *et al.*, 1989; Hounsell *et al.*, 1985, 1989). Core 6 may therefore be produced in the fetal intestine. However, it is possible that core 6 arises from the breakdown of core 2 by cleavage of the Gal residue, although the synthesis of core 6 from GalNAc-R has been reported in human ovarian tissue (Yazawa *et al.*, 1986).

10. CHAIN ELONGATION

O-glycan core structures may be elongated by repeating Gal β1-3/4GlcNAc β1-3 (N-acetyllactosamine) units to form type 1 and 2 chains, respectively, of the carbohydrate backbone. The enzymes involved in poly-N-acetyllactosamine synthesis include families of β3-GlcNAc-transferases, β3-Gal- and β4-Gal-transferases, which may have important roles in determining the overall biological properties of chains (Kataoka *et al.*, 2002). Poly-N-acetyllactosamine chains may be branched to form I antigenic structures Gal β1-4GlcNAc β1-6 (Gal β1-4GlcNAc β1-3) Gal-. The M type core 2 β6-GlcNAc-transferase is involved in the synthesis of these I antigenic structures (Schwientek *et al.*, 1999), as well as the I β6-GlcNAc-transferase that acts on internal GlcNAc residues of Gal β1-4GlcNAc-sequences (Mattila *et al.*, 1998). Colonic tissues, as well as gastric mucosa, have high activities of the β6-GlcNAc-transferase that synthesizes these GlcNAcβ1-6 Gal linkages, as well as β3-GlcNAc-transferases elongating O-glycan core structures 1 and 2 (Brockhausen *et al.*, 1983, 1986, 2001) (Table 1).

In normal colonic mucosa, type 1 and 2 chains are synthesized, but in adenocarcinomas predominantly type 2 is made. Human colon cells synthesize type 1 chains and express β3-Gal-transferase 1 (3GalT1) and 5 (3GalT5) and β3Gal-transferase x (3GalTx). 3GalT1 exhibits lower expression (Bardoni *et al.*, 1999) while 3GalT5 shows little or no expression in adenocarcinoma cells of the colon (Salvini *et al.*, 2001). In contrast, 3GalTx has a higher expression level in adenocarcinoma. Thus the overall activity of 3-Gal-transferase is reduced in colon cancer (Seko *et al.*, 1996).

In contrast, the activity of β4-Gal-transferase appears to be upregulated in colon cancer, but has also been reported to be unaltered (Seko *et al.*, 1996; Yang *et al.*, 1994). Using an antibody against N-acetyllactosamines, it was shown that Gal β1-4GlcNAc β1-3 sequences were weakly expressed in normal mucosa but were present in higher amounts in high grade and advanced human colorectal cancer. Antibody to the stem region of β4Gal-transferase 1 detected the enzyme weakly in normal mucosa, with some increase in low-grade adenoma and marked increase in carcinoma (Ichikawa *et al.*, 1999). It appears therefore, that in colon cancer, there is decreased synthesis of Gal β1-3 GlcNAc structures with relatively higher synthesis of Gal β1-4 GlcNAc structures, and a shift towards type 2 chains. This could promote the synthesis of Lewis^x antigens (Table 1) in cancer cells by providing higher amounts of substrate.

11. SIALYLATION

Glycoproteins from cancer cells, and in particular from metastatic cells (Bresalier *et al.*, 1996) often exhibit a higher degree of sialylation. A large family of sialyltransferases is involved in the addition of a terminal sialic acid to mucin glycoproteins. Sialylation often terminates chains by preventing other glycosyltransferases from acting on sialylated substrates. Sialylation of the Gal residue of O-glycan core 1 prevents the conversion of core 1 to core 2. The core 1 α3-sialyltransferase (ST3Gal1) competes with core 2 β6-GlcNAc-transferase (C2GnT1) for the core 1 substrate. Transfection studies with ST3Gal1 and C2GnT1 have shown that the relative activities of these two enzymes control the overall antigenicity of cell surface mucin, that is whether O-glycan chains remain short and sialylated or become large and complex (Dalziel *et al.*, 2001).

Many sialyltransferase activities are regulated in a cell type and differentiation-specific fashion (Harduin-Lepers *et al.*, 1995). α3-Sialyltransferase activities that synthesize sialyl-Lewis^a (type 1 chains) have been found increased in colon cancer tissues (Ito *et al.*, 1997). The core 1 α3-sialyltransferase activity is increased in several types of cancer cells, including colon cancer (Brockhausen *et al.*, 1995b; Yang *et al.*, 1994). Schneider *et al.* (2001) showed that mRNA levels for α3-sialyltransferases ST3Gal I were higher in colorectal carcinomas. ST3Gal I expression was particularly increased in cases showing invasion of lymph vessels. The expression of another α3-sialyltransferase with a preference for glycolipid substrates and acting on extended type 1 and 2 chains (ST3Gal II) is also upregulated in all colon cancer samples studied (Kudo *et al.*, 1998; Petretti *et al.*, 2000).

Gastric cancer glycoproteins appear to express large amounts of sialylated T antigens (Sotozono *et al.*, 1994). RT-PCR studies showed that ST3Gal-IV expression was increased in gastric cancer tissues (Petretti *et al.*, 1999) and in poorly differentiated colorectal carcinomas (Kudo *et al.*, 1998). This enzyme can act on core 1, as well as on N-acetyllactosamine (type 2) chains. It is not clear if this increased sialyltransferase activity alone would be responsible for the increased T antigen reactivity in gastric cancer.

In the colon, a relatively high activity of α6-sialyl-transferase can convert GalNAc residues of O-glycans to the sialyl-Tn or sialyl-T or other core structures. Thus R-GalNAc can be acted on by specific members of the α6-sialyltransferase (ST6GalNAc) family. mRNA levels of ST6GalNAc-II are increased in cases of colorectal cancer metastases to lymph nodes (Schneider, 2001). This correlates with poor survival and invasive potential.

Changes in sialo–mucin antigen expression in gastrointestinal cancer can be due to altered differentiation status or metaplasia, or changed core proteins. Alterations in the backbone structure, sulfation, O-acetylation or expression of different members of the sialyltransferase family with different substrate specificity and their Golgi localization affect the sialylation status of cells. Although it is not yet possible to predict the exact biological functions of altered mucins, they are useful for diagnosis and prognosis.

12. BLOOD GROUP AND LEWIS ANTIGENS

O-glycans of gastric and intestinal mucins are rich in blood group and tissue antigens (Table 1). Mucins and other glycoproteins produced by colon cancer tissue often express terminal, internal and extended Lewis and sialyl-Lewis antigens, including Lewis^{y}, Lewis^{x} and Lewis^{a} (Table 1) (Shi *et al.*, 1984). The expression of blood group antigens (Table 1), as well as of the enzymes involved in their synthesis, is often region- and cell type-specific (Brockhausen and Kuhns, 1997a).

Aberrant expression of blood group epitopes as well as altered activities of ABO transferases have been found in gastric and colon cancer tissues (Fujitani *et al.*, 2000). Blood groups of cancer glycoproteins may be incompatible, for example, blood group A determinant and A transferase activity may be present in the adenocarcinomas of a blood group O person (David *et al.*, 1993). The blood group O (or H) determinant is the precursor structure for the synthesis of AB blood groups. Thus, the α2-Fuc-transferase synthesizing the H determinant is an important factor regulating blood group ABO expression, and the basis for abnormal ABO blood groups could be a mutation causing a shift in the reading frame of the H transferase gene. Blood group A and B transferases (Yamamoto *et al.*, 1990) have been found either increased, reduced or aberrantly expressed in cancer tissue. Transfection of human colonic and gastric cancer cells with A and B transferase genes resulted in the expression of the corresponding blood groups and abolition of the H antigen (Ichikawa *et al.*, 1997). The A and B expressing cells showed considerably lower motility in culture suggesting that the expression of A and B antigens may regulate the motility and thus invasiveness of tumors.

The expression of the H-gene encoded α2-Fuc-transferase (FUT1) is associated with MUC6 expression in the normal gastric mucosa, while the expression of the secretor gene-encoded FUT2 is associated with MUC5AC. In gastric tumors, this co-regulation is lost (Lopez-Ferrer *et al.*, 2000). In samples from colon cancer patients, α2-Fuc-transferase activity is increased compared to normal (Orntoft *et al.*, 1991; Yang *et al.*, 1994) and is associated with tumor progression (Sun *et al.*, 1995). If core 1 structures are fucosylated in colon cancer, further processing to core 2 structures would be blocked.

The Sd blood group epitope (Table 1) attached to O-glycan core 3 is a major epitope expressed in intestinal tissues. In some colon cancer cells, the Sd epitope is absent (Capon *et al.*, 2001). Accordingly, the mRNA levels of the β4-GalNAc-transferase synthesizing the Sd epitope have been shown to be reduced in the cancer tissues in human gastric mucosa (Dohi *et al.*, 1996).

The appearance of Lewis antigens is regulated by the tissue- and cancer-specific expression of various members of the α3-Fuc-transferase family. The interaction of sialyl-Lewis^x on cancer cells with endothelial cells (Mannori *et al.*, 1995; Sawada *et al.*, 1994) has been proposed to be a factor in the invasion and metastasis of cancer cells. Consistent with this idea is that the occurrence of the sialyl-Lewis^x epitope in colon cancer patients is associated with poor survival (Nakamori *et al.*, 1993).

Gastric cancer tissues aberrantly express Lewis antigens and preferentially express sialyl-Lewis^x and sialyl-Lewis^a. These antigens are present at very low concentrations in normal gastric mucosa. The dimeric sialyl-Lewis^x appears to be related to a poor prognosis (Kim *et al.*, 2002). The activities of the Lewis gene-encoded enzyme α3/4-Fuc-transferase III as well as α3-Fuc-transferase synthesizing sialyl-Lewis^x are high in gastric cancer tissues, and sialyl-Lewis^x antigen was seen in most patients (Ikehara *et al.*, 1998). Often Golgi membrane-bound enzymes are cleaved and released into the serum. Increased α3-Fuc-transferase activity has been observed in the sera of patients with gastric cancer (Yazawa *et al.*, 1989), and this has been correlated with the clinical stage, and appears to reflect a higher activity in tumor tissues.

Most colorectal cancer cells express sialyl-Lewis^x and sialyl-Lewis^a antigens (Table 1), and their levels have been correlated with the progression of the disease. Lewis^b antigens are expressed in a gradient from high expression in the proximal colon to low expression in the distal colon. During cancer development, in the adenoma stage there is an increase in sialyl-Lewis^b expression which increases further during progression to cancer (Nishihara *et al.*, 1999). The molecular basis for this appears to be the regulation of H gene-encoded and especially the Se gene-encoded α2-Fuc-transferases.

Many of the individual enzymes synthesizing Lewis antigens have been shown to be abnormally expressed in colon cancer tissues. In colonic cancer, not only α2- but also α3- and α4-Fuc-transferase activities are increased (Dohi *et al.*, 1994). The mRNA levels for α3-Fuc-transferases IV have been found to be significantly elevated in colorectal carcinoma (Hanski *et al.*, 1996; Ito *et al.*, 1997; Petretti *et al.*, 2000). α3-Fuc-transferase III expression was less abundant in carcinomas with distant metastases (Petretti *et al.*, 2000), but both α3-Fuc-transferases III and VI are increased in poorly differentiated colorectal carcinomas (Kudo *et al.*, 1998). α3-Fuc-transferases III and VI, as well as α3-sialyltransferase ST3Gal IV synthesize sialyl-Lewis^x in normal mucosa, but the same enzymes may not be responsible for the up-regulated sialyl-Lewis^x expression in colorectal cancer.

The synthesis of the backbone structures also controls the abundance of Lewis antigens. The sialyl-Lewis^x structures of colon cancer cells CX-1.1 (Klopocki *et al.*, 1998) are mainly attached to O-glycans, and O-glycan core 2 is particularly important in providing a branch carrying selectin ligands (Li *et al.*, 1996). Both the L and the M enzyme C2GnT in the colon synthesize core 2 and are expected to contribute to the complexity of O-glycans in cancer. However, core 2 synthesis is controlled by high activities of α2-Fuc-transferases and α3-sialyltransferases in colon cancer, which reduce further processing of core 1 to core 2.

13. SULFATION

Colonic and gastric mucin O-glycans carry sulfate esters mainly at the 3-hydroxyl of Gal and the 6-hydroxyl of GlcNAc residues. Families of Gal- and GlcNAc-sulfotransferases are involved in their synthesis (Brockhausen, 2003; Brockhausen and Kuhns, 1997b; Carter *et al.*, 1988; Kuhns *et al.*, 1995). Sulfation of O-glycans at the 3- or 6-position of Gal and the 6-position of GlcNAc plays a role in cell adhesion by providing tight binding to selectins. Sulfated sugars also mediate bacterial binding and are acted on by bacterial sulfatases (Brockhausen, 2003; Brockhausen and Kuhns, 1997b; Hemmerich and Rosen, 1994; Tsuboi *et al.*, 1996). Sulfation often blocks further processing of glycan chains. Thus O-glycan core 1 cannot be converted to core 2 if it is sulfated (Kuhns *et al.*, 1995). However, if terminal GlcNAc is 6-sulfated, a Galβ1-4 residue can still be transferred to GlcNAc.

Sulfated mucins are commonly found in gastric carcinoma (Huang *et al.*, 1986). However, in colon cancer, a decrease in sulfomucin expression is observed, and sialomucins prevail instead of sulfomucins (Yamori *et al.*, 1987; Yang *et al.*, 1994). In colon cancer patients, core 1 Gal-3-sulfotransferase (Gal3ST-4) is decreased in mRNA expression (Seko *et al.*, 2002) and activity (Yang *et al.*, 1994). The core 1 3-sulfotransferase (Gal3ST-4) activity is ubiquitous (Kuhns *et al.*, 1995; Seko *et al.*, 2001), while another member of this sulfotransferase family, Gal3ST-2, is expressed in the intestine. A lower expression of Gal3ST-2 is seen in non mucinous adenocarcinoma (Seko *et al.*, 2002b). During progression of polyposis cells to tumorigenic cells, mucin sulfation decreases and this has been correlated with decreased core 1 3-Gal sulfotransferase activity (Vavasseur *et al.*, 1994).

One of the GlcNAc-6-sulfotransferases (GlcNAc6ST-1) shows a low expression in the colon, and has a broad specificity towards GlcNAc-containing substrates (Uchimura *et al.*, 2002). In contrast, GlcNAc6ST-2 (SulT b) is not expressed in normal mucosa or nonmucinous adenocarcinoma, but is expressed in high endothelial venules, and its main role is probably to synthesize selectin ligands. However, colonic mucinous adenocarcinoma tissue and human colon cancer LS174T and LS180 cells do express GlcNAc6ST-2 (Seko *et al.*, 2002a; Uchimura *et al.*, 2002). In spite of the expression of the GlcNAc6ST-2, core 2 O-glycans of mucin isolated from LS174T-HM7 cells appear to have Lewis^x determinants with sulfate esters only at the 3-position of the terminal Gal (but not GlcNAc) residues (Capon *et al.*, 1997), and sulfotransferase activity in LSB cells (derived from the LS174T cell line) was not detectable towards GlcNAc β1-3Gal β-methyl substrate (Brockhausen *et al.*, 1998b). GlcNAc6ST-2 has been shown to act on a number of GlcNAc-terminating oligosaccharides, including O-glycan cores 2 and 3 (Seko *et al.*, 2002a; Uchimura *et al.*, 2002), while GlcNAc6ST-3 cannot act on core 3 but utilizes core 2 substrates (Lee *et al.*, 1999; Seko *et al.*, 2000; Uchimura *et al.*, 2002).

GlcNAc6ST-3 (SulT a, hIGlcNAc6ST) is preferentially expressed in the intestine and is highly active in the colon (Lee *et al.*, 1999). Thus different members of transferase families with different properties contribute to the tissue-specific range of oligosaccharide structures. In cancer, the balance of these individual enzymes is disturbed. It remains to be elucidated how the alterations in mucin sulfation in cancer cells relate to their biological properties.

14. CONCLUSIONS AND FUTURE DIRECTIONS

The mucin secreting cells of the gastrointestinal tract produce a variety of mucins with complex O-glycan and N-glycan chains. Both the structures of the peptide backbones and the oligosaccharides can change in cancer. The basic question is what directs the glycodynamics in gastrointestinal cancer? Who conducts the symphony?

A number of recurring themes have emerged, such as the increased expression of the sialyl-Tn, T and sialyl-Lewis^x antigens in gastrointestinal cancers. Although some of these changes appear to be valuable diagnostic and prognostic factors, they are heterogeneous in nature and it is difficult to predict the structures arising from a specific cancer cell. In a selected number of cases, the alterations have been correlated with the expression and activities of specific enzymes synthesizing these antigens. Glycosyltransferases building the backbone structure for the attachment of antigens may also play important roles in directing the biosynthesis towards cancer-associated antigens.

When determining the mechanisms of glycan alterations in cancer, technical difficulties are one of many reasons for apparent discrepancies. Were the maximal initial velocities measured and were the appropriate specific substrates used? Different cells have a characteristic glycosylation potential, and glycosylation can be altered under the influence of external factors. In light of natural large variations in the normal tissues, what is a significant difference in cancer? How large does a difference in enzyme activity have to be in order to cause a significant structural variation and the cancer glycotype *in vivo*? Why are so many glycosyltransferases and sulfotransferases abnormal in their expression and activities? We know that glycosylation can change during the cell cycle, and during cell proliferation and differentiation. Are the changes observed secondary to altered growth and differentiation of cancer cells or abnormal cell death? What intracellular organization exists that can efficiently glycosylate hundreds of Ser/Thr residues on the same molecule? How can we mimic these intracellular conditions for glycosylation *in vitro*? Specific patterns have emerged in cancer indicating that selected enzymes are affected. This suggests that signaling events affect the expression of selected transferase genes. In addition, other secondary factors in a tumor cell can influence enzyme activitities. Once the glycosylation has changed, cell adhesion or other extracellular events are affected, and this will allow the cell to invade or metastasize, or to escape the immune system. Cancer antigens, such as the T and Tn antigens, have the potential to stimulate an immune response which gives hope for a natural control of cancer cell growth.

In future we will have to search for defined and better models that reflect the glycodynamics of gastrointestinal cancer cells. The enzymatic mechanisms for alterations in cancer cells should be elucidated by examining the expression, activities and properties of all enzymes involved in the pathways to cancer glycosylation. The intracellular organization of the glycosyltransferase assembly line controlling glycosylation needs to be elucidated. We also have to define the molecular basis for cancer progression and metastasis, and the functional roles glycan chains play in these processes. Finally, we hope to understand how the immune system can deal naturally with cancer cells, or with the help of immune therapy and stimulation using mucin epitopes.

15. ACKNOWLEDGMENTS

The financial support from the Canadian Cystic Fibrosis Foundation and the Natural Sciences and Engineering Research Council of Canada is gratefully acknowledged. The author is a Research Scientist of The Arthritis Society.

16. REFERENCES

Adachi, T., Hinoda, Y., Nishimori, I., Adachi, M., and Imai, K., 1997, Increased sensitivity of gastric cancer cells to natural killer and lymphokine-activated killer cells by antisense suppression of N-acetylgalactosaminyltransferase, *J Immunol.* 159:2645–2651.

Bardoni, A., Valli, M., and Trinchera, M., 1999, Differential expression of β1,3 galactosyltransferases in human colon cells derived from adenocarcinomas or normal mucosa, *FEBS.* 451:75–80.

Blottiere, H.M., Burg, C., Zennadi, R., Perrin, P., Blanchardie, P., Bara, J., Meflah, K., and LePendu, J., 1992, Involvement of histo-blood-group antigens in the susceptibility of colon carcinoma cells to natural killer-mediated cytotoxicity, *Int J Cancer.* 52:609–618.

Boland, C.R. and Clapp, N.K., 1987, Glycoconjugates in the colons of New World Monkeys with spontaneous colitis. Association between inflammation and neoplasia, *Gastroenterology.* 92:625–634.

Bresalier, R.S., Boland, C.R., and Kim, Y.S., 1984, Characteristics of colorectal carcinoma cells with high metastatic potential, *Gastroenterology.* 87:115–122.

Bresalier, R.S., Niv, Y., Byrd, J.C., Duh, Q., Toribara, N., Rockwell, R., Dahiya, R., and Kim, Y.S., 1991, Mucin production by human colonic carcinoma cells correlates with their metastatic potential in animal models of colon cancer metastasis, *J Clin Invest.* 87:1037–1045.

Bresalier, R., Ho, S., Schoeppner, H., Kim, Y., Sleisenger, M., Brodt, P., and Byrd, J., 1996, Enhanced siaylation of mucin-associated carbohydrate structures in human colon cancer metastasis, *Gastroenterology.* 110:1354–1367.

Brockhausen, I., Williams, D., Matta, K.L., Orr, J., and Schachter, H., 1983, Mucin synthesis. III. UDP-GlcNAc : Galβ1-3(GlcNAcβ1-6)GalNAc-R (GlcNAc to Gal) β3-N-acetylglucosaminyltransferase, an enzyme in porcine gastric mucosa involved in the elongation of mucin-type oligosaccharides, *Can J Biochem Cell Biol.* 61:1322–1333.

Brockhausen, I., Matta, K.L., Orr, J., and Schachter, H., 1985, Mucin synthesis. VI. UDP-GlcNAc : GalNAc-R β3-N-acetylglucosaminyltransferase and UDP-GlcNAc : GlcNAcβ1-3GalNAc-R (GlcNAc to GalNAc) β6-N-acetylglucosaminyltransferase from pig and rat colon mucosa, *Biochemistry.* 24:1866–1874.

Brockhausen, I., Matta, K.L., Orr, J., Schachter, H., Koenderman, A.H.L., and van den Eijnden, D.H., 1986, Mucin Synthesis VII. Conversion of R1-β1-3Gal-R2 to R1-β1-3(GlcNAcβ1-6)Gal-R2 and of R1-β1-3 GalNAc-R2 to R1-β1-3(GlcNAcβ1-6)GalNAc-R2 by a β6-N-acetylglucosaminyltransferase in pig gastric mucosa, *Eur J Biochem.* 157:463–474.

Brockhausen, I., Möller, G., Merz, G., Adermann, K., and Paulsen, H., 1990, Control of glycoprotein synthesis: The peptide portion of synthetic O-glycopeptide substrates influences the activity of O-glycan core 1 uridine 5′-diphospho-galactose: N-acetylgalactosamineα-R β3-galactosyl-transferase, *Biochemistry.* 29:10206–10212.

Brockhausen, I., Romero, P., and Herscovics, A., 1991, Glycosyltransferase changes upon differentiation of CaCo-2 human colonic adenocarcinoma cells, *Cancer Res.* 5:3136–3142.

Brockhausen, I., 1995a, The biosynthesis of O-glycosylproteins, in: *Glycoproteins.* J. Montreuil, J. Vliegenthart, and H. Schachter, eds., Vol. 29A, New Comprehensive Biochemistry, Elsevier Pub., New York, New York. pp. 201–259.

Brockhausen, I., Yang, J., Burchell, J., Whitehouse, C., and Taylor-Papadimitriou, J., 1995b, Mechanism underlying aberrant glycosylation of the MUC1 mucin in breast cancer cells, *Eur J Biochem.* 233:607–617.

Brockhausen, I., Toki, D., Brockhausen, J., Peters, S., Bielfeldt, T., Kleen, A., Paulsen, H., Meldal, M., Hagen, F., and Tabak, L., 1996, Specificity of O-glycosylation by bovine colostrum UDP-GalNAc: polypeptide α-N-acetylgalactosaminyltransferase using synthetic glycopeptide substrates, *Glycoconj J.* 13:849–856.

Brockhausen, I. and Kuhns, W., 1997a, *Glycoproteins and human disease.* Medical Intelligence Unit, CRC Press and Mosby Year Book, Chapman & Hall, NY.

Brockhausen, I. and Kuhns, W., 1997b, Role and metabolism of glycoconjugate sulfation. *Trends Glycosci Glycotechnol.* 9:379–398.

Brockhausen, I. and Schachter, H., 1997, Glycosyltransferases involved in N-and O-glycan biosynthesis, in: *Glycosciences: status and perspectives.* H.J. and S. Gabius, eds., Chapman & Hall, Weinheim, pp. 78–113.

Brockhausen, I., Yang, J., Dickinson, N., Ogata, S., and Itzkowitz, S., 1998a, Enzymatic basis for sialyl-Tn expression in human colon cancer cells, *Glycoconj J.* 15:595–603.

Brockhausen, I., Schutzbach, J., and Kuhns, W., 1998b, Glycoproteins and their relationship to human disease, *Acta Anat.* 161:36–78.

Brockhausen, I., 1999, Pathways of O-glycan biosynthesis in cancer cells, *Biochim Biophys Acta.* 1473:67–95.

Brockhausen, I., Yang, J., Lehotay, M., Ogata, S., and Itzkowitz, S., 2001, Pathways of mucin O-glycosylation in normal and malignant rat colonic epithelial cells reveal a mechanism for cancer-associated Sialyl-Tn antigen expression, *Biol Chemistry.* 382:219–232.

Brockhausen, I., Lehotay, M., Yang, J., Qin, W., Young, D., Lucien, J., Coles, J., and Paulsen, H., 2002, Glycoprotein biosynthesis in porcine aortic endothelial cells and changes in the apoptotic cell population, *Glycobiology.* 12:33–45.

Brockhausen, I., 2003, Sulfotransferases involved in glycoprotein synthesis, *Biochem Soc Trans.* 31:318–325.

Campbell, B., Finnie, I.A., Hounsell, E.F., and Rhodes, J.M., 1995, Direct demonstration of increased expression of Thomsen-Friedenreich (TF) antigen in colonic adenocarcinoma and ulcerative colitis mucin and its concealment in normal mucin, *J Clin Invest.* 95:571–576.

Cao, Y., Stosiek, P., Springer, G.F., and Karsten, U., 1996, Thomsen-Friedenreich-related carbohydrate antigens in normal adult human tissues: a systematic and comparative study, *Histochem Cell Biol.* 106:197–207.

Capon, C., Leroy, Y., Wieruszeski, J.M., Ricart, G., Strecker, G., Montreuil, J., and Fournet, B., 1989, Structures of O-glycosidically linked oligosaccharides isolated from human meconium glycoproteins, *Eur J Biochem.* 182:139–152.

Capon, C., Wieruszeski, J.M., Lemoine, J., Byrd, J.C., Leffler, H., and Kim, Y.S., 1997, Sulfated Lewis X determinants as a major structural motif in glycans from LS174T-HM7 human colon carcinoma mucin, *J Biol Chem.* 272:31957–31968.

Capon, C., Maes, E., Michalski, J.C., Leffler, H., and Kim, Y.S., 2001, Sd (a)-antigen-like structures carried on core 3 are prominent features of glycans from the mucin of normal human descending colon, *Biochem J.* 358:657–664.

Carraway, K.L., Fregien, N., Carraway, K.L., 3rd, Carraway, C.A., 1992, Tumor sialomucin complexes as tumor antigens and modulators of cellular interactions and proliferation, *J Cell Sci.* 103:299–307.

Carter, S., Slomiany, A., Gwozdzinski, K., Liau, Y., and Slomiany, B., 1988, Enzymatic sulfation of mucus glycoprotein in gastric mucosa, *J Biol Chem.* 263:11977–11984.

Clausen, H. and Bennett, E.P., 1996, A family of UDP-GalNAc: polypeptide N-acetylgalactosaminyl-transferases control the initiation of mucin-type O-linked glycosylation. *Glycobiology.* 6:635–646.

Corfield, A.P., Carroll, D., Myerscough, N., and Probert, C.S., 2001, Mucins in the gastrointestinal tract in health and disease, *Front Biosci.* 6:D1321–1357.

Dall'Olio, F., Chiricolo, M., Ceccarelli, C., Minni, F., Marrano, D., and Santini, D., 2000, Beta-galactoside alpha2,6 sialyltransferase in human colon cancer: contribution of multiple transcripts to regulation of enzyme activity and reactivity with Sambucus nigra agglutinin, *Int J Cancer.* 88:58–65.

Dall'Olio, F., Chiricolo, M., Mariani, E., and Facchini, A., 2001, Biosynthesis of the cancer-related sialyl-alpha 2,6-lactosaminyl epitope in colon cancer cell lines expressing beta-galactoside alpha 2,6-sialyltransferase under a constitutive promoter, *Eur J Biochem.* 268:5876–5884.

Dalziel, M., Whitehouse, C., McFarlane, I., Brockhausen, I., Gschmeissner, S., Schwientek, T., Clausen, H., Burchell, J., and Taylor-Papadimitriou, J., 2001, The relative activities of C2GnT1 and ST3Gal-I glycosyltransferases determine O-glycan structure and expression of a tumour-associated epitope on MUC1, *J Biol Chem.* 276:11007–11015.

Dang, C.T., Shapiro, C.L., and Hudis, C.A., 2002, Potential role of selective COX-2 inhibitors in cancer management, *Oncology*. 16:30–36.

David, L., Nesland, J.M., Clausen, H., Carneiro, F., and Sobrinho-Simões, M., 1992, Simple mucin-type carbohydrate antigens (Tn, Sialosyl-Tn and T) in gastric mucosa, carcinomas and metastases, *APMIS Suppl*. 27, 100:162–172.

David, L., Leitao, D., Sobrinho-Simoes, M., Bennett, E.P., White, T., Mandel, U., Dabelsteen, E., and Clausen, H., 1993, Biosynthetic basis of incompatible histo-blood group A antigen expression: anti-A transferase antibodies reactive with gastric cancer tissue of type O individuals, *Cancer Res*. 53:5494–5500.

De Bolos, C., Real, F.X., and Lopez-Ferrer, A., 2001, Regulation of mucin and Glycoconjugate expression: from normal epithelium to gastric tumors, *Front Biosci*. 6:D1256–D1263.

Dimitroff, C.J., Pera, P., Dall'Olio, F., Matta, K.L., Chandrasekaran, E.V., Lau, J.T., and Bernacki, R.J., 1999, Cell surface n-acetylneuraminic acid alpha2,3-galactoside-dependant intercellular adhesion of human colon cancer cells, *Biochem Biophys Res Commun*. 256:631–636.

Dohi, T., Hashiguchi, M., Yamamoto, S., Morita, H., and Oshima, M., 1994, Fucosyltransferase-producing sialyl Le (x) carbohydrate antigen in benign and malignant gastrointestinal mucosa, *Cancer*. 73:1552–1561.

Dohi, T., Yuyama, Y., Natori, Y., Smith, P.L., Lowe, J.B., and Oshima, M., 1996, detection of N-acetylgalactosaminyltransferase mRNA which determines expression of Sda blood group carbohydrate structure in human gastrointestinal mucosa and cancer, *Int J Cancer*. 67:626–631.

Egea, G., Francí, C., Gambús, G., Lesuffleur, T., Zweibaum, A., and Real, F.X., 1993, Cis-golgi resident proteins and O-glycans are abnormally compartmentalized in the RER of colon cancer cells, *J Cell Sci*. 105:819–830.

Fujitani, N., Liu, Y., Toda, S., Shirouzu, K., Okamura, T., and Kimura, H., 2000, expression of H type 1 antigen of ABO histo-blood group in normal colon and aberrant expressions of H type 2 and H type 1 antigens in colon cancer, *Glycoconj J*. 17:331–338.

Fung, P.Y.S., Madej, M., Koganty, R.R., and Longenecker, B.M., 1990, Active specific immunotherapy of a murine mammary adenocarcinoma using a synthetic tumor-associated glycoconjugate, *Cancer Res*. 50:4308–4314.

Goupille, C., Hallouin, F., Meflah, K., and Le Pendu, J., 1997, Increase of rat colon carcinoma cells tumorigenicity by α(1-2) fucosyltransferase gene transfection, *Glycoconj J*. 7:221–229.

Goupille, C., Marionneau, S., Bureau, V., Hallouin, F., Meichenin, M., Rocher, J., and Le Pendu, J., 2000, alpha1,2Fucosyltransferase increases resistance to apoptosis of rat colon carcinoma cells, *Glycobiology*. 10:375–382.

Granovsky, M., Bielfeldt, T., Peters, S., Paulsen, H., Meldal, M., Brockhausen, J., and Brockhausen, I., 1994, O-glycan core 1 UDP-Gal: GalNAc β3-galactosyltransferase is controlled by the amino acid sequence and glycosylation of glycopeptide substrates. *Eur J Biochem*. 221:1039–1046.

Gum, J.R., Jr., Crawley, S.C., Hicks, J.W., Szymkowski, D.E., and Kim, Y.S., 2002, MUC17, a novel membrane-tethered mucin. *Biochem Biophys Res Commun*. 291:466–475.

Hanski, C., Hanski, M.L., Zimmer, T., Ogerek, D., Devine, P., and Riecken, E.O., 1995, Characterization of the major sialyl-Le^x-positive mucins present in colon, colon carcinoma, and sera of patients with colorectal cancer, *Cancer Res*. 55:928–933.

Hanski, C., Klußmann, E., Wang, J., Böhm, C., Ogorek, D., Hanski, M.L., Krüger-Krasagakes, S., Eberle, J., Schmitt-Gräff, A., and Riecken, E.O., 1996, Fucosyltransferase III and sialyl-Le(x) expression correlate in cultured colon carcinoma cells but not in colon carcinoma tissue, *Glycoconj J*. 13:727–733.

Harduin-Lepers, A., Recchi, M., and Delannoy, P., 1995, 1994, the year of sialyltransferases. *Glycobiology*. 5:741–758.

Hemmerich, S. and Rosen, S.D., 1994, 6′-sulfated sialyl Lewis x is a major capping group of GlyCAM-1, *Biochemistry*. 33:4830–4835.

Hiller, K.M., Mayben, J.P., Bendt, K.M., Manousos, G.A., Senger, K., Cameron, H.S., and Weston, B.W., 2000, Transfection of alpha (1,3) fucosyltransferase antisense sequences impairs the proliferative and tumorigenic ability of human colon carcinoma cells, *Mol Carcinog*. 27:280–288.

Hoff, S., Matsushita, Y., Ota, D.M., Cleary, K.R., Yamori, T., Hakomori, S., and Irimura, T., 1989, Increased expression of sialyl-dimeric Le^x antigen in liver metastases of human colorectal carcinoma, *Cancer Res*. 49:6883–6888.

Hounsell, E.F., Lawson, A.M., Feeney, J., Gooi, H.C., Pickering, N.J., Stoll, M.S., Lui, S.C., and Feizi, T., 1985, Structural analysis of the O-glycosidically linked core region oligosaccharides of human meconium glycoproteins which express oncofetal antigens, *Eur J Biochem*. 148:367–377.

Hounsell, E.F., Lawson, A.M., Stoll, M.S., Kane, D.P., Chasmore, G.C., Carruthers, R.A., Feeney, J., and Feizi, T., 1989, Characterization by mass spectrometry and 500 MHz proton nuclear magnetic resonance spectroscopy of penta- and hexasaccharide chains of human foetal gastrointestinal mucins (meconium glycoproteins), *Eur J Biochem.* 186:597–610.

Hsueh, C.T., Chiu, C.F., Kelsen, D.P., and Schwartz, G.K., 2000, Selective inhibition of cyclooxygenase-2 enhances mitomycin-C-induced apoptosis, *Cancer Chemother Pharmacol.* 45:389–396.

Huang, C.B., Xu, J., Huang, J.F., and Meng, X.Y., 1986, Sulphomucin colonic type intestinal metaplasia and carcinoma in the stomach. A histochemical study of 115 cases obtained by biopsy, *Cancer.* 57:1370–1375.

Huang, J., Byrd, J.C., Yoon, W.H., and Kim, Y.S., 1992, Effect of benzyl-alpha-GalNAc, an inhibitor of mucin glycosylation, on cancer-associated antigens in human colon cancer cells, *Oncology Res.* 4:507–515.

Ichikawa, T., Nakayama, J., Sakura, N., Hashimoto, T., Fukuda, M., Fukuda, M.N., and Taki, T., 1999, Expression of N-acetyllactosamine and beta1,4-galactosyltransferase (beta4GalT-I) during adenoma-carcinoma sequence in the human colorectum, *J Histochem Cytochem.* 47:1593–1602.

Ikehara, Y., Nishihara, S., Kudo, T., Hiraga, T., Morozumi, K., Hattori, T., and Narimatsu, H., 1998, The aberrant expression of Lewis a antigen in intestinal metaplastic cells of gastric mucosa is caused by augmentation of Lewis enzyme expression, *Glycoconj J.* 15:799–807.

Inoue, M., Takahashi, S., Yamashina, I., Kaibori, M., Okumura, T., Kamiyama, Y., Vichier-Guerre, S., Cantacuzene, D., and Nakada, H., 2001, High density O-glycosylation of the MUC2 tandem repeat unit by N-acetylgalactosaminyltransferase-3 in colonic adenocarcinoma extracts, *Cancer Res.* 61:950–956.

Ito, H., Hiraiwa, N., Sawada-Kasugai, M., Akamatsu, S., Tachikawa, T., Kasai, Y., Akiyama, S., Ito, K., Takagi, H., and Kannagi, R., 1997, Altered mRNA expression of specific molecular species of fucosyl- and sialyl-transferases in human colorectal cancer tissues, *Int J Cancer.* 71:556–564.

Itzkowitz, S.H., Yuan, M., Montgomery, C.K., Kjeldsen, T., Takahashi, H.K., Bigbee, W.L., and Kim, Y.S., 1989, Expression of Tn. sialosyl-Tn and T antigens in human colon cancer, *Cancer Res.* 49:197–204.

Itzkowitz, S.H., Bloom, E.J., Kokal, W.A., Modin, G., Hakomori, S.I., and Kim, Y.S., 1990, Sialosyl-Tn: a novel mucin antigen associated with prognosis in colorectal cancer patients, *Cancer* 66:1960–1966.

Itzkowitz, S.H., Bloom, E.J., Lau, T.S., and Kim, Y.S., 1992, Mucin associated Tn and sialosyl-Tn antigen expression in colorectal polyps, *Gut.* 33:518–523.

Iwai,T., Inaba, N., Naundorf, A., Zhang, Y., Gotoh, M., Iwasaki, H., Kudo, T., Togayachi, A., Ishizuka, Y., Nakanishi, H., and Narimatsu, H., 2002, Molecular cloning and characetrization of a novel UDP-GlcNAc: GalNAc-peptide β1,3-N-acetylglucosaminyltransferase (β3Gn-T6), an enzyme synthesizing the core 3 structure of O-glycans, *J Biol Chem.* 277:12802–12809.

Jang, K.T., Chae, S.W., Sohn, J.H., Park, H.R., and Shin, H.S., 2002, Coexpression of MUC1 with p53 or MUC2 correlates with lymph node metastasis in colorectal carcinomas, *J Korean Med Sci.* 17:29–33.

Jass, J.R. and Walsh, M.D., 2001, Altered mucin expression in the gastrointestinal tract: a review, *J Cell Mol Med.* 5:327–351.

Kakeji, Y., Maehara, Y., Morita, M., Matsukuma, A., Furusama, M., Takahashi, I., Kusumoto, T., Ohno, S., and Sugimachi, K., 1995, Correlation between sialyl Tn antigen and lymphatic metastasis in patients with Borrmann type IV gastric carcinoma, *Brit J Cancer.* 71:191–195.

Kakiuchi, Y., Tsuji, S., Tsuji, M., Murata, H., Kawai, N., Yasumaru, M., Kimura, A., Komori, M., Irie, T., Miyoshi, E., Sasaki, Y., Hayashi, N., Kawano, S., and Hori, M., 2000, Cyclooxygenase-2 activity altered the cell-surface carbohydrate antigens on colon cancer cells and enhanced liver metastasis, *Cancer Res.* 62:1567–1572.

Kang, R., Ikeda, Y., Miyoshi, E., Wang, W., Li, W., Ihara, Y., Sheng, Y., and Taniguchi, N., 2000, Cell cycle-dependant regulation of N-acetylglucosaminyltransferase-III in a human colon cancer cell line, Colo201, *Arch Biochem Biophys.* 374:52–58.

Kataoka, K. and Huh Nh, N.H., 2002, A novel beta1,3-N-acetlglucosaminyltransferase involved in invasion of cancer cells as assayed in vitro, *Biochem Biophys Res Commun.* 294:843–848.

Keppler, O.T., Peter, M.E., Hinderlich, S., Moldenhauer, G., Stehling, P., Schmitz, I., Schwartz-Albiez, R., Reutter, W., and Pawlita, M., 1997, Differential sialylation of cell surface glycoconjugates in a human B lymphoma cell line regulates susceptibility for CD95 (APO-1/Fas)-mediated apoptosis and for infection by a lymphotropic virus, *Glycobiology.* 9:557–569.

Kijima-Suda, I., Miyamoto, Y., Toyoshima, S., Itoh, M., and Osawa, T., 1986, Inhibition of experimental pulmonary metastasis of mouse colon adenocarcinoma 26 sublines by a sialic acid:nucleoside conjugate having sialyltransferase inhibiting activity, *Cancer Res.* 46:858–862.

Kim, Y.S., Gum, J., and Brockhausen, I., 1996, Mucin glycoproteins in neoplasia, *Glycoconj J.* 13:693–707.

Kim, Y.S., 1998, Mucin glycoproteins in colonic neoplasia, *Keio J Med.* 47:10–18.

Kim, M.J., Kim, H.S., Song, K.S., Noh, S.H., Kim, H.G., Paik, Y.K., and Kim, H.O., 2002, Altered expression of Lewis antigen on tissue and erythrocytes in gastric cancer patients, *Yonsei Med J.* 43:427–434.

King, M.J., Chan, A., Roe, R., Warren, B.F., Dell, A., Morris, H.R., Bartolo, C.C., Durdey, P., and Corfield, A.P., 1994, Two different glycosyltransferase defects that result in GalNAcα-O-peptide (Tn) expression, *Glycobiology.* 4:267–269.

Klopocki, A.G., Laskowska, A., Antoniewicz-Papis, J., Duk, M., Lisowska, E., and Ugorski, M., 1998, Role of sialosyl Lewis (a) in adhesion of colon cancer cells—the antisense approach, *Eur J Biochem.* 253:309–318.

Kohsaki, T., Nishimori, I., Nakayama, H., Miyazaki, E., Enzan, H., Nomoto, M., Hollingsworth, M.A., and Onishi, S., 2000, Expression of UDP-GalNAc: polypeptide N-acetylgalactosaminyltransferase isozymes T1 and T2 in human colorectal cancer, *J Gastroenterol.* 35:840–848.

Kojima, N., Handa, K., Newman, W., and Hakomori, S., 1992, Inhibition of selectin-dependent tumor cell adhesion to endothelial cells and platelets by blocking O-glycosylation of these cells, *Biochem Biophys Res Commun.* 182:1288–1295.

Kuan, S.F., Byrd, J.C., Basbaum, C.B., and Kim, Y.S., 1987, Characterization of quantitiative mucin variants from a human colon cancer cell line, *Cancer Res.* 47:5715–5724.

Kuan, S.F., Byrd, J.C., Basbaum, C., and Kim, Y.S., 1989, Inhibition of mucin glycosylation by aryl-N-acetyl-α-galactosaminides in human colon cancer cells, *J Biol Chem.* 264:19271–19277.

Kudo, T., Ikehara, Y., Togayachi, A., Morozumi, K., Watanabe, M., Nakamura, M., Nishihara, S., and Narimatsu, H., 1998, Up-regulation of a set of glycosyltransferase genes in human colorectal cancer, *Lab Investig.* 78:797–811.

Kuhns, W., Rutz, V., Paulsen, H., Matta, K.L., Baker, M.A., Barner, M., Granovsky, M., and Brockhausen, I., 1993, Processing O-glycan core 1, Galβ1-3GalNAcα-R. Specificities of core 2 UDP-GlcNAc:Galβ1-3GalNAc-R β6-N-acetylglucosaminyltransferase and CMP-SA: Galβ1-3GalNAc-R α3-sialyltransferase, *Glycoconj J.* 10:381–394.

Kuhns, W., Jain, R., Matta, K.L., Paulsen, H., Baker, M.A., Geyer, R., and Brockhausen, I., 1995, Characterization of a novel mucin sulfotransferase activity synthesizing sulfated O-glycan core 1, 3-sulfate-Gal-beta1-3GalNAc-alpha-R, *Glycobiology.* 5:689–697.

Kurosaka, A., Nakajima, H., Funakoshi, J., Matsuyama, M., Nagayo, T., and Yamashina, I., 1983, Structures of the major oligosaccharides from a human rectal adenocarcinoma glycoprotein, *J Biol Chem.* 258:11594–11598.

Kurosaka, A., Funakoshi, I., Matsuyama, M., Nagayo, T., and Yamashina, I., 1985, UDP-GalNAc: GalNAc-mucin alpha-N-acetylgalactosamine transferase activity in human intestinal cancerous tissue, *FEBS Lett.* 190:259–262.

Lee, J.K., Bhakta, S., Rosen, S.D., and Hemmerich, S., 1999, Cloning and characterization of a mammalian N-acetylglucosamine-6-sulfotransferase that is highly restricted to intestinal tissue, *Biochem Biophys Res Commun.* 263:543–549.

Leppanen, A., Zhu, Y., Maaheimo, H., Helin, J., Lehtonen, E., and Renkonen, O., 1998, Biosynthesis of branched polylactosaminoglycans. Embryonal carcinoma cells express midchain beta 1,6-N-acetylglucosaminyltransferase activity that generates branches to performed linear backbones. *J Biol Chem.* 273:17399–17405.

Li, F., Wilkins, P.P., Crawley, S., Weinstein, J., Cummings, R.D., and McEver, R.P., 1996, Post-translational modifications of recombinant P-selectin glycoprotein ligand-1 required for binding to P-and E-selectin, *J Biol Chem.* 271:3255–3264.

Lin, S., Kemmner, W., Grigull, S., and Schlag, P.M., 2002, Cell surface alpha 2,6 sialylation affects adhesion of breast carcinoma cells, *Exp Cell Res.* 276:101–110.

Lise, M., Belluco, C., Perera, S.P., Patel, R., Thomas, P., and Ganguly, A., 2000, Clinical correlations of alpha2,6-sialyltransferase expression in colorectal cancer patients, *Hybridoma.* 19:281–286.

Lopez-Ferrer, A., de Bolos, C., Barranco, C., Garrido, M., Isern, J., Carlstedt, I., Reis, C.A., Torrado, J., and Real, F.X., 2000, Role of fucosyltransferases in the association between apomucin and Lewis antigen expression in normal and malignant gastric epithelium, *Gut.* 47:349–356.

Mannori, G., Crottet, P., Cecconi, O., Hanasaki, K., Aruffo, A., Varki, N., and Bevilacqua, M.P., 1995, Differential colon cancer cell adhesion to E-, P-, and L-selectin: role of mucin-type glycoproteins, *Cancer Res.* 55:4425–4431.

Marionneau, S., Bureau, V., Goupille, C., Hallouin, F., Rocher, J., Vaydie, B., and Le Pendu, J., 2000, Susceptibility of rat colon carcinoma cells to lymphokine activated killer-mediated cytotoxicity is decreased by alpha1,2-fucosylation, *Int J Cancer*. 86:713–717.

Mattila, P., Salminen, H., Hirvas, L., Niittymäki, J., Salo, H., Niemelä, Fukuda, M., Renkonen, O., and Renkonen, R., 1998, The centrally acting β1,6N-acetylglucosaminyltransferase (GlcNAc to Gal). Functional expression, purification, and acceptor specificity of a human enzyme involved in midchain branching of linear poly-N-acetyllactosamines, *J Biol Chem*. 273:27633–27639.

Moniaux, N., Escande, F., Porchet, N., Aubert, J., and Batra, S.K., 2001, Structural organization and classification of the human mucin genes, *Front Biosci*. 6:1192–1206.

Muchmore, E.A., Varki, N.M., Fukuda, M., and Varki, A., 1987, Developmental regulation of sialic acid modifications in rat and human colon, *FASEB J*. 1:229–235.

Murata, K., Miyoshi, E., Kameyama, M., Ishikawa, O., Kabuto, T. Sasaki, Y., Hiratsuka, M., Ohigashi, H., Ishiguro, S., Ito, S., Honda, H., Takemura, F., Taniguchi, N., and Imaoka, S., 2000, Expression of N-acetylglucosaminyltransferase V in colorectal cancer correlates with metastasis and poor prognosis, *Clin Cancer Res*. 6:1772–1777.

Nakamori, S., Kameyama, M., Imaoka, S., Furukawa, H., Ishikawa, O., Sasaki, Y., Kabuto, T., Iwanaga, T., Matsushita, Y., and Irimura, T., 1993, Increased expression of sialyl Lewis^x antigen correlates with poor survival in patients with colorectal carcinoma: clinicopathological and immunohistochemical study, *Cancer Res*. 53:3632–3637.

Nishihara, S., Hiraga, T., Ikehara, Y., Kudo, T., Iwasaki, H., Morozumi, K., Akamatsu, S., and Tachikawa, T., 1999, Molecular mechanisms of expression of Lewis b antigen and other type I Lewis antigens in human colorectal cancer, *Glycobiology*. 9:607–616.

Oates, M.D.G., Rosbottom, A.C., and Schrager, J., 1974, Further investigations into the structure of human gastric mucin: The structural configuration of the oligosaccharide chains, *Carbohydr Res*. 34:115–137.

Ogata, S., Maimonis, P.J., and Itzkowitz, S.H., 1992, Mucins bearing the cancer-associated sialosyl-Tn antigen mediate inhibition of natural killer cell cytotoxicity, *Cancer Res*. 52:4741–4746.

Ogata, S., Koganty, R., Reddish, M., Longenecker, B.M., Chen, A., Perez, C., and Itzkowitz, S.H., 1998, Different modes of sialyl-Tn expression during malignant transformation of human colonic mucosa, *Glycoconj J*. 15:29–35.

Orntoft, T., Harving, N., and Langkilde, N., 1990, O-linked mucin-type glycoproteins in normal and malignant colon mucosa: lack of T antigen expression and accumulation of Tn and sialosyl-Tn antigens in carcinomas, *Int J Cancer*. 45:666–672.

Orntoft, T.F., Greenwell, P., Clausen, H., and Watkins, W.M., 1991, Regulation of the oncodevelopmental expression of type 1 chain ABH and Lewis^b blood group antigens in human colon by α-2-L-fucosylation, *Gut*. 32:287–293.

Passaniti, A. and Hart, G., 1988, Cell surface sialylation and tumor metastasis, *J Biol Chem*. 263:7591–7603.

Petretti, T., Schulze, B., Schlag, P.M., and Kemmner, W., 1999, Altered mRNA expression of glycosyltransferases in human gastric carcinomas, *Biochim Biophys Acta*. 1428:209–218.

Petretti, T., Kemmner, W., Schulze, B., and Schlag, P.M., 2000, Altered mRNA expression of glycosyltransferases in human colorectal carcinomas and liver metastases, *Gut*. 46:359–366.

Piller, F., Cartron, J.P., Maranduba, A., Veyrieres, A., Leroy, Y., and Fournet, B., 1984, Biosynthesis of blood group I antigens. identification of a UDP-GlcNAc:GlcNAc beta1-3Gal(-R)beta1-6(GlcNAc to Gal) N-acetylglucosaminyltransferase in hog gastric mucosa, *J Biol Chem*. 259:13385–13390.

Podolsky, D.K., 1985, Oligiosaccharide structures of human colonic mucin, *J Biol Chem*. 260:8262–8271.

Podolsky, D.K., 1985, Oligosaccharide structures of isolated human colonic mucin species, *J Biol Chem*. 260:15510–15515.

Rhodes, J.M., 1996, Unifying hypothesis for inflammatory bowel disease and associated colon cancer: sticking the pieces together with sugar, *The Lancet*. 347:40–44.

Roth, J., Wang, Y., Eckhardt, A.E., and Hill, R.L., 1994, Subcellular localization of the UDP-N-acetyl-D-galactosamine: polypeptide N-acetylgalactosaminyltransferase-mediated O-glycosylation reaction in the submaxillary gland, *Proc Natl Acad Sci USA*. 91:8935–8939.

Röttger, S., White, J., Wandall, H.H., Olivo, J.C., Stark, A., Bennett, E.P., Whitehouse, C., Berger, E.G., Clausen, H., and Nilsson, T., 1998, Localization of three human polypeptide GalNAc-transferases in HeLa cells suggests initiation of O-linked glycosylation throughout the Golgi apparatus, *J Cell Sci*. 111:45–60.

Salvini, R., Bardoni, A., Valli, M., and Trinchera, M., 2001, Beta 1,3-Galactosyltransferase beta 3Gal-T5 acts on the GlcNAcbeta 1→3Galbeta1→4GlcNAcbeta 1→R sugar chains of carcinoembryonic antigen and other N-linked glycoproteins and is down-regulated in colon adenocarcinomas, *J Biol Chem.* 276:3564–3573.

Sawada, R., Tsuboi, S., and Fukuda, M., 1994, Differential E-selectin-dependent adhesion efficiency in sublines of a human colon cancer exhibiting distinct metastatic potentials, *J Biol Chem.* 269:1425–1431.

Schachter, H. and Brockhausen, I., 1992, in: *Glycoconjugates: composition, structure and function*. H.J. Allen and E.C. Kisailus, eds., Marcel Dekker, Inc., New York, NY, pp. 263–332.

Schneider, F., Kemmer, W., Haensch, W., Franke, G., Gretschel, S., Karsten, U., and Schlag, P.M., 2001, Overexpression of sialyltransferase CMP-sialic acid: Galbeta1,3GalNAc-R alpha6-sialyltransferase is related to poor patient survival in human colorectal carcinomas, *Cancer Res.* 61:4605–4611.

Schwartz, B., Breslier, R.S., and Kim, Y.S., 1992, The role of mucin in colon-cancer metastasis, *Int J Cancer.* 52:60–65.

Schwientek, T., Nomoto, M., Levery, S.B., Merkx, G., van Kessel, A.G., Bennett, E.P., Hollingsworth, M.A., and Clausen, H., 1999, Control of O-glycan branch formation. Molecular cloning of human cDNA encoding a novel beta1,6-N-acetylglucosaminyltransferase forming core 2 and core 4, *J Biol Chem.* 274:4504–4512.

Seko, A., Ohkura, T., Kitamura, H., Yonezawa, S., Sato, E., and Yamashita, K., 1996, Quantitative differences in GlcNAc:beta1→3 and GlcNAc:beta1→4 galactosyltransferase activities between human colonic adenocarcinomas and normal colonic mucosa, *Cancer Res.* 56:3468–3473.

Seko, A., Sumiya, J., Yonezawa, S., Nagata, K., and Yamashita, K., 2000, Biochemical differences between two types of N-acetylglucosamine:→6sulfotransferases in human colonic adenocarcinomas and the adjacent normal mucosa: specific expression of a GlcNAc:→6sulfotransferase in mucinous adenocarcinoma, *Glycobiology.* 10:919–929.

Seko, A., Hara-Kuge, S., and Yamashita, K., 2001, Molecular cloning and characterization of a novel human galactose 3-O-sulfotransferase that transfers sulfate to Gal β1-3GalNAc residue in O-glycans, *J Biol Chem.* 276:25697–25704.

Seko, A., Nagata, K., Yonezawa, S., and Yamashita, K., 2002a, Ectopic expression of a GlcNAc 6-O-sulfotransferase, GlcNAc6ST-2, in colonic mucinous adenocarcinoma, *Glycobiology.* 12:379–388.

Seko, A., Nagata, K., Yonezawa, S., and Yamashita, K., 2002b, Down-regulation of Gal 3-0-sulfotransferase-2 (Gal3ST-2) Expression in Human Colonic Non-mucinous Adenocarcinoma, *Jpn J Cancer Res.* 93:507–515.

Shi, Z.R., McIntyre, J., Knowles, B.B., Solter, D., and Kim, Y.S., 1984, Expression of a carbohydrate differentiation antigen, stage-specific embryonic antigen 1, in human colonic adenocarcinoma, *Cancer Res.* 44:1142–1147.

Shibao, K., Izumi, H., Nakayama, Y., Ohta, R., Nagata, N., Nomoto, M., Matsuo, K., Yamada, Y., Kitazato, K., Itoh, H., and Kohno, K., 2002, Expression of UDP-N-acetyl-alpha-D-galactosamine-polypeptide GalNAc N-acetylgalactosaminyl transferase-3 in relation to differentiation and prognosis in patients with colorectoral carcinoma, *Cancer.* 94:1939–1946.

Shimodaira, K., Nakayama, J., Nakamura, N., Hasebe, O., Katsuyama, T., and Fukuda, M., 1997, Carcinoma-associated expression of core 2 beta-1,6-N-acetylglucosaminyltransferase gene in human colorectal cancer: role of O-glycans in tumor progression, *Cancer Res.* 57:5201–5206.

Singhal, A., Fohn, M., and Hakomori, S., 1991, Induction of alpha-N-acetylgalactosamine-O-serine (Tn) antigen-mediated cellular immune response for active immunotherapy in mice, *Cancer Res.* 51:1406–1411.

Slomiany, B.L., Zdebska, E., and Slomiany, A., 1984, Structural characterization of neutral oligosaccharides of human H+Le b+ gastric mucin, *J Biol Chem.* 259:2863–2869.

Slomiany, B.L., Murty, V.L., Slomiany, A., 1980, Isolation and characterization of oligosaccharides from rat colonic mucus glycoprotein, *J Biol Chem.* 255:9719–9723.

Slomski, C.A., Durham, J.P., and Watne, A.L., 1986, Glycosyltransferase levels in familial polyposis coli, *J Surgical Res.* 40:406–410.

Sotozono, M.A., Okada, Y., and Tsuji, T., 1994, The Thomsen-Friedenreich antigen-related carbohydrate antigens in human gastric intestinal metaplasia and cancer, *J Histochem Cytochem.* 42:1575–1584.

Springer, G., Desai, P., Wise, W., Carlstedt, S., Tegtmeyer, H., Stein, R., and Scanlon E., 1990, Pancarcinoma T and Tn epitopes: autoimmunogens and diagnostic markers that reveal incipient carcimomas and help establish prognosis, in: *Immunodiagnosis of cancer*. R. Herbermann, ed., 2nd edn, Marcel Dekker, New York. pp 587–612.

Springer, G.F., Desai, P.R., Spencer, B.D., Tegtmeyer, H., Carlstedt, S.C., and Scanlon, E.F., 1995, T/Tn antigen vaccine is effective and safe in preventing recurrence of advanced breast carcinoma, *Cancer Detec Preven*. 19:374–380.

Sugita, Y., Fujiwara, Y., Hoon, D.S., Miyamoto, A., Sakon, M., Kuo, C.T., and Monden, M., 2002, Overexpression of beta 1,4N-acetylgalactosaminyl-transferase mRNA as a molecular marker for various types of cancers, *Oncology*. 62:149–156.

Sun, J., Thurin, J., Cooper, H.S., Wang, P., Mackiewicz, M., Steplewski, Z., and Blaszczyk-Thurin, M., 1995, Elevated expression of H type GDP-L-fucose: β-D-galactoside α-2-L-fucosyltransferase is associated with human colon adenocarcinoma progression, *Proc Natl Acad Sci USA*. 92:5724–5728.

Sylvester, P.A., Myerscough, N., Warren, B.F., Carlstedt, I., Corfield, A.P., Durdey, P., and Thomas, M.G., 2001, Differential expression of the chromosome 11 mucin genes in colorectal cancer, *J Pathol*. 195:327–335.

Takada, A., Ohmori, K., Yoneda, T., Tsuyoka, K., Hasegawa, A., Kiso, M., and Kannagi, R., 1993, Contribution of carbohydrate antigens sialyl Lewis a and sialyl Lewis x to adhesion of human cancer cells to vascular endothelium, *Cancer Res*. 53:354–361.

Takahashi, I., Maehara, I., Kusumoto, T., Yoshida, M., Kakej, Y., Kusumoto, H., Furusawa, M., and Sugimachi, K., 1993, Predictive value of preoperative serum sialyl Tn antigen levels in prognosis of patients with gastric cancer, *Cancer*. 72:1836–1840.

Takano, R., Muchmore, E., and Dennis, J.W., 1994, Sialylation and malignant potential in tumour cell glycosylation mutants, *Glycobiology*. 4:665–674.

Tang, X., Sun, Y.J., Half, E., Kuo, M.T., and Sinicrope, F., 2002, Cyclooxygenase-2 overexpression inhibits death receptor 5 expression and confers resistance to Tumor Necrosis Factor-related apoptosis-including ligand-induced apoptosis in human colon cancer cells, *Cancer Res*. 62:4903–4908.

Tsuboi, S., Isogai, Y., Hada, N., King, J.K., Hindsgaul, O., and Fukuda, M., 1996, 6′-Sulfo sialyl Le^x but not 6-sulfo sialyl Le^x expressed on the cell surface supports L-selectin-mediated adhesion, *J Biol Chem*. 271:27213–27216.

Uchimura, K., El-Fasakhany, F.M., Hori, M., Hemmerich, S., Blink, S.E., Kansas, G.S., Kanamori, A., Kumamoto, K., Kannagi, R., and Muramatsu, T., 2002, Specificities of N-acetylglucosamine-6-O-sulfotransferases in relation to L-selectin ligand synthesis and tumor-associated enzyme expression, *J Biol Chem*. 277:3979–3984.

van Halbeek, H., Dorland, L., Vliegenthart, J.F.G., Kochetkov, N.K., Arbatsky, N.P., and Derevitskaya, V.A., 1982, Characterization of the primary structure and the microheterogeneity of the carbohydrate chains of porcine blood-group substance by 500 MHz 1H-NMR spectroscopy, *Eur J Biochem*. 127:21–29.

Van Klinken, B.J.W., De Bolos, C., Bueller, H.A., Dekker, J., and Einerhand, A.W.C., 1997, Biosynthesis of mucins (MUC2–6) along the longitudinal axis of the gastrointestinal tract, *Am J Physiol*. 273:G296–G302.

Varki, A., 1993, Biological roles of oligosaccharides: all of the theories are correct. *Glycobiology*. 3:97–130.

Vavasseur, F., Dole, K., Yang, J., Matta, K., Myerscough, N., Corfield, A., Paraskeva, C., and Brockhausen, I., 1994, O-glycan biosynthesis in human colorectal adenoma cells during progression to cancer, *Eur J Biochem*. 222:415–424.

Vavasseur, F., Yang, J., Dole, K., Paulsen, H., and Brockhausen, I., 1995, Synthesis of core 3: Characterization of UDP-GlcNAc: GalNAc β3-N-acetylglucosaminyl-transferase activity from colonic tissues. Loss of the activity in human cancer cell lines, *Glycobiology*. 5:351–357.

Weston, B.W., Hiller, K.M., Mayben, J.P., Manousos, G.A., Bendt, K.M., Liu, R., and Cusack, J.C. Jr., 1999, Expression of human alpha(1,3)fucosyltransferase antisense sequences inhibits selectin-mediated adhesion and liver metastasis of colon carcinoma cells, *Cancer Res*. 59:2127–2135.

Wood, E., Hounsell, E.F., and Feizi, T., 1981, Preparative affinity chromatography of sheep gastric mucins having blood-group Ii activity, and release of antigenically active oligosaccharides by alkaline-borohydride degradation, *Carbohydr Res*. 90:269–282.

Yamamoto, F.I., Clausen, H., White, T., Marken, J., and Hakomori, S.I., 1990, Molecular genetic basis of the histo-blood group ABO system, *Nature*. 345:229–233.

Yamori, T., Kimura, H., Stewart, K., Ota, D., Cleary, K., and Irimura, T., 1987, Differential production of high molecular weight sulfated glycoproteins in normal colonic mucosa, primary colon carcinoma, and metastases, *Cancer Res*. 47:2741–2747.

Yang, J., Byrd, J., Siddiki, B., Chung, Y., Okuno, M., Sowa, M., Kim, Y., Matta, K., and Brockhausen, I., 1994, Alterations of O-glycan biosynthesis in human colon cancer tissue, *Glycobiology*. 4:873–884.

Yazawa, S., Asao, T., Nagamachi, Y., Abbas, S.A., and Matta, K.L., 1989, Tumor-related elevation of serum (alpha——3)-L-fucosyltransferase activity in gastric cancer, *J Cancer Res Clin Oncol*. 115:451–455.

Yazawa, S., Abbas, S.A., Madiyalakan, R., Barlow, J.J., and Matta, K.L., 1986, N-acetyl-β-D-glucosaminyltransferases related to the synthesis of mucin-type glycoproteins in human ovarian tissue, *Carboh Res.* 149:241–252.

Yeh, J.C., Ong, E., Fukuda, M., 1999, Molecular cloning and expression of a novel beta-1,6-N-acetylglucosaminyltransferase that forms core 2, core 4, and I branches. *J Biol Chem.* 274:3215–3221.

Zhu, Y., Srivatana, U., Ullah, A., Gagneja, H., Berenson, C.S., and Lance, P., 2001, Suppression of a sialyltransferase by antisense DNA reduces invasiveness of human colon cancer cells *in vitro*, *Biochim Biophys Acta.* 1536:148–160.

12

O-GlcNAc GLYCOSYLATION AND NEUROLOGICAL DISORDERS

Tony Lefebvre[1], Marie-Laure Caillet-Boudin[2], Luc Buée[2], André Delacourte[2], and Jean-Claude Michalski[1]

[1]UMR 8576
UGSF/CNRS
cité scientifique
59655 Villeneuve d'Ascq
France
[2]U422
INSERM
place de Verdun, 59021 Lille cedex, France

1. INTRODUCTION

O-GlcNAc: a glycosylation type analogous to phosphorylation—the Yin-Yang hypothesis.

O-GlcNAc simply consists in the attachment of a single residue of N-acetylglucosamine on serine or threonine of nuclear and cytosolic proteins. It has been discovered in 1984 by Gerald Hart's group that used galactosyltransferase as an impermeant probe for accessible GlcNAc residues on living lymphocytes (Torres and Hart, 1984). It was soon recognised as a cell surface protein modification but it is clear now that this glycosylation interesses particularly intracellular proteins and moreover cytosolic and nuclear proteins.

The most outstanding feature of this glycosylation is that it is not static, but highly dynamic. The first demonstration of this phenomenon was done on cytokeratins 8 and 18: the biosynthetic and degradation rates of the carbohydrate moiety were faster than the cytokeratins core as determined by metabolic radiolabeling or pulse-chase experiments (Chou *et al.*, 1992). Similar results were observed with small heat shock protein alpha B-crystallin (Roquemore *et al.*, 1996).

In the same way it has been demonstrated that phosphorylation and *O*-GlcNAc posttranslational modifications could compete for the same site or for sites neighboring the peptidic backbone. Use of kinase and phosphatase inhibitors argued for such a process between phosphorylation and *O*-GlcNAc. In fact, a decrease of phosphorylation using

Glycobiology and Medicine, edited by John S. Axford
Kluwer Academic / Plenum Publishers, New York, 2003

kinase inhibitors results in an increase of *O*-GlcNAc glycosylation whereas an increase of phosphorylation by inhibition of phosphatase activities results in a decrease of *O*-GlcNAc incorporation (Griffith and Schmitz, 1999; Lefebvre *et al.*, 1999). The reciprocity between phosphorylation and *O*-GlcNAc has been well documented on the carboxyl terminal domain of RNA polymerase II (CTD) (Comer and Hart, 2001), which has multiple essential roles in transcription initiation, promoter clearance, transcript elongation, and recruitment of the RNA processing machinery. Specific phosphorylation events are associated with the spatial and temporal coordination of these different activities. Using synthetic CTD substrates, the authors showed that *O*-GlcNAc and phosphate modification of the CTD were mutually exclusive at the level of the enzymes responsible for their addition. Another well-known example of *O*-GlcNAc and phosphorylation competition is on the threonine-58 c-Myc proto-oncogene (Kamemura *et al.*, 2002).

The competition between these two posttranslational modifications on the same protein should result in differences in the behavior of this protein. *O*-GlcNAc is a neutral sugar in contrast to phosphorylation that is acid, and then negatively charged at physiological pH. This leads to differences in the conformation of the protein, and then in the assembly with partners and activity (Figure 1).

O-GlcNAc is a relatively new glycosylation type, but there are numerous hypotheses showing that it could be a determinant in cancer, diabetes and neurological disorders.

In this review, to determine if modifications in the *O*-GlcNAc glycosylation process could be related to neurological pathologies, we attempt to collect the maximum of links between *O*-GlcNAc modification and neuropathologies. The accent will be put also on the role of phosphorylation in such diseases.

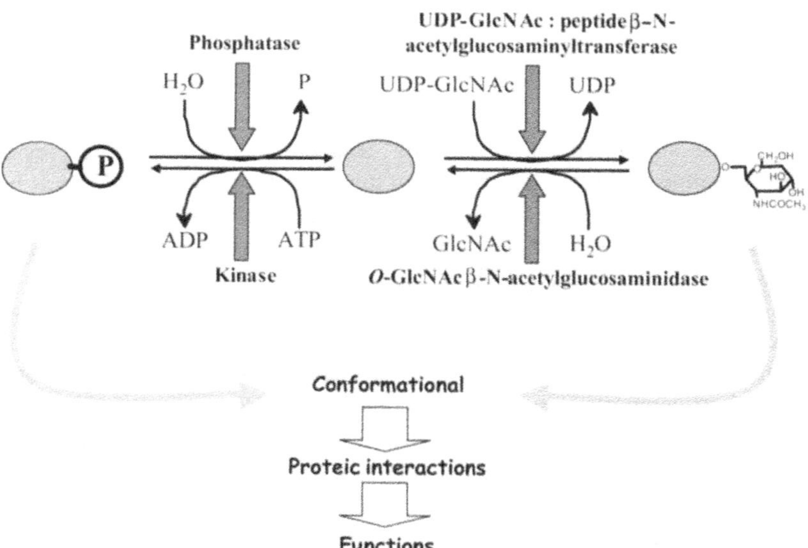

Figure 1. Reciprocity between phosphate and *O*-GlcNAc on a same protein should lead to differences in its biological activities.

2. *O*-GlcNAc IN THE BRAIN—CORRELATION WITH NEUROLOGICAL DISEASES

Whereas the exact role of *O*-GlcNAc is not known, there are several evidences that it is important in brain and that a dysfunction of the regulation of the glyco-deglyco process could contribute to brain disorders.

O-GlcNAc modification has important functional roles in physiological processes of neural cell throughout development, in adulthood and aging (Rex-Mathes *et al.*, 2001). These authors have investigated the *O*-GlcNAc expression of early post-natal cerebellar neurons from mouse brain of different ages. They concluded that the glycosylation was ubiquitously expressed from embryo (day 10) until adulthood, with no significant difference in expression of subcellular fractions from brains of mice with age at an accelerated rate compared to normal mice. Similarly, Yao and Coleman have compared *O*-GlcNAc levels in autopsied normal human brain and in Alzheimer's patients brains using galactosyltransferase assay as a probe (Yao and Coleman, 1998). They concluded that the number of *O*-GlcNAc-containing proteins and the overall *O*-GlcNAc level do not appear to be different between Alzheimer's disease and control brains (normal age-matched controls). However, Yao and Coleman found a significant change with a marked reduction on a 160 kDa *O*-GlcNAc-bearing protein in Alzheimer's disease, that has been identified as AP-3 (as discussed below).

In a different study, Griffith and Schmitz demonstrated that the expression of the proteins modified with *O*-GlcNAc is significantly upregulated in Alzheimer brains compared to age matched control brains (Griffith and Schmitz, 1995).

Interestingly and even with controversial data, numerous neuronal proteins known to be phosphorylated were recently reported to be *O*-GlcNAc modified. Most of them are implicated in neurological disorders. In the following paragraphs, the different data concerning these proteins are presented. The eventual relationship between phosphorylation, *O*-GlcNAc glycosylation and disease is discussed. Another evidence of the importance of *O*-GlcNAc in neurons is the fact that the enzymes responsible for the *O*-GlcNAc dynamism are themselves enriched in the brain.

2.1. Neuronal *O*-GlcNAc-Proteins

The list of neuronal proteins modified with *O*-GlcNAc is growing without cessation for twenty years now. *O*-GlcNAc could modify proteins of the transcriptional machinery (RNA pol II and transcription factors), structural proteins (cytokeratins, talin, vincristin, 4.1 band...), nuclear pore proteins and numerous neuronal proteins we have listed in Table 1. Of these proteins some play a critical role in neurodegenerative diseases: Tau and the beta-amyloid being the best example in Alzheimer's disease and neurofilaments in lateral amyotrophic sclerosis and Alzheimer's disease.

2.1.1. Microtubule-Associated Proteins.

2.1.1.1. Tau Proteins. Tau (Tubulin associated unit) proteins belong to the family of microtubule-associated proteins (low molecular weight). Tau proteins arise from alternative splicing of a single gene located on the long arm of chromosome 17, at band position 17q21. They are mainly expressed in neurons where they play an important role in the

Table 1. An overview of the identified *O*-GlcNAc neuronal proteins.

O-GlcNAc neuronal proteins	References
Beta-amyloid precursor	Griffith *et al.*, 1995, *J Neurosc Res*
Microtubule-associated proteins	
• MAP1, 2 and 4 (HMW)	Ding and Vandre, 1996, *J Biol Chem*
• Tau protein	Arnold *et al.*, 1996, *J Biol Chem*
	Lefebvre *et al.*, 2003, *Biochim Biophys Acta*
Neurofilament H, M and L (nerve conduction)	Dong *et al.*, 1993, *J Biol Chem*
Synapsin I (synaptic vesicles anchor to the cytoskeleton)	Cole and Hart, 1999, *J Neurochem*
Ankyrin (node of Ranvier)	Zhang and Bennett, 1996, *J Biol Chem*
Clathrin assembly protein AP-3	Murphy *et al.*, 1994, *J Biol Chem*

assembly, the stability and the orientation of tubulin monomers into microtubules to constitute the axonal microtubules network. They are phosphoproteins of molecular mass ranging from 45 to 70 kDa according to the isoform and the phosphorylation degree. Their degree of phosphorylation is a good marker of cell integrity. It is heavily disturbed in numerous neurodegenerative disorders, leading to a collapse of the microtubule network and the presence of intraneuronal lesions resulting from Tau aggregation (for review see Delacourte and Buee, 2000). These aggregations can differ by their structure (PHF; Paired Helical Filaments or Straight Filaments), and by their composition in Tau isoforms.

Structurally Tau proteins could be divided in three parts: an acid N-terminal region which contains a projection domain to cytoskeletal components, a central region which contains numerous proline residues and which is a target for protein kinases, and a basic C-terminal region which contains 3 (3R) or 4 (4R) tubulin-binding repeats. The phosphorylation sites are mainly located on each side of Tubulin-binding repeats except a few sites inside this domain. Then, the phosphorylation state of Tau proteins is determinant in microtubule assembly.

Other posttranslational modifications of Tau include glycation (non-enzymatic glycosylation), ubiquitination, proteolysis, N-acetylation, *N*- and *O*-glycosylation, *O*-GlcNAc glycosylation etc. The *O*-GlcNAc glycosylation of Tau proteins has been demonstrated on bovine Tau, with 12 or more sites (Arnold *et al.*, 1996) and on human Tau (Lefebvre *et al.*, 2003). In this last report, it was shown that okadaic acid-induced Tau-hyperphosphorylation is accompanied (i) by a decrease in *O*-GlcNAc of Tau proteins and (ii) by a decrease in the nuclear transport of these proteins. This was demonstrated using lectin blotting with wheat germ agglutinin (WGA), galactosyltransferase and western blotting using phosphorylation-dependent antibodies. *O*-GlcNAc glycosylation occurs on the less phosphorylated molecules (Lefebvre *et al.*, 2003). Initial site mapping on bovine Tau indicate that one major attachment site for *O*-GlcNAc is localized to the microtubule-binding domain on bovine Tau (Arnold *et al.*, 1996). We know very little about *O*-GlcNAc function. As described above, the competition between *O*-GlcNAc and phosphorylation on Tau proteins could contribute to their nuclear localization. The exact localization of the *O*-GlcNAc sites on Tau proteins could help us to understand how these residues could influence the function of this microtubule associated unit protein. As Tau proteins are hyperphosphorylated in many neurodegenerative diseases, this competition *O*-GlcNAc/phosphorylation may be disrupted in the course of such pathologies (Lefebvre *et al.*, 2003).

The existence of a balance between phosphorylation and *O*-GlcNAc glycosylation could explain why, *in vivo*, the hyperphosphorylated Tau proteins that aggregate in paired helical filaments during neurodegenerative diseases are devoid of *O*-GlcNAc residues. Hence, *O*-GlcNAc glycosylation may be considered to be a marker of healthy brain Tau.

Figure 2 represents known Tau phosphorylation sites with their associated protein kinase. According to the existence of a balance between *O*-GlcNAc and phosphorylation we could imagine that these phosphorylation sites are as many as *O*-GlcNAc sites.

Competition between *O*-GlcNAc and phosphorylation is not restricted to Tau proteins but occurs on various proteins as demonstrated in a neuroblastoma cell line, Kelly cells (Lefebvre *et al.*, 1999). Interestingly, the effect of okadaic acid, a protein phosphatases 1 and 2A inhibitor, on *O*-GlcNAc level was more pronounced in the neuroblastoma cell line than other cell lines such as HeLa and COS7, probably depicting the importance of *O*-GlcNAc in brain (Lefebvre *et al.*, 1999).

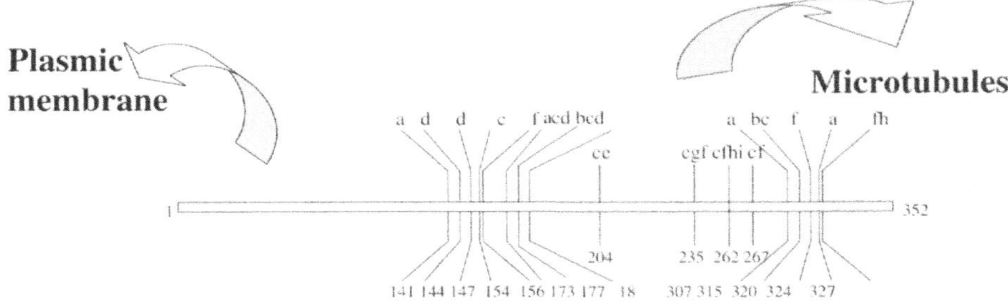

a : Tau protein kinase I / glycogen synthase kinase 3β
b : glycogen synthase kinase 3α / protein kinase FA
c : glycogen synthase kinase 3α / protein kinase FA (after heparin activation)
d : cyclin-dependent kinases ($p34^{cdc2}/p58^{cyclin\ A}$)
e : 35/41 kDa kinase
f : protein kinase A (cAMP dependant)
g : protein kinase C
h : Ca^{2+}/calmodulin-dependent protein kinase II
i : p110-MARK (*Microtubule-Affinity Regulating Kinase*)
k : SAP K (*Stress-Activated Protein kinases*)

Figure 2. Known phosphorylation sites on Tau proteins. Putative sites for *O*-GlcNAc modification?

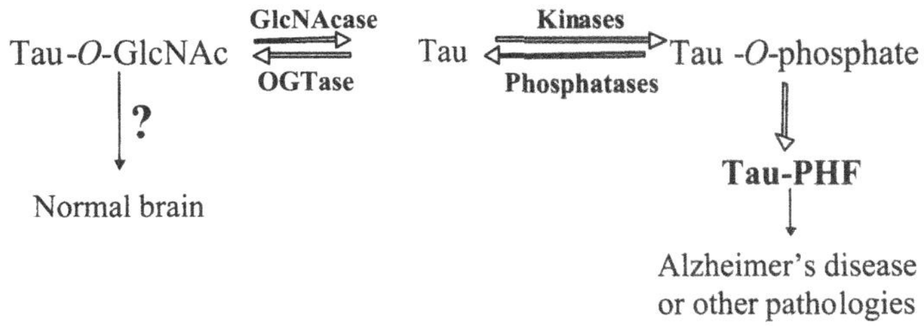

Figure 3. What's the exact role of *O*-GlcNAc in Alzheimer's disease? The importance of Tau phosphorylation in Alzheimer's disease is well known. The role of *O*-GlcNAc on Tau remains to be established.

2.1.1.2. High Molecular Weight Microtubule-Associated Proteins. MAP2 and MAP4 belong to the high molecular weight Microtubule Associated Proteins. MAP2 is a rod-like molecule composed of a C-terminal binding domain and a 200 kDa N-terminal projection domain. Phosphorylation occurs within both the projection and binding domains and extensive phosphorylation on MAP2 reduces its ability to promote microtubule assembly or to bind to preformed microtubules. *O*-GlcNAc glycosylation occurred on at least three sites in the projection domain (Ding and Vandre, 1996). Stoichiometric analysis indicated that nearly 10% of the MAP2 isolated from rat brain is modified by *O*-GlcNAc.

MAP4 was also reported as an *O*-GlcNAc glycosylated molecule (Ding and Vandre, 1996). Phosphorylation of MAP4 within its microtubule-binding domain affects microtubule polymerization and stability and cell cycle progression (Chang *et al.*, 2001). Interestingly, the level of the *O*-GlcNAc modification is increased during mitosis as compared to interphasis (Ding and Vandre, 1996).

2.1.2. Neurofilaments. Neurofilaments are the major neuronal intermediate filaments in adult neurons in central and peripheral nervous system. They belong to the class IV intermediate filament proteins. Mammalian neurofilaments are composed of three polypeptide subunits, designed as NF-L (Low molecular weight, 62 kDa), NF-M (Medium molecular weight, 160 kDa) and NF-H (High molecular weight, 200 kDa) and assemble as heteropolymers. All of the neurofilament subunits are phosphorylated but most of the phosphate on NF-M and NF-H are present in relatively small regions of the molecule in the carboxyterminal tail region rich in KSP sequences. These proteins play an important role in the growth and maintenance of large myelinated axons. Accumulation of neurofilaments are a pathological feature of several human neurodegenerative diseases like lateral amyotrophic sclerosis or infantile spinal muscular atrophy. This suggests that neurofilament transport is disrupted in disease states. However, it is not clear whether this accumulation is a consequence or a main cause of motor neuron dysfunction. Increasing evidence suggests that the phosphorylation of neurofilaments is a mechanism for regulating their transport properties (Miller *et al.*, 2002).

In a first report, only the low and the medium molecular weight subunits were demonstrated to be modified with *O*-linked N-acetylglucosamine (Dong *et al.*, 1993). The predominant *O*-GlcNAc sites were identified: on Thr 21 and Ser 27, 34 and 48 for NF-L, in the N-terminal head domain and one major site on Thr 19 and 48 and Ser 34 for NF-M within the NH2-terminal head domain, and another on Thr 431 located at the tail domain. In a second report it was shown that NF-H was extensively modified by *O*-GlcNAc at Thr 53, Ser 54, and Ser 56 in the head domain (Dong *et al.*, 1996). Curiously somewhat *O*-GlcNAc occurred at multiple sites within the Lys-Ser-Pro repeat motif in the tail domain, a region in assembled neurofilaments known to be nearly stoichiometrically phosphorylated on each of the approximately 50 KSP repeats. The proximity of *O*-GlcNAc and phosphorylation sites in both head and tail domains of each subunit indicates that these modifications may influence one another and play a role in filament assembly and network formation. Up to date, no data reports an effect of the *O*-GlcNAc residues on neurofilaments function or a relationship with disease.

2.1.3. Beta-Amyloid Protein. One of the characteristic features of Alzheimer's disease is the progressive accumulation of amyloid fibrils in senile plaques composed of the amyloid beta-protein (42 amino acid). Amyloid beta-protein is derived from the sequential intracellular cleavage of APP (beta-amyloid precursor protein) (Wolfe, 2002).

It has been demonstrated that the beta-amyloid precursor protein (APP) is modified with *O*-linked N-acetylglucosamine (Griffith and Schmitz, 1995). This case is particularly interesting since it is the first report of a plasma protein that bears *O*-GlcNAc residues. APP is a single transmembrane domain protein and is expressed ubiquitously. Alternative splicing generates multiple transcripts. Of these, three variants are mainly expressed in the brain: APP 695, APP 751, APP 770. APP 695 is the isoform most highly expressed in neurons. Cleavage of APP occurs via either a nonamyloidogenic (generating soluble APPα and a truncated Aβ fragment) or an amyloidogenic pathway (generating Aβ and soluble APPβ). The relationship between APP *O*-GlcNAc glycosylation and APP processing or Alzheimer's disease is unknown.

2.1.4. Assembly Protein-3. Synaptic vesicles (SVs) assemble at the presynaptic compartment through a clathrin-dependent mechanism that involves one or more assembly proteins (APs). Assembly protein-3 (AP-3) is one of the clathrin assembly proteins. It is a synapse-specific protein, which was identified by different laboratories with different names: pp155, AP180, NP185 and F1–20. AP-3 plays a role in promoting clathrin assembly in clathrin-coated vesicle. It is probably involved in synaptic vesicle recycling too. Interactions between AP-3 and AP-2 could be important in vesicle dynamics. AP-3/AP-2 complex is more efficient at assembling clathrin under physiological conditions than is either protein alone (Hao *et al.*, 1999). AP-3 is phosphorylated *in vivo*. Its phosphorylation weakens both the binding of AP-2 by AP-3 and the cooperative clathrin assembly activity of these proteins. Phosphorylation of AP-3, by modulating the affinity of AP-3 for AP-2, may contribute to the regulation of clathrin assembly *in vivo*. AP-3 is also known to be *O*-GlcNAc modified (Murphy *et al.*, 1994). Murphy *et al.* have demonstrated using pulse-chase experiment that the mature form of AP-3 was also *O*-GlcNAc modified. They also showed that AP-3 in its phosphorylated form bound WGA-sepharose, indicating that both modifications, that is *O*-GlcNAc and phosphorylation could be present together on the molecule. The two posttranslational modifications have been mapped on the central 50 kDa structural domain. Capping of the *O*-GlcNAc residues by galactosylation did not affect interaction between AP-3 and clathrin; these authors concluded that *O*-GlcNAc did not play a role in interaction between AP-3 and clathrin.

Yao and Coleman (Yao and Coleman, 1998) found a significant change with a marked reduction on AP-3 in Alzheimer's disease (the reduction was more evident in brain neocortical regions). The reduction was negatively correlated with the density of neurofibrillary tangles (i.e., the intraneuronal aggregates constituted by hyperphosphorylated Tau proteins). In fact, the *O*-GlcNAc/AP-3 ratio is not changed, but it is the level of AP-3 protein that decreases in Alzheimer's disease in agreement with the synapse loss occurring in Alzheimer's disease. The conclusion is that the loss of glycosylated AP-3 may be an earlier event in the pathological cascade of synapse in Alzheimer's disease.

2.1.5. Synapsin I. Synapsin I belongs to a family of five related neuron-specific phosphoproteins, synapsins, associated with the membranes of synaptic vesicles that have been implicated in the regulation of neurotransmitter release (Ferreira and Rapoport, 2002). They tether synaptic vesicles to actin filaments in a phosphorylation-dependent manner, controlling the number of vesicles available for release at the nerve terminus. A growing body of evidence suggests that synapsins play a broad role during neuronal development. They participate in the formation and maintenance of synaptic contacts among central neurons.

Each synapsin has a specific role during the elongation of undifferentiated processes and their posterior differentiation into axons and dendrites. Synapsin anchors synaptic vesicles to the cytoskeleton and ensures a steady supply of fusion-competent synaptic vesicles. It is phosphorylated and it is well known that phosphorylation plays a role in synapsin interaction with synapses vesicles and cytoskeleton. This protein is also *O*-GlcNAc glycosylated on seven sites which are present in the B and D domains: Ser 55, Thr 56, Thr 87, Ser 516, Thr 524, Thr 562, and Ser 576 (Cole and Hart, 1999). These glycosylation sites are around the five phosphorylation sites suggesting a competition between *O*-GlcNAc and phosphate.

2.1.6. Ankyrin. Ankyrin G of 270 and 480 kDa are proteins, belonging to the spectrin-binding proteins that are localized at nodes of Ranvier regions within myelinated axons where ion channels are enriched and the ion fluxes of action potential occur. They may be involved in coupling the voltage-dependent sodium channel and in bridging the membrane to the spectrin/actin network. The 480 kDa protein has been demonstrated to be modified with *O*-GlcNAc in the 46 kDa serine-rich domain which distinguishes these ankyrins from other members of the ankyrin family (Zhang and Bennett, 1986). The role of *O*-GlcNAc on ankyrins is unknown. Nevertheless, a relationship between phosphorylation and spectrin–ankyrin association has been shown (Ghosh and Cox, 2001). Treatment of erythroid cells with Ser/Thr phosphatase inhibitors stimulates the hyperphosphorylation of ankyrin in chicken and dissociated the bulk of ankyrins from cytoskeletal spectrin. This observation demonstrates that the cytoskeletal association of ankyrin is regulated by phosphorylation.

2.2. Glucose, the Hexosamine Pathway and *O*-GlcNAc

Recently it was shown that *O*-GlcNAc modifying proteins could proceed from cellular glucose through the hexosamine biosynthetic pathway, resulting in elevated UDP-N-acetylglucosamine (GlcNAc) concentrations in particular in pancreatic beta-cells (Konrad *et al.*, 2000; for review see Konrad and Kudlow, 2002). This hexosamine pathway is directly linked to the glucose metabolism and interestingly numerous proteins involved in the metabolism of glucose are themselves *O*-GlcNAc, such as casein-kinase II, glycogen synthase-kinase 3 (Lubas and Hanover, 2000) and insulin receptor substrate-1 and 2 (Patti *et al.*, 1999). The use of streptozotocin, a diabetogenic product, in conjunction with glucose could up regulate glycosylated proteins including p135, identified as *O*-linked N-acetyltransferase (OGT) (Konrad *et al.*, 2001). Glutamine:fructose-6-phosphate amido transferase is the rate-limiting enzyme acting in the hexosamine biosynthetic pathway (Sayeski and Kudlow, 1996) and could be considered as the key-enzyme. It catalyzes the conversion of fructose-6-phosphate to glucosamine-6-phosphate.

Glucose metabolism is essential for neural function that influences many normal cellular processes, from neurotransmitter synthesis to ATP production. The glucose, and by consequence energy metabolism, is changed in aging brain, both in the insulin and in the acetylcholine signal transduction. Cumulative evidences suggest that during aging and Alzheimer's disease, the brain actively adapts its glucose metabolism. Glucose is preserved for anabolism and the oxidative utilization of ketone bodies is enhanced (Heiniger, 2000). It suggests that in Alzheimer's disease, brain may not follow a suicide but a rescue program. Changes in brain glucose levels have been shown in numerous neurodegenerative diseases and they may be involved in pathological processes (Hoyer, 1998). Thus, brain is likely to be susceptible to lowered glucose metabolism for its *O*-GlcNAc glycosylation processes (Figure 4).

Interestingly, diabetic neuropathies are a complication of diabetes mellitus and have been linked to problems in the glycemic control. Interestingly, several models have described a relationship between glucose level and phosphorylation, implicating specific protein kinases. The hypothesis that mitogen-activated kinases (MAPK) form transducers for the damaging effects of high glucose has been tested (Purves *et al.*, 2001). MAPK are implicated in the etiology of diabetic neuropathies both via direct effects of glucose and via glucose-induced oxidative stress.

A link was established between diabetes and phosphorylation on neurofilaments. Changes in the phosphorylation status of neurofilaments could lead to severe impairments in axon structure and function that may be found in diabetics. Using two animal models of type 1 diabetes it was demonstrated that in diabetic animals an increase in neurofilaments phosphorylation occurs (Fernyhough *et al.*, 1999). This is related to an increase in the phosphorylation of an isoform of the jun N-terminal kinase (JNK) and extracellular signal-regulated kinases (ERK) which in turn hyperphosphorylate neurofilaments. In neurons of diabetic rats, the hyperphosphorylation of neurofilaments may contribute to the distal sensory axonopathy observed in diabetes. In the same focus it has been proposed that hyperglycemia-induced activation of stress-activated protein kinases (SAPKs), that are neurofilament kinases, may be a primary etiological event in diabetic neuropathy (Fernyhough and Schmidt, 2002).

Studies performed on starving mice have shown an hyperphosphorylation of Tau with a similar regional selectivity as those observed in Alzheimer's disease (Planel *et al.*, 2001). As for neurofilament, kinases specific for Tau and Ser/Thr phosphatases are involved. During starvation Tau phosphorylation was accompanied by a decrease in the activity of

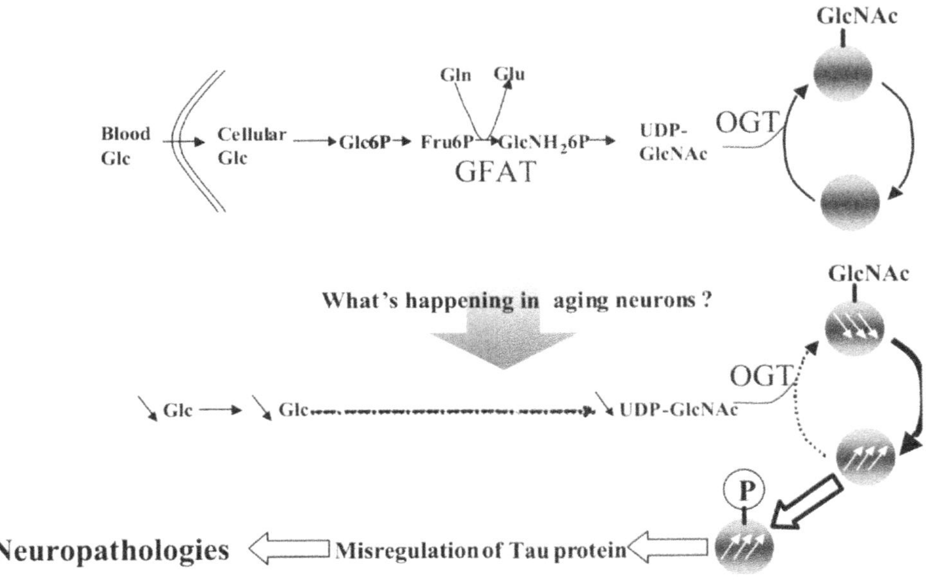

Figure 4. The hexosamine biosynthetic pathway leads to the synthesis of *O*-GlcNAc. Misregulation in aging neurons?

Tau-specific kinases: Tau protein kinase I/glycogen synthase kinase 3 (TPKI/GSK3), and cyclin-dependent kinase 5 (cdk5). In this process of hyperphosphorylation of Tau, this study also suggested that the inhibition of PP2A is predominant. These two examples strongly suggest that there is a competitive relationship between phosphorylation and *O*-GlcNAc: starvation may induce hypoglycosylation via the hexosamine pathway and be a consequence of hyperphosphorylation of neuronal proteins.

2.3. The Enzymes of the Cycling *O*-GlcNAc

The enzymes responsible for the versatility of *O*-GlcNAc are known and cloned: the *O*-linked N-acetylglucosaminyl transferase (OGT) and the *O*-linked N-acetyl-glucosaminidase (*O*-GlcNAcase). These enzymes are nucleocytoplasmic proteins that are interestingly enriched in the brain (Cole and Hart, 2001), were particularly enriched in the cytosol of synaptosomes and both map regions associated with neurodegenerative diseases.

2.3.1. The OGT. This enzyme (Uridine diphospho-N-acetylglucosamine: Polypeptide β-N-acetylglucosaminyltransferase; EC 2.4.1.94) has been characterized for the first time in rat cytosolic liver and in extracts of rabbit reticulocyte membranes (Haltiwanger *et al.*, 1990, 1992). In SDS-PAGE it gives two bands of 110 and 78 kDa: the 78 kDa unit could be a proteolytic form of the 110 kDa one. The catalytic activity is supported by the 110 kDa unit that has been cloned (Lubas and Hanover, 2000).

OGT is highly conserved since we found 80% homology between *C. elegans* and human.

The enzyme contains at N-terminus 9–13 tetratricopeptides (34 amino-acids) that should play a role in interaction with other proteins and for the trimerization of OGT. It is tyrosine phosphorylated and interestingly *O*-GlcNAc glycosylated itself. The enzyme has no consensus sequence for glycosylation and it is suggested that perhaps it exists not as one but many OGT in eukaryota.

It is enriched in pancreas (beta-cells) and brain (Nolte and Muller, 2002).

OGT is also found in plants, where it plays a role in gibberellins pathway (plant hormones with diverse roles in plant growth and development), it is named SPINDLY (Swain *et al.*, 2001). No DNA related sequences have been found in *E. coli* or in *Saccharomyces cerevisiae*.

OGT is reported to have homology with a large group of diverse sugar processing enzymes that include glycogen phosphorylase, UDP-GlcNAc-2-epimerase and MurG, a glycosyl transferase (Wrabl and Grishin, 2001). Proteins homologous to the OGT form a large superfamily termed GPGTF (Glycogen Phosphorylase/Glycogen Transferase).

Recently, alloxan a beta-cell toxin, has been described as an inhibitor of OGT (Konrad *et al.*, 2002). This uracil analog could block both glucosamine and streptozotocin induced protein glycosylation. Nevertheless, this inhibitor is also able to block beta-cell glucokinase and may also inhibit other enzymes recognizing uracil moieties.

OGT is necessary for the viability of ES cells (using Cre-loxP recombination) and is located at the Xp13 region a region known to contain genes implicated in X-linked Parkinson's disease (Nolte and Muller, 2002; Shafi *et al.*, 2000).

Lastly, studying the molecular mechanism by which growth factors regulate hematopoietic cell viability or apoptosis, OGT has been demonstrated to be involved in the prevention of apoptotic mechanism (Fletcher *et al.*, 2002).

Table 2. Summary of the main features of cycling enzymes of *O*-GlcNAc.

	OGT	*O*-GlcNAcase
Chromosomal location	Xp13	10q24
Molecular weight (kDa)	110 and 78	103
Tissue expression	All tissues (enriched in pancreas and brain)	All tissues (enriched in placenta and brain)
Subcellular distribution	Nucleus and cytosol	Nucleus and cytosol
Inhibitor	Alloxan	PUGNAc and STZ
Posttranslational modifications	Phosphate (on tyrosine) and *O*-GlcNAc	Substrate for caspase-3

2.3.2. The O-GlcNAcase. *O*-N-acetylglucosaminidase has been purified for the first time from rat spleen (Dong and Hart, 1994) and has been recently cloned (Gao *et al.*, 2001). It is repertoried as EC 3.2.1.52. The protein contains 916 amino-acid and has a molecular weight of 103 kDa. As OGT *O*-GlcNAcase is a nucleocytoplasmic enzyme which is distributed in all tissues but is particularly abundant in brain, in skeletal muscle and pancreas. It was originally known as hyaluronidase and it is also named hexosaminidase C with a near neutral pH optimum (hexosaminidase A and B are lysosomal and have acidic pH optima). It is not inhibited by GalNAc. Two inhibitors are known for *O*-GlcNAcase: the streptozotocin, an irreversible inhibitor (Roos *et al.*, 1998) and PUGNAc, *O*-(2-acetamido-2-deoxy D-glucopyranosylidene)amino-N-phenylcarbamate (Haltiwanger *et al.*, 1998). The features of OGT and *O*-GlcNAse are summarized in Table 2.

O-GlcNAcase did not share protein motifs with other components. Heat shock proteins, HSP110 and HSC70, and intracellular signal transducers, amphiphysin, DRP-2, and calcineurin, appear to be in complex with *O*-GlcNAcase. We must note that deregulation of amphiphysin, DRP-2, and calcineurin have all been associated with neurological diseases (Wells *et al.*, 2002).

As for OGT, the gene for *O*-GlcNAcase maps a region associated with Alzheimer's disease and other neurological disorders, the chromosomal location 10q24 (Bertram *et al.*, 2000; Myers *et al.*, 2000).

3. CONCLUDING REMARKS

This review summarizes links that have been drawn between *O*-GlcNAc glycosylation and neuronal metabolism, and tries to correlate the effect of this posttranslational modification on the development of neurological diseases (Figure 5). By comparison to phosphorylation we wanted to show that little is known about *O*-GlcNAc and neuropathologies: phosphorylation has been well studied on numerous neuronal proteins, Tau and neurofilaments being the best example. In contradiction, *O*-GlcNAc is a relatively new posttranslational modification and studies performed on proteins are more a description of the glycosylation than a study of its functional significance. So up to date, its impact in pathologies has not allowed us to clearly understand its importance in brain disorders even if there are evidences of its link with such diseases. The first parallel is that enzymes that regulate *O*-GlcNAc are mapped to chromosomal locations linked to neuropathologies. Nevertheless to confirm that these enzymes are implicated in such diseases, more works and experiments have to be done. The function of *O*-GlcNAc in neuronal proteins had to be determined: for instance according to the existence of the *O*-GlcNAc-phosphorylation

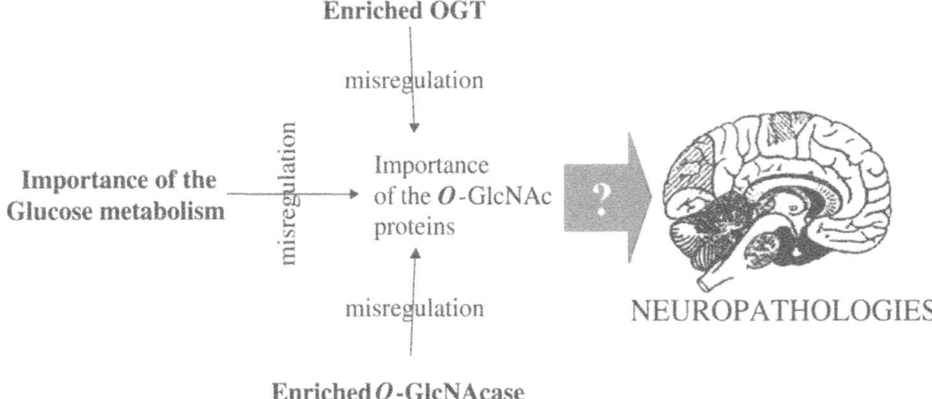

Figure 5. In the brain—summary of links drawn between *O*-GlcNAc glycosylation and neuronal diseases. The impact of this glycosylation in neuropathologies seems evident but remains to be confirmed.

balance we can just suspect that a protein in its glycosylated form could have a different role than that played by the phosphorylated form. An existing finding is that glycosylated Tau could be a marker for healthy brain in contradiction with hyperphosphorylated Tau. Another thing is the importance of *O*-GlcNAc in brain and the link drawn between glucose metabolism, phosphorylation and *O*-GlcNAc.

The next years will be rich in comprehending *O*-GlcNAc functions: by localization of the exact sites of glycosylation, by knowing the interacting partners in the different forms and by studying the conformational changes induced by the sugar on neuronal proteins. Perhaps these findings will highlight the contribution of *O*-GlcNAc to neuronal diseases.

4. ACKNOWLEDGMENTS

The authors thank the Genepole of Lille.

5. REFERENCES

Arnold, C.S., Johnson, G.V., Cole, R.N., Dong, D.L., Lee, M., and Hart, G.W., 1996, The microtubule-associated protein Tau is extensively modified with *O*-linked N-acetylglucosamine, *J Biol Chem.* 271:28741–28744.

Bertram, L., Blacker, D., Mullin, K., Keeney, D., Jones, J., Basu, S., Yhu, S., McInnis, M.G., Go, R.C.P., Vekrellis, K., Selkoe, D.J., Saunders, A.J., and Tanzi, R.E., 2000, Evidence for genetic linkage of Alzheimer's disease to chromosome 10q, *Science.* 290: 2302–2303.

Chang, W., Gruber, D., Chari, S., Kitazawa, H., Hamazumi, Y., Hisanaga, S., and Bulinski, J.C., 2001, Phosphorylation of MAP4 affects microtubule properties and cell cycle progression, *J Cell Sci.* 114:2879–2887.

Chou, C.F., Smith, A.J., and Omary, M.B., 1992, Characterization and dynamics of *O*-linked glycosylation of human cytokeratin 8 and 18, *J Biol Chem.* 267:3901–3906.

Cole, R.N. and Hart, G.W., 1999, Glycosylation sites flank phosphorylation sites on synapsin I: *O*-linked N-acetylglucosamine residues are localized within domains mediating synapsin I interactions, *J Neurochem.* 73:418–428.

Cole, R.N. and Hart, G.W., 2001, Cytosolic *O*-glycosylation is abundant in nerve terminals, *J Neurochem.* 79:1080–1089.

Comer, F.I. and Hart, G.W., 2001, Reciprocity between *O*-GlcNAc and *O*-phosphate on the carboxyl terminal domain of RNA polymerase II, *Biochemistry.* 40:7845–7852.

Delacourte, A. and Buee, L., 2000, Tau pathology: a marker of neurodegenerative disorders, *Curr Opin Neurol.* 13:371–376.

Ding, M. and Vandre, D.D., 1996, High molecular weight microtubule-associated proteins contain *O*-linked-N-acetylglucosamine. *J Biol Chem.* 271:12555–12561.

Dong, D.L., Xu, Z.S., Chevrier, M.R., Cotter, R.J., Cleveland, D.W., and Hart, G.W., 1993, Glycosylation of mammalian neurofilaments. Localization of multiple *O*-linked N-acetylglucosamine moieties on neurofilament polypeptides L and M, *J Biol Chem.* 268:16679–16687.

Dong, D.L. and Hart, G.W., 1994, Purification and characterization of an *O*-GlcNAc selective N-acetyl-beta-D-glucosaminidase from rat spleen cytosol, *J Biol Chem.* 269:19321–19330.

Dong, D.L., Xu, Z.S., Hart, G.W., and Cleveland, D.W., 1996, Cytoplasmic *O*-GlcNAc modification of the head domain and the KSP repeat motif of the neurofilament protein neurofilament-H, *J Biol Chem.* 271:20845–20852.

Fernyhough, P., Gallagher, A., Averill, S.A., Priestley, J.V., Hounsom, L., Patel, J., and Tomlinson, D.R., 1999, Aberrant neurofilaments phosphorylation in sensory neurons of rats with diabetic neuropathy, *Diabetes.* 48:881–889.

Fernyhough, P. and Schmidt, R.E., 2002, Neurofilaments in diabetic neuropathy, *Int Rev Neurobiol.* 50:115–144.

Ferreira, A. and Rapoport, M., 2002, The synapsins: beyond the regulation of neurotransmitter release, *Cell Mol Life Sci.* 59:589–595.

Fletcher, B.S., Dragstedt, C., Notterpek, L., and Nolan, G.P., 2002, Functional cloning of SPIN-2, a nuclear anti-apoptotic protein with roles in cell cycle progression, *Leukemia*, 16:1507–1518.

Gao, Y., Wells, L., Comer, F.I., Parker, G.J., and Hart, G.W., 2001, Dynamic *O*-glycosylation of nuclear and cytosolic proteins: cloning and characterization of a neutral, cytosolic beta-N-acetylglucosaminidase from human brain, *J Biol Chem.* 276:9838–9845.

Ghosh, S. and Cox, J.V., 2001, Dynamics of ankyrin-containing complexes in chicken embryonic erythroid cells: role of phosphorylation, *Mol Biol Cell.* 12:3864–3874.

Griffith, L.S. and Schmitz, B., 1995, *O*-linked N-acetylglucosamine is upregulated in Alzheimer brains, *Biochem Biophys Res Commun.* 213:424–431.

Griffith, L.S. and Schmitz, B., 1999, *O*-linked N-acetylglucosamine levels in cerebellar neurons respond reciprocally to pertubations of phosphorylation, *Eur J Biochem.* 262:824–831.

Hao, W., Luo, Z., Zheng, L., Prasad, K., and Lafer, E.M., 1999, AP180 and AP-2 interact directly in a complex that cooperatively assembles clathrin, *J Biol Chem.* 274:22785–22794.

Haltiwanger, R.S., Holt, G.D., and Hart, G.W., 1990, Enzymatic addition of *O*-GlcNAc to nuclear and cytoplasmic proteins. Identification of a uridine diphospho-N-acetylglucosamine:peptide beta-N-acetylglucosaminyltransferase, *J Biol Chem.* 265:2563–2568.

Haltiwanger, R.S., Blomberg, M.A., and Hart, G.W., 1992, Glycosylation of nuclear and cytoplasmic proteins. Purification and characterization of a uridine diphospho-N-acetylglucosamine : polypeptide beta-N-acetylglucosaminyltransferase, *J Biol Chem.* 267:9005–9013.

Haltiwanger, R.S., Grove, K., and Philipsberg, G.A., 1998, Modulation of *O*-linked N-acetylglucosamine levels on nuclear and cytoplasmic proteins *in vivo* using the peptide *O*-GlcNAc-beta-N-acetylglucosaminidase inhibitor *O*-(2-acetamido-2-deoxy-D-glucopyranosylidene)amino-N-phenylcarbamate, *J Biol Chem.* 273:3611–3617.

Heininger, K., 2000, A unifying hypotheis of Alzheimer's disease. IV. Causation and sequence of events, *Rev Neurosci.* 11: 213–328.

Hoyez, S., 1998, Risk factors for Alzheimer's disease during aging. Impacts of glucose/energy metabolism, *J Neural Transm Suppl.* 54:187–194.

Kamemura, K., Hayes, B.K., Comer, F.I., and Hart, G.W., 2002, Dynamic interplay between *O*-glycosylation and *O*-phosphorylation of nucleocytoplasmic proteins: alternative glycosylation/phosphorylation of THR-58, a known mutational hot spot of c-Myc in lymphomas, is regulated by mitogens, *J Biol Chem.* 277:19229–19235.

Konrad, R.J., Janowski, K.M., and Kudlow, J.E., 2000, Glucose and streptozotocin stimulate p135 *O*-glycosylation in pancreatic islets, *Biochem Biophys Res Commun.* 267:26–32.

Konrad, R.J., Tolar, J.F., Hale, J.E., Knierman, M.D., Becker, G.W., and Kudlow, J.E, 2001, Purification of the *O*-glycosylated protein p135 and identification as *O*-GlcNAc transferase, *Biochem Biophys Res Commun.* 288:1136–1140.

Konrad, R.J., Zhang, F., Hale, J.E., Knierman, M.D., Becker, G.W., and Kudlow, J.E., 2002, Alloxan is an inhibitor of the enzyme *O*-linked N-acetylglucosamine transferase, *Biochem Biophys Res Commun.* 293:207–212.

Konrad, R.J. and Kudlow, J.E., 2002, The role of *O*-linked protein glycosylation in beta-cell dysfunction, *Int J Mol Med.* 10:535–539.

Lefebvre, T., Alonso, C., Mahboub, S., Dupire, M.J., Zanetta, J.P., Caillet-Boudin, M.L., and Michalski, J.C., 1999, Effect of okadaic acid on *O*-linked N-acetylglucosamine levels in a neuroblastoma cell line, *Biochim Biophys Acta.* 1472:71–81.

Lefebvre, T., Ferreira, S., Dupond-Wallois, L., Bussière, T., Dupire, M.-J., Delacourte, A., Michalski, J.-C., and Caillet-Boudin, M.-L., 2003, Evidence of a balance between phosphorylation and *O*-GlcNAc glycosylation of Tau proteins—A role in nuclear localization, *Biochim Biophys Acta.* 1619(2): 167–176.

Lubas, W.A. and Hanover, J.A., 2000, Functional expression of *O*-linked GlcNAc transferase. Domain structure and substrate specificity, *J Biol Chem.* 275:10983–10988.

Miller, C.C., Ackerley, S., Brownlees, J., Grierson, A.J., Jacobsen, N.J., and Thornhill, P., 2002, Axonal transport of neurofilaments in normal and disease states, *Cell Mol Life Sci.* 59:323–330.

Murphy, J.E., Hanover, J.A., Froehlich, M., DuBois, G., and Keen, J.H., 1994, Clathrin assembly protein AP-3 is phosphorylated and glycosylated on the 50-kDa structural domain, *J Biol Chem.* 269:21346–21352.

Myers, A., Holmans, P., Marshall, H., Kwon, J., Meyer, D., Ramic, D., Shears, S., Booth, J., Wavrant DeVrieze, F., Crook, R., Hamshere, M., Abraham, R., Tunstall, N., Rice, F., Carty, S., Lillystone, S., Kehoe, P., Rudrasingham, V., Jones, L., Lovestone, S., Perez-Tur, J., Williams, J., Owen, M.J., Hardy, J., and Goate, A.M., 2000, Susceptibility locus for Alzheimer disease on chromosome 10, *Science.* 290:2304–2305.

Nolte, D. and Muller, U., 2002, Human *O*-GlcNAc transferase (OGT): genomic structure, analysis of splice variants, fine mapping in Xq13.1, *Mamm Genome.* 13:62–64.

Patti, M.E., Virkamaki, A., Landaker, E.J., Kahn, C.R., and Yki-Jarvinen, H., 1999, Activation of the hexosamine pathway by glucosamine *in vivo* induces insulin resistance of early postreceptor insulin signaling events in skeletal muscle, *Diabetes.* 48:1562–1571.

Planel, E., Yasutake, K., Fujita, S.C., and Ishiguro, K., 2001, Inhibition of protein phosphatases 2A overrides Tau protein kinase I/glycogen synthase kinase 3 beta and cyclin-dependent kinase 5 inhibition and results in Tau hyperphosphorylation in the hippocampus of starved mouse, *J Biol Chem.* 276:34298–34306.

Purves, T., Middlemas, A., Agthong, S., Jude, E.B., Boulton, A.J., Fernyhough, P., and Tomlinson, D.R., 2001, A role for mitogen-activated protein kinases in the etiology of diabetic neuropathy, *FASEB J.* 15:2508–2514.

Rex-Mathes, M., Werner, S., Strutas, D., Griffith, L.S., Viebahn, C., Thelen, K., and Schmitz, B. 2001, *O*-GlcNAc expression in developing and aging mouse brain, *Biochimie.* 83:583–590.

Roos, M.D., Xie, W., Su, K., Clark, J.A., Yang, X., Chin, E., Paterson, A.J., and Kudlow, J.E., 1998, Streptozotocin, an analog of N-acetylglucosamine, blocks the removal of *O*-GlcNAc from intracellular proteins, *Proc Assoc Am Physicians.* 110:422–432.

Roquemore, E.P., Chevrier, M.R., Cotter, R.J., and Hart, G.W., 1996, Dynamic *O*-GlcNAcylation of the small heat shock protein alpha B-crystallin, *Biochemistry.* 35:3578–3586.

Sayeski, P.P. and Kudlow, J.E., 1996, Glucose metabolism to glucosamine is necessary for glucose stimulation of transforming growth factor-alpha gene transcription, *J Biol Chem.* 271:15237–15243.

Shafi, R., Iyer, S.P., Ellies, L.G., O'Donnell, N., Marek, K.W., Chui, D., Hart, G.W., and Marth, J.D., 2000, The *O*-GlcNAc transferase gene resides on the X chromosome and is essential for embryonic stem cell viability and mouse ontogeny, *Proc Natl Acad Sci USA.* 97:5735–5739.

Swain, S.M., Tseng, T.S., and Olszewski, N.E., 2001, Altered expression of SPINDLY affects gibberellin response and plant development, *Plant Physiol.* 126: 1174–1185.

Torres, C.R. and Hart, G.W., 1984, Topography and polypeptide distribution of terminal N-acetylglucosamine residues on the surfaces of intact lymphocytes, Evidence for *O*-linked GlcNAc, *J Biol Chem.* 259:3308–3317.

Wells, L., Gao, Y., Mahoney, J.A., Vosseller, K., Chen, C., Rosen, A., and Hart, G.W., 2002, Dynamic *O*-glycosylation of nuclear and cytosolic proteins: further characterization of the nucleocytoplasmic beta-N-acetylglucosaminidase, *O*-GlcNAcase, *J Biol Chem.* 277:1755–1761.

Wrabl, J.O. and Grishin, N.V., 2001, Homology between *O*-linked GlcNAc transferases and proteins of the glycogen phosphorylase superfamily, *J Mol Biol.* 314:365–374.

Wolfe, M.S., 2002, Secretase as a target for Alzheimer's disease, *Curr Top Med Chem.* 2:371–383.

Yao, P.J. and Coleman, P.D., 1998, Reduced *O*-glycosylated clathrin assembly protein AP180: implication for synaptic vesicle recycling dysfunction in Alzheimer's disease, *Neurosci Lett.* 252:33–36.

Zhang, X. and Bennett, V., 1986, Identification of *O*-linked N-acetylglucosamine modification of ankyrinG isoforms targeted to nodes of Ranvier, *J Biol Chem.* 271:31391–31398.

CONGENITAL DEFECTS IN GLYCOSYLATION

13

THE CARBOHYDRATE EPITOPE OF THE NEUTRALIZING ANTI-HIV-1 ANTIBODY 2G12

Christopher N. Scanlan[1,2], Ralph Pantophlet[2], Mark R. Wormald[1], Erica Ollmann Saphire[2,3], Daniel Calarese[3], Robyn Stanfield[3], Ian A. Wilson[3,4], Hermann Katinger[5], Raymond A. Dwek[1], Dennis R. Burton[2], and Pauline M. Rudd[1]

[1]The Glycobiology Institute, Department of Biochemistry, University of Oxford, Oxford OX1 3QU, United Kingdom
[2]Department of Immunology
[3]Department of Molecular Biology
[4]Skaggs Institute for Chemical Biology
The Scripps Research Institute, La Jolla, California 92037
[5]Institute of Applied Microbiology, University of Agriculture
1190 Vienna, Austria

2G12 is a broadly neutralizing human monoclonal antibody against human immunodeficiency virus type-1 (HIV-1) that has previously been shown to bind to a carbohydrate-dependent epitope on gp120. Here, site-directed mutagenesis and carbohydrate analysis were used to define further the 2G12 epitope. Alanine scanning mutagenesis showed that elimination of the N-linked carbohydrate attachment sequences associated with residues N295, N332, N339, N386, and N392 by N $\rightarrow$ A substitution produced significant decreases in 2G12 binding affinity to $gp120_{JR\text{-}CSF}$. The mutagenesis studies provided no convincing evidence for the involvement of gp120 amino acid side chains in 2G12 binding. Antibody binding was inhibited when gp120 was treated with *Aspergillus saitoi* mannosidase, Jack Bean mannosidase, or endoglycosidase H, indicating that Manα1 $\rightarrow$ 2Man-linked sugars of oligomannose glycans on gp120 are required for 2G12 binding. Consistent with this finding, the binding of 2G12 to gp120 could be inhibited by monomeric mannose but not by other hexoses. The data presented here suggests that the most likely epitope for 2G12 is formed from a specific cluster of mannose residues on the outer face of gp120, with the other glycans playing an indirect role in maintaining epitope conformation. [*Journal of Virology*, July 2002, pp. 7306–7321, Vol. 76, No. 14.]

Glycobiology and Medicine, edited by John S. Axford
Kluwer Academic / Plenum Publishers, New York, 2003

1. INTRODUCTION

According to the World Health Organisation there are over 40 million individuals worldwide living with HIV, all of whom are expected to develop AIDS. The development of an effective vaccine, providing sterilizing immunity, remains the primary goal in efforts to control the epidemic.

1.1. Neutralizing Antibodies against HIV-1

Most vaccine initiatives, aimed at eliciting a humoral response, have been targeted to the external HIV glycoproteins: gp120 and gp41 (Burton, 1997). Although the immune response to HIV and its structural proteins is generally characterized by low levels of neutralizing antibodies (Burton and Montefiori, 1997; Connor *et al.*, 1998; Kostrikis *et al.*, 1996; Moog *et al.*, 1997; Moore *et al.*, 1996; Parren *et al.*, 1999), a few such antibodies, cross reactive to many different isolates of the virus have been characterized (Burton *et al.*, 1994; Conley *et al.*, 1994; D'Souza *et al.*, 1994; Thomas *et al.*, 2002; Trkola *et al.*, 1996). The epitopes of these neutralizing antibodies may act as a template for improved vaccine design. Three of these MAbs bind to the surface glycoprotein, gp120, which is the viral receptor for CD4 and chemokine receptors CCR5 and CXCR4. The most potent of these MAbs are b12, which recognizes an epitope overlapping the CD4 receptor site (Burton *et al.*, 1994; Roben *et al.*, 1994), and 2G12 (Kunert *et al.*, 1998; Trkola *et al.*, 1996), which recognizes an epitope based around the C4/V4 region of gp120 and is highly sensitive to the presence of N-linked glycans in this region. Another Fab with broad neutralizing ability, X5, recognizes a region close to the coreceptor binding site on gp120 and overlapping the epitope recognized by CD4-induced MAbs, such as 17b (Moulard *et al.*, 2002). GP41 is also a target for broadly neutralising antibodies: one MAb, 2F5, binds to an epitope involving a linear motif (ELDKWA) on the membrane proximal region of the transmembrane envelope protein gp41 (Conley *et al.*, 1994; Kunert *et al.*, 1998; Parker *et al.*, 2001; Zwick *et al.*, 2001). Recently, two MAbs, Z13 and 4E10, have been described which recognize a region close to the C terminus of the 2F5 epitope (Stiegler *et al.*, 2001; Zwick *et al.*, 2001).

1.2. 2G12 Binds to the "Silent" Face of gp120

The MAb 2G12 has been shown, *in vitro*, to neutralize a wide spectrum of different HIV-1 isolates, including those from different clades, with the notable exception of clade E (Trkola *et al.*, 1995, 1996). *In vivo*, the MAb protects macaques against vaginal challenge with the chimeric virus SHIV 89.6P (Mascola *et al.*, 2000). The antibody recognizes a unique epitope that does not compete with any of the large panel of MAbs to gp120 that have been produced (Moore and Sodroski, 1996). The binding of 2G12 to gp120 had previously been shown to be inhibited by a number of mutations that disrupted sequences encoding attachment of N-linked carbohydrates. These sequences were located in the C2 and C3 regions around the base of the V3 loop, the C4 region, and the V4 loop (Trkola *et al.*, 1996) and probably clade C (Sanders *et al.*, 2002). The crystal structure of the core of gp120 suggests that these carbohydrate attachment sites are clustered together on a region of the gp120 known as the "silent face" (Kwong *et al.*, 1998, 2000). This solvent-accessible face is largely covered by carbohydrate and expected to be relatively weakly immunogenic and, hence, is described as immunologically silent (Wyatt *et al.*, 1993, 1998).

1.3. Glycosylation and Antibody Recognition

Glycosylation can limit the immunogenicity of glycoproteins such as gp120. This may be for a number of reasons. Firstly, glycoproteins exhibit micro-heterogeneity: a single protein sequence might display multiple antigenic glycoforms leading to the dilution of any single response (Rudd and Dwek, 1997). Secondly, the carbohydrates added to potential antigens are the same as those added to "self"-proteins and therefore immunological tolerance may come into play. Thirdly, the affinities of protein–carbohydrate interactions are generally weaker than those between proteins so that high affinity antibodies to carbohydrates do not generally develop. Furthermore large, dynamic glycans can cover an otherwise immunogenic protein epitope (Wilson and Stanfield, 1995; Woods *et al.*, 1994).

Despite large variations in gp120 sequence the 25 consensus *N*-linked glycosylation sites are well-conserved between strains. Mass spectrometric analysis of gp120 has indicated that oligomannose-type glycans are clustered around the C3/C4 base, whereas complex glycans are found on the V1/V2, V3, V4, and V5 domains (Zhu *et al.*, 2000).

1.4. Neutralizing Antibody 2G12

The antibody 2G12 is thought to bind to the silent outer domain of gp120. A previous study showed that loss of *N*-linked glycans around the C3/V4 stem prevented 2G12 binding (Trkola *et al.*, 1996). Additionally, mutations deleting the *N*-link coding sequences have been observed in escape mutants (Poignard *et al.*, 1999). These data were, however, ambiguous. 2G12 might either bind directly to gp120 carbohydrates, or it may bind to a protein epitope with a glycosylation dependent conformation. In this study, the potential contributions of both carbohydrate and protein to the 2G12 epitope were investigated. Site directed alanine-scanning mutagenesis showed that protein side chains contributed little to the specificity of the 2G12 epitope. Glycosidase digestion of gp120 glycans, mutation of *N*-linked coding sequences and alteration of the *N*-linked glycosylation pathway all reduced or abrogated the affinity of 2G12 for gp120. The binding of 2G12 to its epitope can be specifically inhibited by monosaccharides. From the data reported here, it appears that the 2G12 epitope requires $\alpha 1 \rightarrow 2$ linked D-mannose structures present within a cluster of *N*-linked oligomannose glycans on the silent outer face of gp120 (Scanlan *et al.*, 2002).

2. RESULTS

2.1. Alanine Scanning Mutagenesis of gp120$_{JR\text{-}CSF}$ to Identify Residues Important for 2G12 Binding

Several mutations which disrupt attachment sites for N-linked carbohydrates on gp120 have previously been shown to reduce 2G12 binding to monomeric gp120 (Trkola *et al.*, 1996) significantly. However, no systematic mutagenesis analysis has been carried out, and the involvement of extensive regions of gp120 protein surface in the interaction with 2G12 cannot be excluded. Therefore, we decided to carry out extensive alanine scanning mutagenesis. We targeted amino acids predicted to be accessible to antibody from the

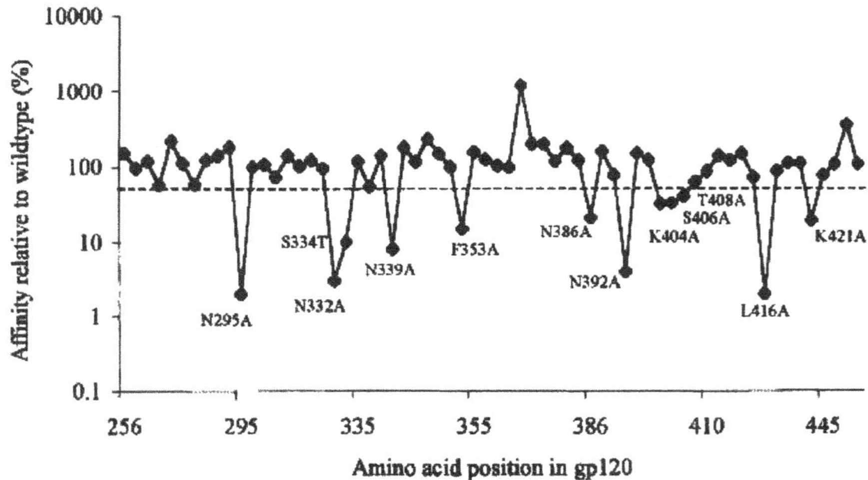

Figure 1. Apparent affinity of 2G12 for alanine mutants of gp $120_{JR\text{-}CSF}$ relative to that of parent gp $120_{JR\text{-}CSF}$. HxB2 sequence numbering is used (Korber *et al.*, HIV Sequence Database, 2001 [http://hiv-web.lanl.gov]). The substitutions that caused a more than 50% (dashed line) reduction in apparent affinity are labeled.

only available structures of gp120: those of core gp120 complexed to CD4 and the Fab fragment 17b (Kwong *et al.*, 1998, 2000).

Mutagenesis was carried out using gp120 from the isolate JR-CSF as the parent. Sixty-three single-amino-acid variants were generated; all of these substituted alanine for the amino acid in the parent gp120 with a small number of exceptions. Where alanine occurs in the parent gp120, it was substituted by lysine. In two cases where threonine and serine form part of an N-linked carbohydrate signal sequence (specifically, T297 and S334), they were substituted by serine and threonine, respectively, to maintain the signal (NXS/T) while altering the residue at these positions. Mutant monomeric gp120s from recombinant pseudovirions were captured onto ELISA wells and probed with various concentrations of 2G12 to generate a binding curve for each mutant. Apparent binding affinities were determined from the concentration of 2G12 at half-maximal binding. The apparent affinity of 2G12 for each mutant gp120 was then related to that for wild-type gp120.

The overwhelming majority of alanine substitutions in gp120 had a limited effect on 2G12 binding. Twelve amino acid substitutions resulted in significant decreases in 2G12 affinity. Five of these substitutions altered triplet sequence motifs (NXS/T) coding for potential N-linked glycosylation sites: N295A, N332A, N339A, N386A, and N392A (Figure 1). Other substitutions producing lowered affinity for 2G12 were on the silent face of gp120 (S334T, L416A), in the V4 loop (K404A, S406A, T408A), in the coreceptor binding site (K421A), and at the junction between the inner and outer domains of gp120 (F353A).

2.2. Conformation of gp120 Measured by IgG b12

For those mutants showing decreased affinities for 2G12, we also investigated binding of the human neutralizing antibody b12, which recognizes an epitope overlapping the CD4 binding site (Burton *et al.*, 1994; Roben *et al.*, 1994). N295A and N332A mutants showed essentially unchanged b12 binding affinities (see also Table 2 in Scanlan *et al.*, 2002). However, N339A, N386A, and N392A mutants all displayed significantly lowered b12 affinities, implying that the substitutions may induce extensive misfolding or conformational perturbation. Therefore, the corresponding carbohydrate chains may not be involved in 2G12 binding. The retention of 2G12 binding by the mutant T388A, in which the carbohydrate signal sequence at position 386–388 is eliminated, suggests that the carbohydrate chain at N386 is indeed not involved in 2G12 binding.

To further investigate the importance of the carbohydrates at N339 and N392, N → Q substitutions were generated. The N339Q mutant bound 2G12 with an affinity similar to that of the parent gp120 and bound b12 with an enhanced affinity. This implies that the carbohydrate chain at position 339 may not be crucial for 2G12 binding but that substitution of Asn with Ala, although not with Gln, disrupts the conformation of the 2G12 (and b12) epitope. In contrast, an N392Q mutant, like the N392A mutant, bound 2G12 with considerably lower affinity than the parent gp120, but bound b12 with unchanged affinity. This is consistent with there being a supporting role for the carbohydrate chain at N392 in 2G12 binding.

Of the remaining mutants that showed decreased affinities for 2G12 relative to the parent gp120, four (S334T, L416A, E353A, and K421A) displayed a similar reduction in affinity for b12. The substitutions involved may, therefore, result in some disruption of global conformation or misfolding. Three substitutions in the V4 loop—K404A, S406A, and T408A—produced very modest decreases in the affinity of 2G12 for gp120 while maintaining b12 affinity.

For two amino acid substitutions, G458A and S365A, significant increases in both 2G12 and b12 binding affinities were observed. Both of these residues are located in the CD4 binding site of gp120. Mutations of these residues thus appears to lead to global conformational changes of gp120.

2.3. Mapping of Mutagenesis Results onto Model of gp120

Data from alanine scanning experiments were mapped onto the crystal structure of the complexed gp120 core of HIV-1_{HxB2} (Figure 2). This approach was thought to be valid as, although the mutagenesis studies used gp$120_{JR\text{-}CSF}$, the structure of the core seems to be highly conserved between isolates. One caveat is that the structure of gp120 is that of the core molecule complexed to CD4 and Fab 17b, and some differences between liganded and unliganded core gp120 have been proposed (Myszka *et al.*, 2000). The carbohydrates attached to N295, N332, N339, N386, and N392 lie on either side of the V3 and V4 loops of the outer face of gp120 (Figure 3). Previous site-specific analysis of N-linked glycosylation of gp120 revealed that oligomannose sugars are attached to these five asparagine residues (Zhu *et al.*, 2000).

The proximity of the carbohydrate chains to the V3 and V4 loops raises the question as to whether these loops could be involved in 2G12 binding. The very modest effect of a V3 deletion on 2G12 binding (67% of wild type affinity) argues against involvement of the V3 loop. Similarly, deletion of the V1 or V1/V2 regions has modest effects, suggesting that the V1/V2 loop does not significantly impact 2G12 binding. On the other hand, the modest decreases in affinity for 2G12 associated with substitutions in the V4 loop described above suggest that the V4 loop may have some role in 2G12 binding. A V4 loop-deleted

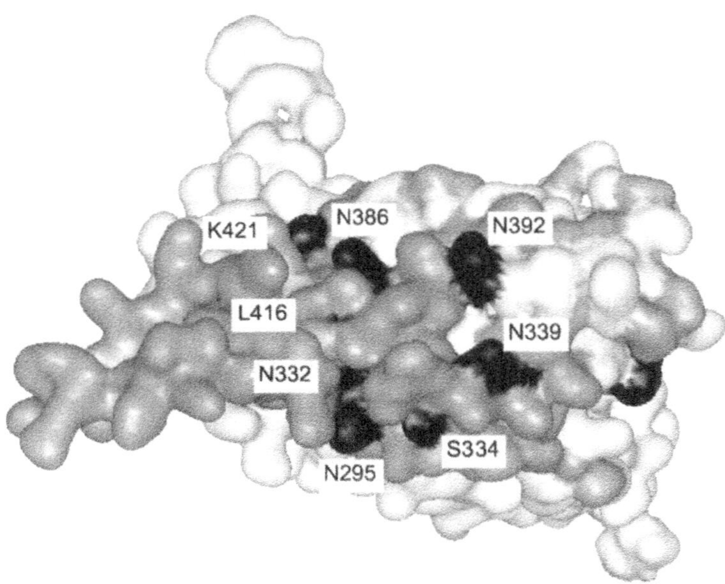

Figure 2. gp120 protein structure with amino acids colored to denote the effects of alanine substitutions on 2G12 affinity. The view shows the surface of the C4-V4 face of gp120. Coordinates were taken from the structure of the CD4-liganded core of $gp120_{HXB2}$ (Kwong *et al.*, 2000). Mutations that did not cause a significant decrease in affinity are shown in gray, and those that caused a decrease in relative affinity are indicated in black. For clarity, the V4 loop has been omitted.

mutant was not generated, as previous studies revealed that the envelope of such a mutant is not processed and only gp160 can be immunoprecipitated from transfected cells. However, alignments of primary sequences of isolates that are effectively neutralized by 2G12 indicated little conservation of amino acid type or loop length between strains. Therefore, it seems unlikely that V4 loop residues are specifically involved in 2G12 binding.

2.4. Glycosidase Digestion of gp120 N-linked Glycans

Exoglycosidase and endoglycosidase digestion of the oligomannose glycans of monomeric gp120 gave further insight into the carbohydrate structures that might be required for the 2G12 epitope. Removal of mannose residues from gp120 by endoH, leaving only the core asparagine-linked GlcNAc, dramatically reduced the affinity of 2G12 for gp120. In contrast, the affinity of IgG1 b12 for gp120 was unaffected, indicating that endoH treatment does not have global conformational effects on gp120. Removal of either the Manα1 $\rightarrow$ 2Man-linked residues or Manα1 $\rightarrow$ 2,3,6Man-linked residues by linkage-specific mannosidases greatly reduced the affinities of both 2G12 and CVN for gp120, but not that of b12.

From these experiments, it appears that the epitope of 2G12 is either formed mainly from the outer Manα1 $\rightarrow$ 2Man residues of oligomannose chains or also involves Manα1 $\rightarrow$ 2,3,6Man residues in the context of Manα1 $\rightarrow$ 2Man residues (for more details see Figures 5 and 6 in Scanlan *et al.*, 2002).

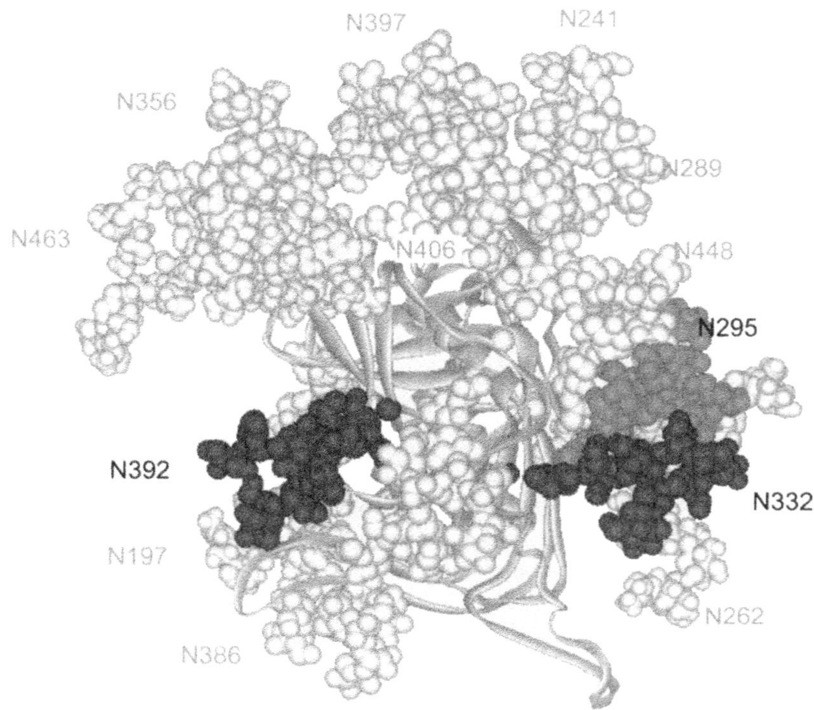

Figure 3. Model of gp120 showing the location of N-linked carbohydrates relevant to this study. The oligomannose-type carbohydrates attached to N332 and to N332 are strongly conserved amongst 2G12-sensitive strains (black). The carbohydrate attached to N295 (dark gray) is strongly implicated by mutagenesis as being essential for the 2G12 epitope. The remaining glycans are of both the complex (N463, N197, N262) and oligomannose (N356, N397, N242, N289, N448, N406, N386) types.

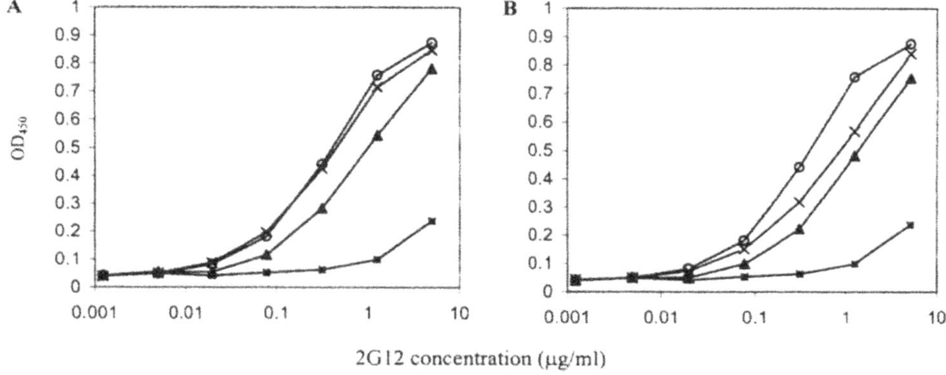

Figure 4. Monosaccharide inhibition of 2G12 binding to gp120. (A) Binding of 2G12 to $gp120_{JR\text{-}FL}$ in the presence of 0(○), 5(×), 50(▲), or 500 mM mannose (■). (B) Binding of 2G12 to $gp120_{JR\text{-}FL}$ in the presence of 500 mM mannose (■), galactose (▲), *N*-acetylglucosamine (×), or buffer (○).

2.5. Carbohydrate Inhibition of the Interaction of 2G12 and gp120

The results from mannosidase digestion strongly suggest that mannose residues are involved in the 2G12 epitope. Consistent with this, high concentrations of D-mannose were able to inhibit the interaction of 2G12 and gp120. The specificity for the inhibition of 2G12 for gp120 appears to be restricted to mannose structures. A comparison of several monosaccharides shows that only mannose (Figure 4) and fructose (data not shown) significantly inhibits 2G12 binding (Figure 4).

3. DISCUSSION

The human MAb 2G12 is one of the few known broadly neutralizing anti-HIV-1 antibodies. Definition of its epitope at the molecular level may contribute to the design of an immunogen able to elicit 2G12-like antibodies. The antibody has previously been shown to bind an epitope on gp120 that is sensitive to changes in N-linked glycosylation and that does not overlap that of any other known antibody to gp120 (Moore and Sodroski, 1996). Here, site-directed mutagenesis and glycan modification of gp120 have been used to characterize the protein and carbohydrate contributions to the unique 2G12 binding site.

3.1. Mutagenesis of gp120

Alanine scanning mutagenesis showed that elimination of the N-linked carbohydrate attachment sequences associated with residues N295, N332, N339, N386, and N392 by N→A substitution produced significant decreases in 2G12 binding affinity to $gp120_{JR\text{-}CSF}$. The N295A and N332A substitutions had specific effects on 2G12 binding to gp120 in that binding of the anti-CD4 binding site antibody b12 was unaffected. In contrast, the N339A, N386A, and N392A substitutions also affected b12 binding, suggesting that they may produce conformational perturbation or protein misfolding, thus bringing into question the involvement of the carbohydrates at these positions in 2G12 binding. Indeed, the retention of 2G12 binding by a T388A mutant in which the carbohydrate attachment sequence at positions 386 to 388 was eliminated confirmed that the carbohydrate chain at N386 is not involved in 2G12 binding. The retention of 2G12 binding by an N339Q mutant similarly argued against the importance of the carbohydrate at N339 in 2G12 binding. Since an N392Q substitution significantly reduced 2G12 binding with little effect on b12 binding, the carbohydrate chain at N392 is likely to be important for 2G12 binding. We also noted that, whereas previous studies had suggested that the carbohydrate at N448 might be important in 2G12 binding, we found this conclusion unlikely for $gp120_{JR\text{-}CSF}$ since a mutant (T450A) in which the carbohydrate signal sequence was eliminated at this position still bound 2G12. Therefore, the mutagenesis studies implicated carbohydrate chains at N295, N332, and N392 as most likely to be important in 2G12 binding. Previous site-specific analysis of $gp120_{SF2}$ has shown that oligomannose chains are attached at these positions (Zhu *et al.*, 2000).

3.2. Sequence Alignments of 2G12 Sensitive Strains

Primary sequence comparisons of gp120s known to interact with 2G12 could help to illuminate the relative importance of the residues highlighted by the mutagenesis studies. A comparison of the primary sequences (HIV Sequence Database, 2001 [http://hiv-web.lanl.gov]) of gp120 from a panel of isolates efficiently neutralized by 2G12 showed that the N-linked carbohydrate signal sequences associated with N295, N332, and N392 are particularly highly conserved. In one of the isolates in which a carbohydrate signal sequence is lost at position 332, it is replaced by another one immediately adjacent. The conservation of the carbohydrate signal sequence associated with N339 is less pronounced than that at the other positions which is, again, consistent with the notion that the carbohydrate at this position is not as crucial for 2G12 binding.

3.3. Hypervariable Loops of gp120

The carbohydrate chains implicated in 2G12 recognition are close to both the V3 and V4 loops. Involvement of the V3 loop was excluded for $gp120_{JR\text{-}CSF}$, since deletion of the loop had minimal effect on affinity for 2G12. On the other hand, alanine substitutions in the V4 loop produced very modest decreases in 2G12 affinity. The variability of primary sequences in this region, among isolates neutralized by 2G12, argues against a direct role for the V4 loop in 2G12 binding. However, some role for the V4 loop in modulating or maintaining the 2G12 epitope is consistent with all the data. This is further supported by the observation that clade E isolates, most of which have an additional disulfide bond internal to the V4 loop, are not recognized by 2G12.

3.4. Investigation of the Carbohydrate Moiety of the 2G12 Epitope

Cleavage of specific mannose linkages was found to be sufficient to dramatically reduce 2G12 binding to gp120. EndoH treatment leaving only the core asparagine-linked GlcNAc, Jack Bean mannosidase treatment leaving $Man_1GlcNAc_2$, and *A. saitoi* mannosidase treatment leaving $Man_5GlcNAc_2$ structures were all effective in essentially eliminating 2G12 binding. *A. saitoi* mannosidase removes a single mannose from each of the D2 and D3 arms and two mannoses from the D1 arm of $Man_9GlcNAc_2$, suggesting that one or more of these residues is critical for 2G12 recognition.

Mannose inhibited 2G12 binding to gp120. In contrast, other hexoses did not significantly affect the binding of 2G12 to gp120 (Figure 4). This further indicates a direct interaction of the carbohydrate with the antibody combining site. The high concentrations of D-mannose required for the inhibition suggests that multiple mannose sub-units may be accommodated in the 2G12 paratope.

The requirement for specific mannose structures is consistent with the inhibition of 2G12 binding by CVN. CVN binds specifically to the Manα1 → 2Man termini of $Man_8GlcNAc_2$(D1D3) and $Man_9GlcNAc_2$.

Overall, therefore, our studies support a cluster of mannose residues contributed by upto three different oligomannose chains on the outer face of gp120 as being critical for 2G12 binding, while providing no indication of any direct involvement of protein side chains. The positioning of the epitope, a model of trimeric gp120 suggests that the attachment of 2G12 would not be hindered either by CD4, co-receptors or by other 2G12 molecules attached to adjacent monomeric units (Figure 5). The dense clustering of oligmannose residues, surrounding N295, N332, and N392 is evident. As previously noted (Kwong *et al.*, 2000), the only solvent exposed regions of gp120 not occluded by carbohydrate are the receptor binding sites, some portions of the hypervariable loops and the C and N terminus (which would be expected to covered by gp41).

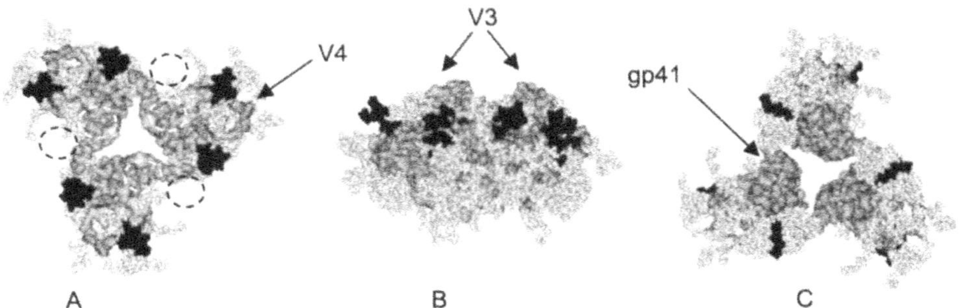

Figure 5. Positioning of the 2G12 epitope on a trimeric model of gp120. The glycans attached to N392, N332, and N295, and which are proposed to form the basis of the 2G12 interaction with gp120, are marked (black). The perspectives shown are from (A) the target cell membrane—looking towards the virus, from (B) the side, and from (C) the viral membrane. The approximate sites of CD4 binding (black broken circle), and the cellular co-receptor (white broken circle) are indicated.

3.5. 2G12 and Carbohydrate Recognition

Exclusive recognition of the carbohydrate of a glycoprotein by an antibody is unusual and raises a number of issues. First, such recognition might be thought to be excluded by tolerance mechanisms, as discussed earlier. However, the cluster of tightly packed oligomannose sugars in the region of the 2G12 epitope is unrepresentative of mammalian glycosylation and may, therefore, invoke an antibody response. There are few, if any, mammalian glycoproteins that are extensively mannosylated. Usually when mannose structures are present, they are mainly associated with one particular site, for example, CD2 (Wyss *et al.*, 1995), Thy-1 (Parekh *et al.*, 1987), and tissue plasminogen activator (Parekh *et al.*, 1998). Even when proteins contain oligomannose sugars at more than one site, a survey of N-glycosylated mammalian proteins in the protein database gave no examples of tightly packed oligomannose clusters such as those on gp120 (A. Petrescu and M. R. Wormald, unpublished data). One factor militating against heavy mannosylation of mammalian proteins is effective clearance by mannose receptors (Taylor and Drickamer, 1993; Weis *et al.*, 1998). Certain pathogens that are heavily mannosylated, such as yeast, are cleared rapidly by this mechanism (Neth *et al.*, 2000). The second unusual aspect of the 2G12–gp120 interaction is that 2G12 has a high affinity for gp120; typically, the K_d for the interaction is in the nanomolar range. However, the affinities of protein–carbohydrate interactions are generally much weaker (Thomas *et al.*, 2000; Wilson and Stanfield, 1995). Surprisingly, the high affinity is also observed for the Fab fragment of 2G12 (R. Pantophlet and D. R. Burton, unpublished data). One way in which higher affinity may be achieved is if multivalency is exercised within a single antibody combining region, that is, if the antibody, for instance, bridges between two or more "subsites" corresponding to mannose residues from different carbohydrate chains. The increase in affinity of proteins for carbohydrates achieved through multiple interactions between the sugar and the protein binding site has been discussed (Cygler *et al.*, 1991; Lee and Lee, 2000; Perkins *et al.*, 1981).

The oligomannose structures located at N295, N332, N339, N386, and N392 are clustered together with the asparagine residues within a 40 by 25 $Å^2$ area on the protein surface (Figure 3). N295 and N332 are located in close proximity to one another, as are N339,

N386, and N392. It appears that the glycan at N332 is likely to be very constrained, and the asparagine residue and the $GlcNAc_2$ (chitobiose) core have extensive interactions with the protein, while the glycan side chain is packed against the V3 loop and the glycan at N295. The mobility of the glycan at N295 is itself expected to be restricted by glycans at N448 and N262. The glycan at N339 is likely to be restricted in its mobility by interactions of the asparagine residue and the chitobiose core with the protein and of the glycan with the V3 and V4 loops. The presence of glycan at N392 may further constrain the glycan at N339. The glycans at N386 and N392 probably have extensive conformational freedom. The reduced dynamic freedom of the glycans on N295, N332, and N339 is not unprecedented (Garcia *et al.*, 1996; Wilson *et al.*, 1981), but it is in contrast to that suggested for a number of other glycoproteins studied (Wormald and Dwek, 1999). The positioning of these carbohydrates and their unusual packing is predicted to form the basis for the 2G12 epitope.

In general, sugar recognition by antibody combining sites involves three to four distinct subsites. The shape and topology of these subsites determine the specificity for a particular 3D arrangement of sugars. In the case of gp120, it is possible to construct a novel sugar epitope (i.e., nonself) from the tight clustering of the different oligomannose structures on the silent face of gp120.

3.6. Immunogenicity of the 2G12 Epitope

2G12 is a neutralizing anti-HIV antibody with a unique epitope requiring specific glycan structures. This requirement for epitope-glycosylation has implications for vaccine design in the light of the generally poor immunogencity of carbohydrate antigens. The suitability of the silent face of gp120 as a template for vaccine design depends on both the ability of the human immune repertoire to form paratopes against glycosylated antigens and, more importantly, the probability of these arising through somatic mutation and clonal selection.

As discussed above, multivalency is required for high affinity antibody–carbohydrate binding. This is because the free energy gained from the accommodation of a hydroxyl group within a receptor is low. Therefore a B-cell, that acquires a mutation allowing the recognition of an additional hydroxyl by its BCR, may not gain a replicative advantage during clonal selection. The evolution of high avidity antibodies whose paratopes contain multiple low affinity sites is an improbable event and might be expected to arise infrequently during B-cell maturation, perhaps explaining why no other known antibodies bind to the "silent" face of HIV.

4. ACKNOWLEDGMENTS

We thank R. Aguilar-Sino, C. Tsang, and G. Reinkensmeier for technical assistance, S. Selvarajah for help with the mutagenesis, and P. Stevens (MRC AIDS Directed Programme Reagent Project) for providing plasmid pEE6HCMVgp120GS. $gp120_{JR\text{-}FL}$ was kindly provided by W. Olson and P. Maddon (Progenics Pharmaceuticals, Tarrytown, N.Y.). CVN was kindly provided by M. Boyd.

This work was supported by NIH grants AI33292 to D.R.B. and GM46192 to I.A.W. E.O.S. is a fellow of the Universitywide AIDS Research Program. C. N. Scanlan is a joint student of the University of Oxford/The Scripps Research Institute graduate programs.

5. REFERENCES

Burton, D.R., 1997, A vaccine for HIV type 1: the antibody perspective, *Proc Natl Acad Sci USA.* 94:10018–10023.

Burton, D.R. and Montefiori, D.C., 1997, The antibody response in HIV-1 infection, *AIDS.* 11 Suppl A:S87–98.

Burton, D.R., Pyati, J., Koduri, R., Sharp, S.J., Thornton, G.B., Parren, P.W., Sawyer, L.S., Hendry, R.M., Dunlop, N., Nara, P.L. *et al.* 1994, Efficient neutralization of primary isolates of HIV-1 by a recombinant human monoclonal antibody, *Science.* 266:1024–1027.

Conley, A.J., Kessler, J.A., 2nd, Boots, L.J., Tung, J.S., Arnold, B.A., Keller, P.M., Shaw, A.R., and Emini, E.A., 1994, Neutralization of divergent human immunodeficiency virus type 1 variants and primary isolates by IAM-41-2F5, an anti-gp41 human monoclonal antibody, *Proc Natl Acad Sci USA.* 91:3348–3352.

Connor, R.I., Korber, B.T., Graham, B.S., Hahn, B.H., Ho, D.D., Walker, B.D., Neumann, A.U., Vermund, S.H., Mestecky, J., Jackson, S., Fenamore, E., Cao, Y., Gao, F., Kalams, S., Kunstman, K.J., McDonald, D., McWilliams, N., Trkola, A., Moore, J.P., and Wolinsky, S.M., 1998, Immunological and virological analyses of persons infected by human immunodeficiency virus type 1 while participating in trials of recombinant gp120 subunit vaccines, *J Virol.* 72:1552–1576.

Cygler, M., Rose, D.R., and Bundle, D.R., 1991, Recognition of a cell-surface oligosaccharide of pathogenic Salmonella by an antibody Fab fragment, *Science.* 253:442–445.

D'Souza, M.P., Geyer, S.J., Hanson, C.V., Hendry, R.M., and Milman, G., 1994, Evaluation of monoclonal antibodies to HIV-1 envelope by neutralization and binding assays: an international collaboration, *AIDS* 8:169–181.

Garcia, K.C., Degano, M., Stanfield, R.L., Brunmark, A., Jackson, M.R., Peterson, P.A., Teyton, L., and Wilson, I.A., 1996, An alphabeta T cell receptor structure at 2.5 A and its orientation in the TCR-MHC complex, *Science.* 274:209–219.

Kostrikis, L.G., Cao, Y., Ngai, H., Moore, J.P., and Ho, D.D., 1996, Quantitative analysis of serum neutralization of human immunodeficiency virus type 1 from subtypes A, B, C, D, E, F, and I: lack of direct correlation between neutralization serotypes and genetic subtypes and evidence for prevalent serum-dependent infectivity enhancement, *J Virol.* 70:445–458.

Kunert, R., Ruker, F., and Katinger, H., 1998, Molecular characterization of five neutralizing anti-HIV type 1 antibodies: identification of nonconventional D segments in the human monoclonal antibodies 2G12 and 2F5, *AIDS Res Hum Retroviruses.* 14:1115–1128.

Kwong, P.D., Wyatt, R., Majeed, S., Robinson, J., Sweet, R.W., Sodroski, J., and Hendrickson, W.A., 2000, Structures of HIV-1 gp120 envelope glycoproteins from laboratory-adapted and primary isolates, *Struct Fold Des.* 8:1329–1339.

Kwong, P.D., Wyatt, R., Robinson, J., Sweet, R.W., Sodroski, J., and Hendrickson, W.A., 1998, Structure of an HIV gp120 envelope glycoprotein in complex with the CD4 receptor and a neutralizing human antibody, *Nature.* 393:648–659.

Kwong, P.D., Wyatt, R., Sattentau, Q.J., Sodroski, J., and Hendrickson, W.A., 2000, Oligomeric modeling and electrostatic analysis of the gp120 envelope glycoprotein of human immunodeficiency virus, *J Virol.* 74:1961–1972.

Lee, R.T. and Lee, Y.C., 2000, Affinity enhancement by multivalent lectin-carbohydrate interaction, *Glycoconj J.* 17:543–551.

Mascola, J.R., Stiegler, G., VanCott, T.C., Katinger, H., Carpenter, C.B., Hanson, C.E., Beary, H., Hayes, D., Frankel, S.S., Birx, D.L., and Lewis, M.G., 2000, Protection of macaques against vaginal transmission of a pathogenic HIV-1/SIV chimeric virus by passive infusion of neutralizing antibodies, *Nat Med.* 6:207–210.

Moog, C., Fleury, H.J., Pellegrin, I., Kirn, A., and Aubertin, A.M., 1997, Autologous and heterologous neutralizing antibody responses following initial seroconversion in human immunodeficiency virus type 1-infected individuals, *J Virol.* 71:3734–3741.

Moore, J.P., Cao, Y., Leu, J., Qin, L., Korber, B., and Ho, D.D., 1996, Inter- and intraclade neutralization of human immunodeficiency virus type 1: genetic clades do not correspond to neutralization serotypes but partially correspond to gp120 antigenic serotypes, *J Virol.* 70:427–444.

Moore, J.P. and Sodroski, J., 1996, Antibody cross-competition analysis of the human immunodeficiency virus type 1 gp120 exterior envelope glycoprotein, *J Virol.* 70:1863–1872.

Moulard, M., Phogat, S.K., Shu, Y., Labrijn, A.F., Xiao, X., Binley, J.M., Zhang, M.Y., Sidorov, I.A., Broder, C.C., Robinson, J., Parren, P.W., Burton, D.R., and Dimitrov, D.S., 2002, Broadly cross-reactive HIV-1-neutralizing human monoclonal Fab selected for binding to gp120-CD4-CCR5 complexes, *Proc Natl Acad Sci USA.* 99:6913–6918.

Myszka, D.G., Sweet, R.W., Hensley, P., Brigham-Burke, M., Kwong, P.D., Hendrickson, W.A., Wyatt, R., Sodroski, J., and Doyle, M.L., 2000, Energetics of the HIV gp120-CD4 binding reaction, *Proc Natl Acad Sci USA.* 97:9026–9031.

Neth, O., Jack, D.L., Dodds, A.W., Holzel, H., Klein, N.J., and Turner, M.W., 2000, Mannose-binding lectin binds to a range of clinically relevant microorganisms and promotes complement deposition, *Infect Immun.* 68:688–693.

Parekh, R.B., Dwek, R.A., Thomas, J.R., Opdenakker, G., Rademacher, T.W., Wittwer, A.J., Howard, S.C., Nelson, R., Siegel, N.R., Jennings, M.G. *et al.* 1989, Cell-type-specific and site-specific N-glycosylation of type I and type II human tissue plasminogen activator, *Biochemistry.* 28:7644–7662.

Parekh, R.B., Tse, A.G., Dwek, R.A., Williams, A.F., and Rademacher, T.W., 1987, Tissue-specific N-glycosylation, site-specific oligosaccharide patterns and lentil lectin recognition of rat Thy-1, *Embo J.* 6:1233–1244.

Parker, C.E., Deterding, L.J., Hager-Braun, C., Binley, J.M., Schulke, N., Katinger, H., Moore, J.P., and Tomer, K.B., 2001, Fine definition of the epitope on the gp41 glycoprotein of human immunodeficiency virus type 1 for the neutralizing monoclonal antibody 2F5, *J Virol.* 75:10906–10911.

Parren, P.W., Moore, J.P., Burton, D.R., and Sattentau, Q.J., 1999, The neutralizing antibody response to HIV-1: viral evasion and escape from humoral immunity, *AIDS.* 13 Suppl A:S137–162.

Perkins, S.J., Johnson, L.N., Phillips, D.C., and Dwek, R.A., 1981, The binding of monosaccharide inhibitors to hen egg-white lysozyme by proton magnetic resonance at 270 MHz and analysis by ring-current calculations, *Biochem J.* 193:553–572.

Poignard, P., Sabbe, R., Picchio, G.R., Wang, M., Gulizia, R.J., Katinger, H., Parren, P.W., Mosier, D.E., and Burton, D.R., 1999, Neutralizing antibodies have limited effects on the control of established HIV-1 infection *in vivo*, *Immunity.* 10:431–438.

Roben, P., Moore, J.P., Thali, M., Sodroski, J., Barbas, C.F., 3rd, and Burton, D.R., 1994, Recognition properties of a panel of human recombinant Fab fragments to the CD4 binding site of gp120 that show differing abilities to neutralize human immunodeficiency virus type 1, *J Virol.* 68:4821–4828.

Rudd, P.M. and Dwek, R.A., 1997, Glycosylation: heterogeneity and the 3D structure of proteins, *Crit Rev Biochem Mol Biol.* 32:1–100.

Sanders, R.W., Venturi, M., Schiffner, L., Kalyanaraman, R., Katinger, H., Lloyd, K.O., Kwong, P.D., and Moore, J.P., 2002, The Mannose-Dependent epitope for neutralizing antibody 2G12 on human immunodeficiency virus type 1 glycoprotein gp120, *J. Virol.* 76:7293–7305.

Scanlan, C.N., Pantophlet, R., Wormald, M.R., Ollmann Saphire, E., Stanfield, R., Wilson, I.A., Katinger, H., Dwek, R.A., Rudd, P.M., and Burton, D.R., 2002, The broadly neutralizing anti-human immunodeficiency virus type 1 antibody 2G12 recognizes a cluster of alpha1 → 2 mannose residues on the outer face of gp120, *J Virol.* 76:7306–7321.

Stiegler, G., Kunert, R., Purtscher, M., Wolbank, S., Voglauer, R., Steindl, F., and Katinger, H., 2001, A potent cross-clade neutralizing human monoclonal antibody against a novel epitope on gp41 of human immunodeficiency virus type 1, *AIDS Res Hum Retroviruses.* 17:1757–1765.

Taylor, M.E. and Drickamer, K., 1993, Structural requirements for high affinity binding of complex ligands by the macrophage mannose receptor, *J Biol Chem.* 268:399–404.

Thomas, R., Patenaude, S.I., MacKenzie, C.R., To, R., Hirama, T., Young, N.M., and Evans, S.V., 2002, Structure of an anti-blood group A Fv and improvement of its binding affinity without loss of specificity, *J Biol Chem.* 277:2059–2064.

Trkola, A., Pomales, A.B., Yuan, H., Korber, B., Maddon, P.J., Allaway, G.P., Katinger, H., Barbas, C.F., 3rd, Burton, D.R., Ho, D.D. *et al.* 1995, Cross-clade neutralization of primary isolates of human immunodeficiency virus type 1 by human monoclonal antibodies and tetrameric CD4-IgG, *J Virol.* 69:6609–6617.

Trkola, A., Purtscher, M., Muster, T., Ballaun, C., Buchacher, A., Sullivan, N., Srinivasan, K., Sodroski, J., Moore, J.P., and Katinger, H., 1996, Human monoclonal antibody 2G12 defines a distinctive neutralization epitope on the gp120 glycoprotein of human immunodeficiency virus type 1, *J Virol.* 70:1100–1108.

Weis, W.I., Taylor, M.E., and Drickamer, K., 1998, The C-type lectin superfamily in the immune system, *Immunol Rev.* 163:19–34.

Wilson, I.A., Skehel, J.J., and Wiley, D.C., 1981, Structure of the haemagglutinin membrane glycoprotein of influenza virus at 3 A resolution, *Nature.* 289:366–373.

Wilson, I.A. and Stanfield, R.L., 1995, A Trojan horse with a sweet tooth, *Nat Struct Biol.* 2:433–436.

Woods, R.J., Edge, C.J., and Dwek, R.A., 1994, Protein surface oligosaccharides and protein function, *Nat Struct Biol.* 1:499–501.

Wormald, M.R. and Dwek, R.A., 1999, Glycoproteins: glycan presentation and protein-fold stability, *Struct Fold Des.* 7:R155–160.

Wyatt, R., Kwong, P.D., Desjardins, E., Sweet, R.W., Robinson, J., Hendrickson, W.A., and Sodroski, J.G., 1998, The antigenic structure of the HIV gp120 envelope glycoprotein, *Nature.* 393:705–711.

Wyatt, R., Sullivan, N., Thali, M., Repke, H., Ho, D., Robinson, J., Posner, M., and Sodroski, J., 1993, Functional and immunologic characterization of human immunodeficiency virus type 1 envelope glycoproteins containing deletions of the major variable regions, *J Virol.* 67:4557–4565.

Wyss, D.F., Choi, J.S., Li, J., Knoppers, M.H., Willis, K.J., Arulanandam, A.R., Smolyar, A., Reinherz, E.L., and Wagner, G., 1995, Conformation and function of the N-linked glycan in the adhesion domain of human CD2, *Science.* 269:1273–1278.

Zhu, X., Borchers, C., Bienstock, R.J., and Tomer, K.B., 2000, Mass spectrometric characterization of the glycosylation pattern of HIV-gp120 expressed in CHO cells, *Biochemistry.* 39:11194–11204.

Zwick, M.B., Labrijn, A.F., Wang, M., Spenlehauer, C., Saphire, E.O., Binley, J.M., Moore, J.P., Stiegler, G., Katinger, H., Burton, D.R., and Parren, P.W., 2001, Broadly neutralizing antibodies targeted to the membrane-proximal external region of human immunodeficiency virus type 1 glycoprotein gp41, *J Virol.* 75:10892–10905.

14

NEW THERAPEUTICS FOR THE TREATMENT OF GLYCOSPHINGOLIPID LYSOSOMAL STORAGE DISEASES

Butters, T. D., Dwek, R. A., and Platt, F. M.

Glycobiology Institute, Department of Biochemistry,
Oxford University, Oxford, UK

1. ABSTRACT

Glycosphingolipid lysosomal storage diseases are a small but challenging group of human disorders to treat. Although these appear to be monogenic disorders where the catalytic activity of enzymes in glycosphingolipid catabolism is impaired, the presentation and severity of disease is heterogeneous. Treatment is often restricted to palliative care, but in some disorders enzyme replacement does offer a significant clinical improvement of disease severity.

An alternative therapeutic approach termed "substrate deprivation" or "substrate reduction therapy" (SRT) aims to reduce cellular glycosphingolipid biosynthesis to match the impairment in catalytic activity seen in lysosomal storage disorders. *N*-Alkylated imino sugars are nitrogen containing polyhydroxylated heterocycles that have inhibitory activity against the first enzyme in the pathway for glucosylating sphingolipid in eukaryotic cells, ceramide-specific glucosyltransferase. The use of *N*-alkylated imino sugars to establish SRT as an alternative therapeutic strategy is described in cell culture and gene knockout mouse disease models. One imino sugar, *N*-butyl-DNJ (*N*B-DNJ) has been used in clinical trials for type 1 Gaucher disease and has shown to be an effective and safe therapy for this disorder. The results of these trials and the prospects of improvement to the design of imino sugar compounds for treating Gaucher and other glycosphingolipid lysosomal storage disorders will be discussed.

2. GLYCOLIPID FUNCTION

Several functions for membrane glycosphingolipids (GSLs) have been described including, membrane stabilization of transmembrane proteins, cell signalling events, and

Glycobiology and Medicine, edited by John S. Axford
Kluwer Academic / Plenum Publishers, New York, 2003

cell–cell adhesion. With the aid of modern oligosaccharide characterization methods and gene cloning of the enzymes that coordinate synthesis and degradation, these molecules are now being seen as potential targets for therapeutic intervention (Dwek *et al.*, 2002; Gagnon and Saragovi, 2002). Recently, GSL participation in detergent insoluble domains, or membrane rafts has emphasized their roles in the trafficking of functionally important molecules to maintain homeostasis (Kobayashi and Hirabayashi, 2000). GSL's also act as ligands for antigen presenting cells and the precise way in which the CD1 molecule coordinates lipid alkyl chains within the structure has been determined following crystallographic analysis (Gadola *et al.*, 2002).

3. GLYCOLIPID METABOLISM—A STATE OF FLUX

We have learned several important lessons from the generation of mouse knockout models although these must be viewed with caution. The flux of glycolipid through the biosynthetic pathway to the membrane, returning back to the cell *via* the lysosome and recycling of the monosaccharide and sphingosine base moieties is continuous (Kolter and Sandhoff, 1999). When this flux is disturbed, in this case by the failure of the lysosome to catabolize the GSLs fully, *de novo* biosynthesis continues and no feedback mechanisms appear to be invoked. The reduced lysosomal efflux leads to GSL storage and a significant expansion of the lysosomes. Other molecules normally destined for degradation are trapped in the insoluble lipid matrix, further restricting access to mutant enzymes. The GSLs may not be exclusively stored in subcellular compartments and in a number of disorders GSL/protein complexes are found in excretions. However, cells that are compromised for lysosomal clearance, follow a pathway that is not well understood (Jeyakumar *et al.*, 2002a), resulting in disease pathogenesis. Thus several mouse models have been generated where the gene for enzyme mediated GSL catabolism has been deleted and these show lysosomal storage of the correct product and emulate some of the disease phenotypes seen in man (Suzuki *et al.*, 1998). By contrast, deletion of the genes encoding glycosyltransferases in the GSL biosynthetic pathway results in variable phenotypes. Complete ablation of ceramide glucosyltransferase (CGT) activity results in embryonic lethality (Yamashita *et al.*, 1999). One conclusion may be that more complex GSLs are required for development but the observation that significant ectodermal apoptosis takes place points to alternative mechanisms. The substrate for CGT is ceramide and a complete block in the synthetic step to glucosylceramide increases the levels of ceramide above the capability of other ceramide specific enzymes such as sphingomyelin synthase, to flux through the pathway. The accumulation of ceramide in metabolically active tissues results in untimely apoptotic cell signalling and death. Experimental disruptions to genes coding for glycosyltransferases involved in synthesizing GSLs further downstream in the pathway appear viable but mice have demonstrable impairment. For example, targeted disruption of two genes that code for glycosyltransferases that synthesize higher gangliosides, results in viable mice that are lethally sensitive to noise and have peripheral nerve degeneration (Allende and Proia, 2002).

4. MANIPULATION OF GLYCOSYLATION

No human mutations in synthesis of GSLs have been described. Either they do not occur because of embryonic lethality or they are so rare that we have yet to discover them.

It is only in the last few years that we have identified mutations at early stages in the *N*-linked glycoprotein biosynthesis pathway that lead to a family of diseases, Congenital Disorders of Glycosylation (CGD) (Freeze, 2002). A complete lack of *N*-linked glycosylation to proteins is not tolerated for multicellular organisms and disruption of the gene coding for a key enzyme coordinating these events, *N*-acetylglucosaminyl-1-phosphotransferase, abruptly stops embryogenesis in mouse, in similar fashion to the CGT knockout (Marek *et al.*, 1999).

Consequently, where intervention during the biosynthesis of glycoproteins and glycolipid is advocated for the treatment of human disease, manipulation of glycosylation must be sufficient to allow biological efficacy without compromising functional cell integrity. In the glycosphingolipidoses, mutations in lysosomal enzymes leads to reduced catalytic activity and storage of GSLs. A number of therapeutic options are available for some of these disorders and the most novel is to reduce the amount of GSL biosynthetic capability to decrease the amount of lysosomal influx (Butters *et al.*, 2000a; Platt and Butters, 1998, 2000). This would permit improved substrate hydrolysis by the mutant enzyme to remove any accumulation in the lysosome.

5. SUBSTRATE REDUCTION THERAPY—*IN VITRO* MODELS OF HUMAN DISEASE

Using a chemically derived model for Gaucher disease, where conduritol β-epoxide was used to irreversibly inactivate the lysosomal β-glucocerebrosidase in macrophages, we were able to demonstrate that a substrate reduction strategy could work in principle using novel inhibitors of ceramide glucosyltransferase (CGT). Two compounds were identified that were able to inhibit this enzyme. The first was an *N*-alkylated glucose mimic, deoxynojirimycin (*N*-butyl-DNJ,. *N*B-DNJ) (Platt *et al.*, 1994a) and subsequent structure/function studies revealed that a galactose derivative, *N*-butyl-DGJ (*N*B-DGJ) was equally effective but more selective (Platt *et al.*, 1994b). Both are competitive, reversible inhibitors for ceramide using *in vitro* assays with isolated enzyme and are non-competitive for UDP-glucose (Butters *et al.*, 2000b). Molecular modelling using the crystal structure of ceramide and the NMR solution structure of *N*B-DNJ indicated homology between several functional groups supporting ceramide mimicry as a plausible mechanism of action for imino sugar inhibition of CGT (Butters *et al.*, 2000b). Gaucher disease is clinically heterogeneous due to different mutations in the glucocerebrosidase gene leading to variable enzyme activities. Where this residual enzyme activity is high, a less severe phenotype is produced, the non-neuronopathic type I disease. The mutation commonly found here is N370S but even for patients with this mutation the level of disease severity is not consistent implicating other epigenetic factors in disease etiology. Where very low levels of residual activity are present, a more severe outcome is predicted involving additional storage and pathology in neural tissue to produce an infantile (type II disease) or juvenile variant (type III disease). Enzyme replacement therapy is used to successfully treat more than 3000 type I Gaucher disease patients (Zimran, 1997) but has little effect on the more severe variants due to the lack of protein access to the brain. A small molecule such as *N*B-DNJ is able to cross the blood brain barrier at concentrations that are able to deplete neuronal storage in mouse models of the human gangliosidoses.

6. MOUSE MODELS FOR HUMAN GANGLIOSIDOSES

Tay-Sachs disease results from a deficiency in β-hexosaminidase A (hex A) activity leading to the accumulation of ganglioside GM2. The accumulation is more pronounced in the brain because of the elevated synthesis of gangliosides in neural tissue. Mouse knockout models for this disorder have no hex A activity and accumulate GM2 in the lysosomes. The presence of a sialidase activity in mouse tissue partially hydrolyzes stored GM2 to asialoGM2 (AGM2) which is a substrate for β-hexosaminidase B. Thus in the mouse, unlike the human disease, the storage levels of glycolipid never exceeds the threshold required to significantly affect normal physiological processes or reduce the normal lifespan. Under conditions of stress however, such as repeated breeding of females, clinical features can be induced, similar to that seen in the late onset variant of Tay-Sachs disease in man (Jeyakumar *et al.*, 2002b).

The oral treatment of the Tay-Sachs mouse with *N*B-DNJ reduces the accumulation of ganglioside in liver and importantly, in brain, as demonstrated by GM2 analysis, histology, and electron microscopy (Platt *et al.*, 1997a).

Further demonstration of the potency of imino sugars, in particular *N*B-DNJ, to dramatically prevent symptom onset was provided by the study of the effects of SRT in a symptomatic model of gangliosidosis, Sandhoff disease. In knock-out models of disease in mouse, where both hex A and hex B activities are null, a severe phenotype is presented in more than one glycolipid species and other *N*-acetylhexosamine containing glycoconjugates. Consequently these transgenic mice show rapid, progressive neurodegeneration and die at 4–5 months of age. Following treatment with *N*B-DNJ from 3–6 weeks of age mice have reduced glycolipid storage in all tissues and increased life expectancy by 40% (Jeyakumar *et al.*, 1999). Complementation of this approach with enzyme delivered by bone marrow transplantation (BMT) provided a significant (25%) synergistic outcome in this severe disease model (Jeyakumar *et al.*, 2001). Disease progression is determined by the amount of residual, or augmented enzyme in the case of BMT, and even small increases in enzyme activity is profoundly effective. For infantile disorders a strategy of enzyme supplementation in addition to SRT should be beneficial, but until storage levels can be affected to reduce the inflammatory component in the brain, this remains an experimental alternative.

7. COMBINATION THERAPY

For those disorders where an existing monotherapy is effective, for example, β-glucocerebrosidase replacement for type I Gaucher patients, there is a potential role for combination with imino sugar therapy. However, the lysosomal β-glucocerebrosidase is also inhibited by *N*B-DNJ with an IC_{50} value of 520 μM using an *in vitro* assay with purified human placental enzyme (Platt *et al.*, 1994b), similar to the Ceredase™ formulation given to Gaucher patients. This is 25 fold higher than the concentration of *N*B-DNJ to inhibit ceramide glucosyltransferase activity (IC_{50}, 20.4 μM) and places the kinetic equilibrium in favor of reduced substrate concentrations, not storage at drug concentrations lower than 50 μM (Platt *et al.*, 1997b). Although theoretically possible, it is difficult to sustain higher doses of imino sugar in mouse in practice. But under conditions where plasma levels are at least 50 μM we have explored the possible interactions between *N*B-DNJ and glucocerebrosidase *in vivo*. These experiments have revealed that β-glucocerebrosidase activity was not inhibited in the presence of *N*B-DNJ. On the contrary, enzyme activity was elevated and the serum half-life of the enzyme was increased (Priestman *et al.*, 2000). This suggests that exposure to concentrations of *N*B-DNJ below the IC_{50} of β-glucocerebrosidase protected the enzyme from inactivation, extending the circulatory half-life. This discovery supports an "enzyme chaperone" strategy where misfolded mutant enzyme could be stabilized in the presence of an active site inhibitor. Longer *N*-alkyl chain DNJ derivatives, for example *N*N-DNJ, have greater affinity for β-glucocerebrosidase and this may lead to therapeutic exploitation for Gaucher disease mutations other than N370S (Sawkar *et al.*, 2002).

One clinical study of combination therapy in type I Gaucher disease has reported limited efficacy in man and is discussed below.

8. CLINICAL STUDIES

A multi-centre Phase I/II study of *N*B-DNJ Zavesca™ was initiated by Oxford GlycoSciences for Gaucher patients eligible for ERT with moderate disease, that is, measurable organomegaly (liver or spleen) and hemoglobin <11.5 g/dl or platelets $<100 \times 10^9$/L. 28 patients who were unable or unwilling to be treated by ERT were recruited with doses of 100 mg three times daily and 22 patients completed the study (Cox *et al.*, 2000). The plasma level of *N*B-DNJ in this study was found to be 6 μM, a concentration that by *in vitro* standards would be sufficient to cause partial depletion of GSLs. The effect on GSL biosynthesis was confirmed following flow cytometry analysis of surface GM1 in circulating leucocytes and revealed that after 12 months levels fell by 38.5%. No effect on the glycosylation of plasma glycoproteins was observed indicating that the concentration of *N*B-DNJ was insufficient to inhibit endoplasmic reticulum resident α-glucosidases. This was not surprising since in cellular studies the Ki value, determined *in vitro*, has to be exceeded by 1000–10,000 times extracellularly to obtain inhibition, probably because access to the ER is restricted. This is not the case for CGT where the active site of the Golgi membrane resident enzyme faces the cytoplasm (Marks *et al.*, 1999).

Clinical improvement in organ volume was found after 12 months, which improved, in those patients who elected to be enrolled in an extension study. After 24 months liver and spleen volumes were reduced 14.5% and 26.4% respectively. A further reduction was observed at 36 months, liver reducing by 17.5% and spleen volume by 29.6%. Hematological parameters were slower to improve or did not achieve a moderate response, suggesting that the kinetics of lysosomal GSL clearance in progenitor cells may be slower to respond. However, quality of life measurements, physical condition, general health perception, and vitality, showed an improvement even after 6 months (Zimran and Elstein, 2003).

A comparative 3-arm study also followed where 36 patients on ERT were recruited and switched to either SRT, maintained on ERT, or a combination of SRT and ERT. After 6 months evaluation, organ volume data and hematological parameters were variable and changes were not significant. However, it is noteworthy that out of the 33 patients that completed the 6 month study, 29 of these continued on SRT monotherapy for a further 6 months and 28 continued beyond the extension phase of the trial (Zimran and Elstein, 2003). Despite no significant changes in clinical markers for disease there was no long-term deteriorating disease, providing some support for maintenance therapy. The rapid clearance of stored glycolipid following ERT would make further improvements in clinical measurements by SRT possible, but only after an extended period of time.

9. FUTURE DEVELOPMENT

*N*B-DNJ has been shown to be an effective therapy for Gaucher disease but has some side effects that may restrict treatment contrary to the mode of action that suggests all other glucosphingolipidoses could be treated. However, *N*B-DNJ (Zavesca™) has recently been approved by the European Commission for the treatment of mild to moderate type 1 Gaucher disease in patients for whom ERT is unsuitable. To reduce side effects, notably following the inhibition of other glucosidases such as the sucrase/isomaltase intestinal enzyme where osmotic diarrhoea is observed, more selective compounds are required. A galactose analogue, *N*B-DGJ, has similar potency to *N*B-DNJ using *in vitro* and cellular studies, but lacks α-glucosidase inhibitory activity. When given orally to mice this is extremely well tolerated and does not induce weight loss that compromises escalated dosing of gangliosidosis mouse models to improve efficacy (Andersson *et al.*, 2000). Phase I clinical studies are now in progress to assess the safety and eventual efficacy of this compound in man.

Brain penetration of small alkyl chain compounds is below 10% of the circulating dose and may not be sufficient at current clinical doses to achieve a reduction in neural GSL storage in man. Longer alkyl chain compounds may be an alternative for the neuronopathic forms of disease since CGT inhibition and tissue uptake improves with increasing chain length (Mellor *et al.*, 2002). Homology modelling using a computer generated molecular model of CGT predicts that longer alkyl chain length compounds fit with more affinity to a large hydrophobic groove that accommodates the ceramide substrate (Butters *et al.*, 2003b). Modifications to the alkyl chain have also reduced the cellular toxicity of these longer alkyl chain length compounds and current interests are now in determining their effects on blood/brain barrier access (Butters *et al.*, 2003a). This would generate a new class of compounds for the treatment of GSL storage disorders that have significant neurodegeneration with usually fatal modalities.

10. CONCLUSIONS

The basic discovery that *N*-alkylated imino sugars are inhibitors of GSL biosynthesis has been developed from *in vitro* demonstration of proof of principle for substrate reduction therapy, to animal models for lysosomal storage of GSLs, to the clinic. Further exploitation of this discovery will see improvements in drug design to treat other disorders where GSLs may contribute to neuropathology, and help to understand the role of GSLs in many physiological processes. This in turn will lead to alternate therapies for other clinical disorders where manipulation of the biosynthetic pathway can be applied to modulate disease phenotype.

11. REFERENCES

Allende, M.L. and Proia, R.L., 2002, Lubricating cell signaling pathways with gangliosides, *Curr Opin Struct Biol.* 12:587–592.

Andersson, U., Butters, T.D., Dwek, R.A., and Platt, F.M., 2000, *N*-butyldeoxygalactonojirimycin: a more selective inhibitor of glycosphingolipid biosynthesis than *N*-butyldeoxynojirimycin, *in vitro* and *in vivo*, *Biochem Pharmacol.* 59:821–829.

Butters, T.D., Dwek, R.A., and Platt, F.M., 2000a, Inhibition of glycosphingolipid biosynthesis: application to lysosomal storage disorders, *Chem Rev.* 100:4683–4696.

Butters, T.D., Dwek, R.A., and Platt, F.M., 2003a, therapeutic applications of imino sugars in lysosomal storage disorders, *Curr Top Med Chem.* 3:561–574.

Butters, T.D., Mellor, H.R., Narita, K., Dwek, R.A., and Platt, F.M., 2003b, Small Molecule Therapeutics for the Treatment of Glycolipid Lysosomal Storage Disorders, *Phil Trans R Soc Lond B* 358:927–945.

Butters, T.D., van den Broek, L.A.G.M., Fleet, G.W.J., Krulle, T.M., Wormald, M.R., Dwek, R.A., and Platt, F.M., 2000b, Molecular requirements of imino sugars for the selective control of N-linked glycosylation and glycosphingolipid biosynthesis, *Tetrahedron:Asymmetry*. 11:113–124.

Cox, T., Lachmann, R., Hollak, C., Aerts, J., van Weely, S., Hrebicek, M., Platt, F., Butters, T., Dwek, R., Moyses, C., Gow, I., Elstein, D., and Zimran, A., 2000, Novel oral treatment of Gaucher's Disease with *N*-butyldeoxynojirimycin (OGT 918) to decrease substrate biosynthesis, *Lancet.* 355:1481–1485.

Dwek, R.A., Butters, T.D., Platt, F.M., and Zitzmann, N., 2002, targeting glycosylation as a therapeutic approach, *Nature Rev Drug Discovery*. 1:65–75.

Freeze, H.H., 2002, Human disorders in N-glycosylation and animal models, *BBA Gen Subjects.* 1573: 388–393.

Gadola, S.D., Zaccai, N.R., Harlos, K., Shepherd, D., CastroPalomino, J.C., Ritter, G., Schmidt, R.R., Jones, E.Y., and Cerundolo, V., 2002, Structure of human CD1b with bound ligands at 2.3 angstrom, a maze for alkyl chains, *Nat Immunol.* 3:721–726.

Gagnon, M. and Saragovi, H.U., 2002, Gangliosides: therapeutic agents or therapeutic targets?, *Expert Opin Ther Patents.* 12:1215–1223.

Jeyakumar, M., Butters, T.D., CortinaBorja, M., Hunnam, V., Proia, R.L., Perry, V.H., Dwek, R.A., and Platt, F.M., 1999, Delayed symptom onset and increased life expectancy in Sandhoff disease mice treated with *N*-butyldeoxynojirimycin, *Proc Nat Acad Sci USA.* 96:6388–6393.

Jeyakumar, M., Butters, T.D., Dwek, R.A., and Platt, F.M., 2002a, Glycosphingolipid lysosomal storage diseases: therapy and pathogenesis, *Neuropathol Appl Neurobiol.* 28:343–357.

Jeyakumar, M., Norflus, F., Tifft, C.J., CortinaBorja, M., Butters, T.D., Proia, R.L., Perry, V.H., Dwek, R.A., and Platt, F.M., 2001, Enhanced survival in Sandhoff disease mice receiving a combination of substrate deprivation therapy and bone marrow transplantation, *Blood.* 97:327–329.

Jeyakumar, M., Smith, D., EliottSmith, E., CortinaBorja, M., Reinkensmeier, G., Butters, T.D., Lemm, T., Sandhoff, K., Perry, V.H., Dwek, R.A., and Platt, F.M., 2002b, An inducible mouse model of late onset Tay-Sachs disease, *Neurobiol Disease.* 10:201–210.

Kobayashi, T. and Hirabayashi, V., 2000, Lipid membrane domains in cell surface and vacuolar systems, *Glycoconjugate J.* 17:163–171.

Kolter, T. and Sandhoff, K., 1999, Sphingolipids—their metabolic pathways and the pathobiochemistry of neurodegenerative diseases, *Angew Chem Int Ed.* 38:1532–1568.

Marek, K.W., Vijay, I.K., and Marth, J.D., 1999, A recessive deletion in the GlcNAc-1-phosphotransferase gene results in peri-implantation embryonic lethality, *Glycobiology*. 9:1263–1271.

Marks, D.L., Wu, K.J., Paul, P., Kamisaka, Y., Watanabe, R., and Pagano, R.E., 1999, Oligomerization and topology of the Golgi membrane protein glucosylceramide synthase, *J Biol Chem.* 274:451–456.

Mellor, H.R., Nolan, J., Pickering, L., Wormald, M.R., Platt, F.M., Dwek, R.A., Fleet, G.W.J., and Butters, T.D., 2002, Preparation, biochemical characterization and biological properties of radiolabelled N-alkylated deoxynojirimycins, *Biochem J.* 366:225–233.

Platt, F.M. and Butters, T.D., 1998, New therapeutic prospects for the glycosphingolipid lysosomal storage diseases, *Biochem Pharmacol.* 56:421–430.

Platt, F.M. and Butters, T.D., 2000, Substrate deprivation: A new therapeutic approach for the glycosphingolipid lysosomal storage diseases, Expert Reviews in Molecular Medicine (http://www-ermm.cbcu.cam.ac.uk) 1–17.

Platt, F.M., Neises, G.R., Dwek, R.A., and Butters, T.D., 1994a, *N*-butyldeoxynojirimycin is a novel inhibitor of glycolipid biosynthesis, *J Biol Chem.* 269:8362–8365.

Platt, F.M., Neises, G.R., Karlsson, G.B., Dwek, R.A., and Butters, T.D., 1994b, *N*-butyldeoxygalactonojirimycin inhibits glycolipid biosynthesis but does not affect N-linked oligosaccharide processing, *J Biol Chem.* 269:27108–27114.

Platt, F.M. Neises, G.R., Reinkensmeier, G., Townsend, M.J., Perry, V.H., Proia, R.L., Winchester, B., Dwek, R.A., and Butters, T.D., 1997a, Prevention of lysosomal storage in Tay-Sachs mice treated with *N*-butyldeoxynojirimycin, *Science.* 276:428–431.

Platt, F.M., Reinkensmeier, G., Dwek, R.A., and Butters, T.D., 1997b, Extensive glycosphingolipid depletion in the liver and lymphoid organs of mice treated with *N*-butyldeoxynojirimycin, *J Biol Chem.* 272:19365–19372.

Priestman, D.A., Platt, F.M., Dwek, R.A., and Butters, T.D., 2000, Imino sugar therapy for type I Gaucher disease, *Glycobiology.* 19:iv–vi.

Sawkar, A.R., Cheng, W.C., Beutler, E., Wong, C.H., Balch, W.E., and Kelly, J.W., 2002, Chemical chaperones increase the cellular activity of N370S beta-glucosidase: a therapeutic strategy for Gaucher disease, *Proc Nat Acad Sci USA.* 99:15428–15433.

Suzuki, K., Proia, R.L., and Suzuki, K., 1998, Mouse models of human lysosomal diseases, *Brain Path.* 8:195–215.

Yamashita, T., Wada, R., Sasaki, T., Deng, C.X., Bierfreund, U., Sandhoff, K., and Proia, R.L., 1999, A vital role for glycosphingolipid synthesis during development and differentiation, *Proc Nat Acad Sci USA.* 96:9142–9147.

Zimran, A., 1997, *Gaucher's Disease*, London, Bailliere Tindall.

Zimran, A. and Elstein, D., 2003, Gaucher disease and the clinical experience with substrate reduction therapy, *Phil Trans R Soc Lond B* 358:961–966.

GLYCOIMMUNOLOGY

15

THE MANNOSE-BINDING LECTIN (MBL) ROUTE FOR ACTIVATION OF COMPLEMENT

M. Kojima, J. S. Presanis, and R. B. Sim

MRC Immunochemistry Unit
Department of Biochemistry
University of Oxford
South Parks Road
Oxford OX1 3QU, UK

1. THE COMPLEMENT SYSTEM

The human complement system consists of more than thirty proteins, which are either soluble in body fluids, especially the blood plasma, or bound to cell membranes (for a recent overview, see Morley and Walport, 2000). These complement components are organized in three activating pathways and one terminal pathway (Figure 1). Activation of the system occurs via the classical, alternative, or lectin pathway, on recognition of particulate materials by binding of complement proteins to charge or carbohydrate arrays (Sim, 1993). This eventually results in the assembly of unstable protease complexes called the C3 convertases, C4b2a and C3bBb, which activate the central complement component C3. The major fragment of activated C3, C3b, binds covalently to complement-activating surfaces, and is responsible for initiating most of the effector mechanisms of the system. C3b and its cleaved form, iC3b, act as opsonins, enhancing phagocytosis of the complement-activating particles, while C3b can also combine with the C3 convertases to form C5 convertases and initiate the assembly of the membrane attack complex (MAC) which can insert into a lipid bilayer on the target cell, causing cell lysis. The minor fragment of activated C3, C3a, is an anaphylatoxin, and like the other anaphylatoxins of the complement system, C4a and C5a, it can mediate inflammatory responses, including smooth muscle contraction and increased vascular permeability. Therefore, the complement system recognizes and targets potentially harmful molecules or cells for destruction by opsonisation, cell lysis, or induction of the inflammatory response. In addition, other

Glycobiology and Medicine, edited by John S. Axford
Kluwer Academic / Plenum Publishers, New York, 2003

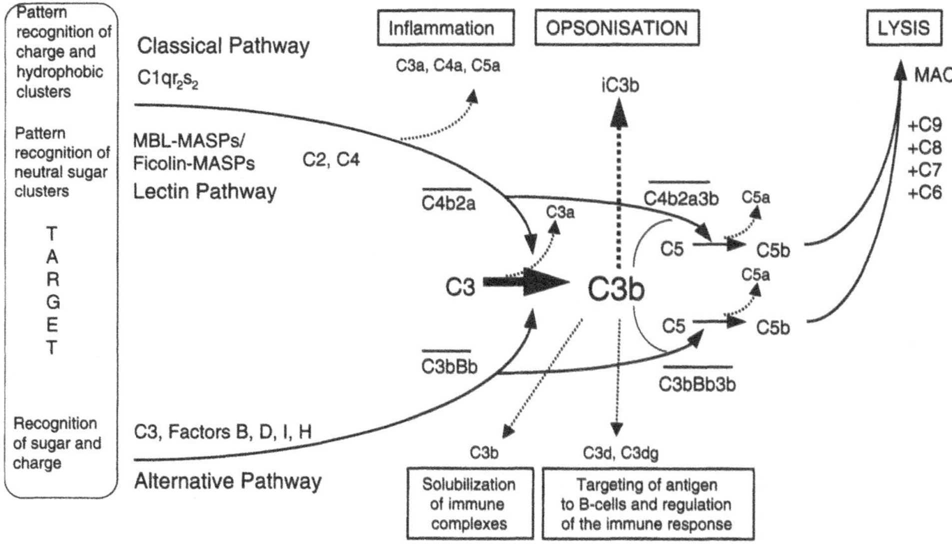

Figure 1. Overview of the complement system. The complement cascade can be activated by three distinct pathways on recognition of a target. All three pathways ultimately lead to the formation of multicomponent serine proteases (C3 convertases) C4b2a and C3bBb, which allows these three routes to converge to the activation of C3. The covalent binding of this protein to activating surfaces is responsible for initiating most of the effects of the system. Figure adapted from Sim and Laich (2000).

effector mechanisms of the system have been reported, which include solubilization of immune complexes and targeting of antigen to B-cells (Dempsey *et al.*, 1996) and to dendritic cells. Excessive or inappropriate activation of complement may damage host tissue, and the system is under elaborate control by regulatory proteins.

1.1. The Classical Pathway

The classical pathway of complement is activated by the binding of the first component of the cascade, the C1 complex, to immune complexes or aggregates containing immunoglobulin G (IgG) or immunoglobulin M (IgM) (Burton and Woof, 1992).

In addition, many other substances also activate the classical pathway, without the need for antibody: these include ligand-bound C-reactive protein (CRP), lipid A of Gram-negative bacteria, capsular polysaccharide of Gram-positive bacteria, mitochondrial cardiolipin, and nucleic acid (Sim, 1993; Sim and Malhotra, 1994). The classical pathway of complement activation needs charge clusters for recognition and such recognition may be antibody-dependent or antibody-independent.

The C1 complex is composed of a recognition molecule, C1q, and the serine protease proenzymes, C1r and C1s, in a stoichiometry of 1 C1q:2 C1r:2 C1s (Figure 2) (Gigli *et al.*, 1976).

When the globular heads of C1q bind to an activator, the associated C1r undergoes autoactivation and activates C1s (Dodds *et al.*, 1978). Activated C1s then cleaves C4 to generate a larger fragment C4b and an anaphylatoxic peptide C4a (Figure 1). Exposed within the C4b is a highly labile internal thiol ester, which is able to react via its acyl group

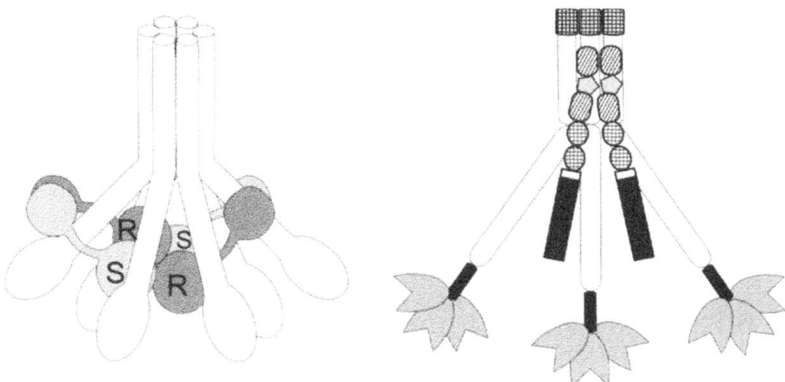

Figure 2. Models for the C1 complex and the MBL-MASPs complex. The C1 complex model is formed by placing the $C1r_2$–$C1s_2$ complex across the C1q arms, then bending each end of the complex around two opposite C1q arms so that both C1s catalytic domains come into contact with the centrally located catalytic domains of C1r. The C1 complex ($C1qr_2s_2$) model is adapted from Arlaud *et al.*, (1987). C1q is shown in white, C1r in dark gray and C1s in light gray. R and S denote the larger catalytic domains of C1r and C1s. The MBL-MASP-2 model is based on an MBL trimer interacting with a MASP-2 dimer where the three N-terminal domains of each MASP bind a separate MBL subunit (Chen and Wallis, 2001).

with nucleophiles, such as hydroxyl groups (hence forming an ester bond) and amino groups (forming an amide bond) (Sim *et al.*, 1981). Only a small proportion (usually <10%) of C4b ends up covalently bound to the activator near the activating C1 complex, while most of the activated C4 reacts with water and diffuses away from the site of complement activation. C2 binds to the C4b, and activated C1s of the nearby C1 complex cleaves C2 to release the small C2b fragment and form the classical pathway C3 convertase, C4b2a. The C2a fragment is a serine protease and, as part of the C4b2a complex, cleaves C3, a homologue of C4 (Figure 1).

As with C4, cleavage of C3 releases the anaphylatoxin C3a and exposes the reactive, labile thiol ester in C3b, now available to bind covalently with any -OH or -NH_2 groups in the vicinity. C3b more readily reacts with OH groups, often on sugars. Specificity of binding by C3b or C4b is therefore conferred by the fact that their thiol ester bonds are short-lived once exposed ($t_{1/2} < 1$ s) and therefore will only react with the target, which has been recognized by C1q. Considerable amplification occurs during activation, with approximately 240 C3b molecules deposited on the activating surface for each C1q molecule (Ollert *et al.*, 1994). If one of these C3b molecules binds covalently to C4b, then a C5 convertase, C4b2a3b, is formed; the C4b and C3b in the complex bind and orient C5 for cleavage by the C2a.

The cleavage of C5 by a C5 convertase initiates the lytic or terminal pathway (Figure 1), which is common to all three activating pathways. It is relevant only if the target has an accessible lipid bilayer. This first step releases C5a, an anaphylatoxin which is also chemotactic. Unlike its homologues C3 and C4 molecules, C5 does not contain an internal thiol ester bond for covalent binding; instead, a binding site with a specificity for C6 is exposed in the freshly activated C5b. Binding of C6 to C5b facilitates interaction with C7 (Morley and Walport, 2000). The resulting C5b-7 complex attaches to the phospholipid bilayer of the target cell membrane. Binding of C8 to the C5b-7 allows the α chain of C8 to insert into the membrane. C5b-8 then binds C9 to generate the mature MAC (membrane attack complex), which may contain 1–18 C9 molecules bound to a C5b-8 complex. The MACs behave as integral membrane components, forming ion-permeable channels (Bhakdi and Tranum-Jensen, 1991) and/or leaky patches (Esser, 1991). Under physiological osmotic pressure, such membrane disruption may result in cell lysis; alternatively, it may trigger a variety of cellular metabolic pathways, resulting in the synthesis and release of inflammatory mediators.

1.2. The Alternative Pathway

The alternative pathway of complement activation relies on the spontaneous hydrolysis of the intramolecular thiol ester bond in C3, which is estimated to occur at 0.5% per hour in the serum (Pangburn and Muller-Eberhard, 1980), as proposed by the "tickover" hypothesis (Nicol and Lachmann, 1973). Such hydrolysis of the thiol ester bond by water induces a conformational change in C3 so that the hydrolyzed C3, designated C3(H_2O), adopts a C3b-like conformation and function (Pangburn and Muller-Eberhard, 1980). The C3(H_2O) can then bind factor B in a Mg^{2+}-dependent manner, and the resulting C3(H_2O)B complex is recognized by factor D. Factor D cleaves a bond in factor B to release the smaller fragment Ba and yield the activated C3(H_2O)Bb complex, an unstable fluid-phase C3 convertase of the alternative pathway. Activation of C3 by this initial convertase releases C3a and the labile C3b component, which may bind covalently to surfaces via its exposed thiol ester bond. If the target surface lacks any membrane-bound down-regulatory factors or does not provide binding sites for soluble regulatory proteins, the surface-bound C3b binds factor B and the surface-bound C3 convertase of the alternative pathway, C3bBb is formed and may be stabilized by properdin.

On non-activating surfaces, a regulatory protein, factor H, may bind to the surface-bound C3b and to sialylated oligosaccharides, or other negative charge clusters and the resulting C3b-H complex is recognized by factor I, which cleaves C3b to generate inactive iC3b and prevents formation of C3bBb and therefore further C3b deposition. The specificity of the pathway, therefore, relies on the carbohydrate distribution (for C3b binding) and charge distribution (for factor H binding) to the target surface. On the host cell surfaces, membrane-bound regulatory factors similar to factor H are present, which prevent alternative pathway activation.

The alternative pathway is activated by a whole spectrum of substances, including IgG, IgA, LPS from Gram-negative bacteria, cell wall teichoic acid from Gram-positive bacteria, viruses, virus-infected cells, yeasts, protozoans, and helminths (Sim, 1993; Sim and Laich, 2000; Sim and Malhotra, 1994). Again the pathway can be activated by both antibody-dependent and -independent mechanisms.

1.3. The Lectin Pathway

The lectin pathway is very similar to the classical pathway and its activation results in the formation of the classical pathway C3 convertase, C4b2a (Figure 1). However, the recognition molecule for the pathway is mannose-binding lectin (MBL), which has a "bunch of tulips" structure like C1q and some of the collectins (Figure 2). MBL binds in a Ca^{2+}-dependent manner to the target via its C-type lectin domains, which recognize neutral sugars (preferentially mannose, *N*-acetylglucosamine, and fucose) on the surfaces of a range of microorganisms, including bacteria, fungi, and viruses (Table 1). In addition, MBL binds to some transformed mammalian cells and some antibodies (Table 2).

In circulation, MBL is associated with its proteases, called MBL-associated serine proteases (MASPs), which are structurally similar to C1r and C1s, and with a small protein of 19 kDa (MAp19). Although the activation mechanisms for the MBL-MASPs complex are yet to be understood, MASP2 becomes activated on binding of MBL to the target and has the capacity to cleave C4 (Thiel *et al.*, 1997) and C2 (Wong *et al.*, 1999); thus the C3 convertase, C4b2a, may be formed on the activator, and the complement cascade can proceed as for the classical pathway. Recently, MASPs have also been found in complex with another family of lectins called ficolins (Matsushita *et al.*, 2000; Matsushita *et al.*, 2001). These are also collagen-containing carbohydrate binding proteins with quaternary structure similar to C1q and MBL. Hence, the lectin pathway may also be activated by ficolin-MASPs complexes.

Table 1. Microorganisms to which MBL can bind

Microbes	References
Bacteria	
Salmonella montevideo	Kuhlman *et al.* (1989)
Escherichia coli	Van Emmerik *et al.* (1994)
Haemophilus influenzae	Van Emmerik *et al.* (1994); Neth *et al.* (2000)
Listeria monocytogenes	Van Emmerik *et al.* (1994)
Neisseria meningitidis	Van Emmerik *et al.* (1994); Neth *et al.* (2000)
Mycobacterium avium	Polotsky *et al.* (1997)
Chlamydia pneumoniae	Swanson *et al.* (1998)
Burkholderia cepacia	Davies *et al.* (2000)
Klebsiella species	Neth *et al.* (2000)
Staphylococcus aureus	Neth *et al.* (2000)
Streptococcus pneumoniae	Neth *et al.* (2000)
Actinomyces israelii	Townsend *et al.* (2001)
Bifidobacterium bifidum	Townsend *et al.* (2001)
Fusobacterium (except F. mortiferum)	Townsend *et al.* (2001)
Leptotrichia buccalis	Townsend *et al.* (2001)
Proprionibacterium acnes	Townsend *et al.* (2001)
Veillonella dispar	Townsend *et al.* (2001)
Viruses	
Influenza A	Malhotra *et al.* (1994a); Kase *et al.* (1999)
Herpes simplex 2	Fischer *et al.* (1994)
HIV-1 and -2	Saifuddin *et al.* (2000)
Fungi	
Saccharomyces cerevisiae	Ikeda *et al.* (1987)
Candida albicans	Tabona *et al.* (1995); Neth *et al.* (2000)
Cryptococcus neoformans	Schelenz *et al.* (1995)
Aspergillus fumigatus	Neth *et al.* (2000)
Protozoa	
Leishmania major and *mexicana*	Green *et al.* (1994)
Trypanosoma cruzi	Kahn *et al.* (1996)
Cryptosporidium parvum	Kelly *et al.* (2000)
Plasmodium falciparum	Klabunde *et al.* (2002)

Table 2. Reported MBL binding to immunoglobulins. Human IgM is not recognized by human MBL (Roos *et al.*, 2003). IgG-G0 refers to the form of IgG in which the N-linked glycans on the CH2 domain terminate in *N*-acetyl glucosamine. In normal human IgG, about 20% of these glycans terminate in *N*-acetyl glucosamine, the rest in galactose. The proportion of *N*-acetyl glucosamine termination is greatly increased in rheumatoid arthritis

Immunoglobulin	References
Agalactosyl IgG (IgG-G0)	Malhotra *et al.* (1995); Rudd *et al.* (1995)
Polymeric IgA	Roos *et al.* (2001)
Human IgM by rat MBL	Koppel and Solomon, (2001)
Human, bovine, and murine IgM by rabbit MBL	Nevens *et al.* (1992)

2. MANNOSE-BINDING LECTIN

Mannose-binding lectin (MBL) is a glycoprotein of the collectin family, members of which all bind to carbohydrates on the surfaces of microorganisms and particulate materials via their C-terminal lectin domains (Malhotra *et al.*, 1992). It is the recognition molecule for the lectin pathway of complement activation, and it is also known as mannan-binding protein (MBP) or mannose-binding lectin, since it binds to yeast mannan (Kawasaki *et al.*, 1978) as well as to mannose coupled to sepharose (Wild *et al.*, 1983). Previous names for MBL include core-specific lectin (CSL), from its interaction with the core motif of N-linked oligosaccharide (Colley *et al.*, 1988), and Ra reactive factor (RaRF), for it binds to Ra chemotype strains of *Salmonella* (Ihara *et al.*, 1982). The term MBL is now widely used, in preference to MBP, to avoid confusion with identically abbreviated proteins (in particular, major basic protein and myelin basic protein) and to emphasize its lectin character.

2.1. MBL in the Group of Collectins

The name collectin is derived from the words "collagen" and "lectin", and the members of this group contain both calcium-dependent carbohydrate-binding (C-type lectin) globular heads and collagen-like regions (Malhotra *et al.*, 1992). Six different collectins have been characterized as extracellular proteins from three major locations (reviewed in Hakansson and Reid, 2000): while only one form of MBL exists in man, two isoforms (MBL-A in serum and MBL-C in serum and in the liver) are found in rodents; lung surfactant proteins A and D (SP-A and SP-D) are found in the pulmonary surfactant on the epithelial lining of the lungs; and both conglutinin (BK) and collectin 43 (CL-43) are bovine serum proteins, more closely related to SP-D than to MBL. In addition, an intracellular collectin of unknown function, collectin liver 1 (CL-L1), was recently discovered in liver cells (Ohtani *et al.*, 1999). The extracellular collectins target invading pathogens—mainly bacteria, but also fungi, viruses, and potential allergens—by binding to their surface carbohydrates. This encounter usually results in aggregation or agglutination of the target particle, followed by opsonisation of the target via direct interaction with receptors on phagocytes (Malhotra *et al.*, 1994b). Unlike other collectins, however, MBL can also activate the complement system, hence foreign materials can be opsonised or lysed via the complement system (Ikeda *et al.*, 1987).

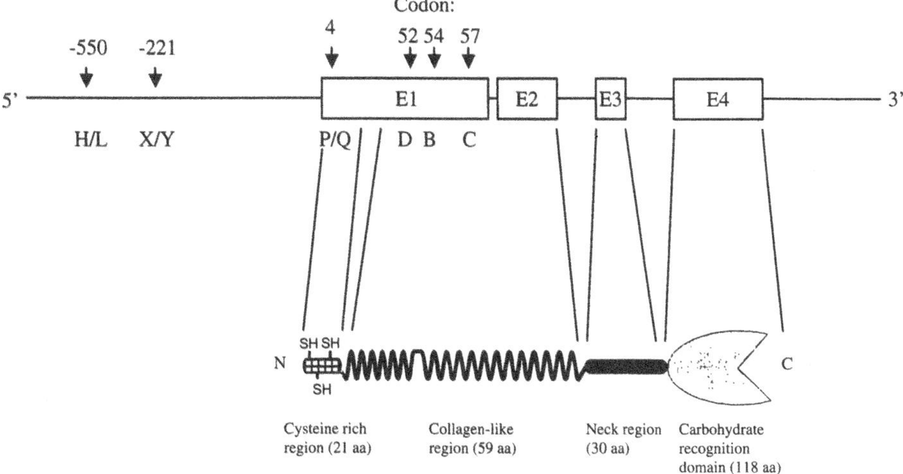

Figure 3. Structure and organization of MBL polypeptide showing structural gene mutations in exon one. MBL consists of an N-terminal cross-linking region, a collagen-like region, a neck region, and a carbohydrate recognition domain. Three polypeptide chains associate to form a trimeric subunit, with a collagen triple helix. Mutations in exon one cause single amino acid substitutions at codons 52, 54, and 57 leading to the D, B, and C variants of MBL (A variant is the wild type). The B and C variants disrupt the Gly-X-Y repeats of the collagen-like region resulting in an altered capacity to form the collagen triple helix. The D variant introduces an additional cysteine residue and so may disrupt oligomer formation by generation of additional disulphide bonds.

Proteins of the collectin family are made of subunits, which are composed of three identical or near-identical polypeptides; each polypeptide has a short N-terminal cross-linking region, containing two or three cysteines, a collagen-like region of variable length, an α-helical coiled-coil neck region, and a carbohydrate recognition domain (CRD) (Holmskov *et al.*, 1994) (Figure 3). In most collectins, these trimeric subunits are further assembled into larger homo-oligomers, which enables them to cross-link several target particles and confers greater avidity.

2.2. MBL is itself a Glycoprotein

Human MBL contains no N-linked glycosylation sites and the total glycosylation of the molecule appears to be low. As in the case of tissue collagen, however, the collagen-like region of MBL undergoes posttranslational modifications, such as hydroxylation of proline residues and hydroxylation and glycosylation of lysine residues: the collagen-like region of human MBL contains eight potential hydroxylation sites and four potential O-glycosylation sites, of which four hydroxyprolines and two hydroxylysines have been identified (Colley *et al.*, 1988) and recombinant human MBL expressed in human hepatoma cell lines has been found to contain glucosylgalactosylhydroxylysines and galactosylhydroxylysines at an approximate ratio of 3:1 (Ma *et al.*, 1997). Since the addition of hydroxylase inhibitor to the expression system resulted in production of lower MBL oligomers, such modifications are believed to be important in the assembly of the structural subunit into higher oligomers (Ma *et al.*, 1997).

2.3. CRD Structure

MBL is a C-type lectin and hence requires calcium for binding to a carbohydrate ligand. This is explained by the direct participation of calcium in the binding by the formation of coordination bonds to the 3- and 4-OH groups of the bound sugar, as determined from the crystal structure of recombinant rat MBL-A lectin domains complexed with an oligosaccharide (Weis *et al.*, 1992). In addition, the structure also reveals the requirement of the 3- and 4-OH groups of the complexed sugar to be in the equatorial plane of the hexose ring-structure in order to allow for hydrogen bonding to amino acid side chains. Hence, MBL can bind to monosaccharides, such as *N*-acetyl-D-glucosamine and mannose, but not D-galactose (Holmskov *et al.*, 1994); the binding specificity, in the order of decreasing affinity, is *N*-acetyl-D-glucosamine > mannose, *N*-acetyl mannosamine and L-fucose > maltose > glucose > D-galactose or *N*-acetyl galactosamine (Haurum *et al.*, 1993a).

The carbohydrate binding site in MBL is situated in a shallow pocket peripherally located on the CRD, hence the carbohydrate affinity of a single CRD is very weak with a dissociation constant K_D of 10^{-3} M (Iobst *et al.*, 1994). But the trimeric organization of a structural subunit and subsequent oligomerization permits multivalent, and hence stronger, interaction between MBL and a carbohydrate-containing surface; using different approaches, the K_D for human MBL binding to mannan or glycosylated bovine serum albumin was estimated to be in the order of 10^{-9} M (Kawasaki *et al.*, 1983; Lee *et al.*, 1992), and a recent crystal structure of trimeric rat MBL-A cross-linked by an oligosaccharide supports such stabilization in higher oligomers (Ng *et al.*, 2002).

The CRDs in a subunit are separated by a cavity on the three-symmetry axis, so that the three carbohydrate-binding sites in a human MBL subunit offer a flat platform with a constant distance of 4.5 nm between the sites; this distance would make it impossible for a single mammalian high-mannose oligosaccharide to interact with more than one CRD, whereas the repeating sugar structures on many microbial surfaces would readily interact with all three domains (Sheriff *et al.*, 1994; Weis and Drickamer, 1994). In addition, sialylation of mammalian cell surface glycoproteins may reduce binding by MBL under normal conditions; however, malignant transformations and viral infections may modify the oligosaccharide structures on the cell surface, and MBL binding to some tumor cells has been reported (Fujita *et al.*, 1995; Muto *et al.*, 1999). In this way, MBL acts as a pattern recognition receptor, recognizing pathogen-associated molecular patterns and distinguishing self from non-self (Janeway, 1989).

2.4. Regulation of Serum Levels of MBL: Structural and Promoter Polymorphisms

The constitutional level of MBL in the circulation is very stable (Nielsen *et al.*, 1995), yet there is very wide inter-individual variation in the serum concentration of MBL: for example, MBL concentrations of 1085 normal Japanese sera have been observed to range from 0.07 to 6.40 μg/ml (Terai *et al.*, 1993). In addition, it is now accepted that MBL deficiency is one of the most common human immunodeficiency states (Turner and Hamvas, 2000). This very large variation between individuals results partly from the occurrence of three structural mutations, giving rise to four MBL variants, and from several polymorphisms in the promoter region.

Three single base mutations in codons 52, 54, and 57 of exon 1 of the *MBL* gene are the major determinants of deficiency; the mutations result in a change of Arg52 to Cys (R52C, D variant) (Madsen *et al.*, 1994), Gly54 to Asp (G54D, B variant) (Sumiya *et al.*, 1991) and Gly57 to Glu (G57E, C variant) (Lipscombe *et al.*, 1992b), respectively, with the A variant indicating the wild-type (Figures 3 and 4). G54D and G57E mutations interrupt the Gly-X-Y repeats of the collagenous region, and studies conducted on recombinant rat MBL-A with homologous mutations indicated that these mutations altered the interchain disulfide bond formation within the N-terminal cross-linking region (Wallis and Cheng, 1999).

The presence of just one of these mutant alleles results in profoundly reduced serum concentrations of MBL: Madsen *et al.* (1994) report that Danish Causcasians with the A/B, A/C, and A/D genotypes have mean MBL concentrations of 207 ng/ml, 225 ng/ml, and 440 ng/ml, respectively, while the homozygotes for the wild-type have 1487 ng/ml. Therefore, individuals who have one or more of these structural mutations possess circulating MBL not only in low amounts, but also in lower oligomer forms than the wild-type MBL (Lipscombe *et al.*, 1995). Together, the mutations lead to defective complement activation.

The frequency of these structural mutations in a population varies between ethnic groups (Table 3) (reviewed in Turner and Hamvas, 2000). The apparently independent origins of the two major variant MBL alleles (B and C) and their high frequency in most populations has been interpreted as evidence for some benefit arising from MBL deficiency.

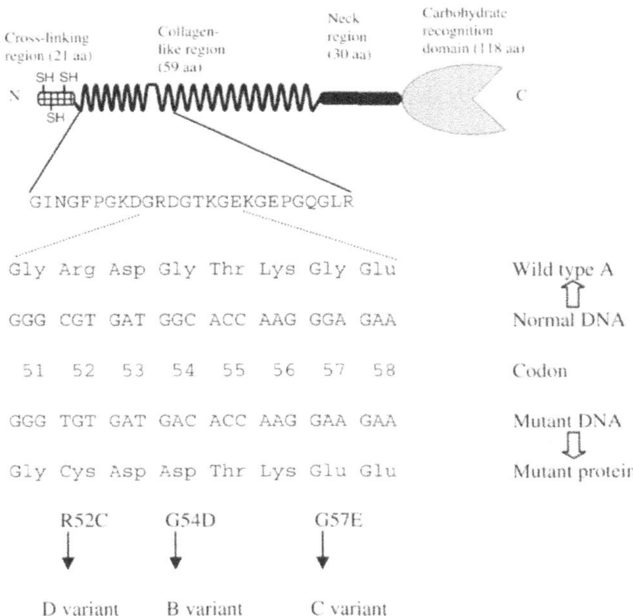

Figure 4. Structure and organization of MBL polypeptide showing structural gene mutations in exon one. Mutations in exon one cause single amino acid substitutions at codons 52, 54, and 57 lead to the D, B, and C variants of MBL (A variant is the wild type). The B and C variants disrupt the Gly-X-Y repeats of the collagen-like region resulting in an altered capacity to form the collagen triple helix. The D variant introduces an additional cysteine residue and so may disrupt oligomer formation by generation of additional disulfide bonds.

Table 3. Frequencies of structural gene mutations in exon one of the human MBL gene in various populations.

Allele	Caucasian Danish	African Kenyan	Asian Chinese	Aboriginal Australian	S. American Indian
A wild type	0.80	0.70	0.89	0.997	0.50
B G54D	0.11	0.02	0.11	0.00	0.46
C G57E	0.03	0.24	0.00	0.00	0.04
D R52C	0.06	0.04	0.00	0.003	0.00

Data on Danish Caucasians, Kenyan Africans, and South American Indians (Mapuche population) are quoted from Madsen *et al.*, (1998a). Data on Chinese Asians are quoted from Lipscombe *et al.* (1992b). Data on Aboriginal Australians (Walpiri population) are quoted from Turner *et al.* (2000).

In addition to the structural gene mutations, there are several polymorphisms within the promoter region of the MBL (Madsen *et al.*, 1995; Madsen *et al.*, 1998a). The promoter polymorphisms largely explain the large variation in MBL concentrations observed in individuals with identical structural genotypes (Garred *et al.*, 1992b; Lipscombe *et al.*, 1992a).

3. FUNCTIONS OF MBL

3.1. Direct Opsonization by MBL

There are two pathways by which MBL may participate in a host defence response; the first by activating and thereby depositing complement components on an activator; the second by acting directly as an opsonin (Hartshorn *et al.*, 1993; Kuhlman *et al.*, 1989). Purified human MBL and recombinant MBL have been shown to effect ingestion of *Salmonella montevideo* by monocytes in a concentration-dependent manner (Kuhlman *et al.*, 1989), and both recombinant wild-type and G54D-mutant human MBLs were observed to mediate the uptake of *S. montevideo* by human neutrophils (Super *et al.*, 1992). In addition, the two allelic forms of MBL have been observed to bind influenza A viruses and enhance the H_2O_2 production of human neutrophils (Hartshorn *et al.*, 1993). Both the wild-type and mutant forms of MBL, therefore, opsonized the target micro-organisms directly for uptake by monocytes and neutrophils; hence, they must be interacting with a receptor on these phagocytic cells (Figure 5).

The main candidate for the MBL receptor is calreticulin (Sim *et al.*, 1998) which also binds C1q and SP-A (Malhotra *et al.*, 1990) and possibly also SP-D (Ogden *et al.*, 2001) Calreticulin does not have a transmembrane segment or lipid anchor, but appears to be bound to the outer surface of a wide-range of cell types via either HLA Class I heavy-chain (Arosa *et al.*, 1999) or via the endocytic receptor protein CD91, and such a calreticulin-CD91 complex has been reported to mediate C1q- and MBL-enhanced engulfment of apoptotic cells (Ogden *et al.*, 2001). This receptor is sometimes referred to as collectin receptor or cC1qR (c indicates it binds to the collagen region of C1q).

Another candidate receptor C1qRp ("p" for phagocytic), was implicated in C1q/SP-A/MBL-enhanced monocyte phagocytosis of IgG- or C4b/C3b-coated erythrocytes; antibodies against C1qRp weakly inhibit such MBL-enhanced phagocytosis (Tenner *et al.*, 1995).

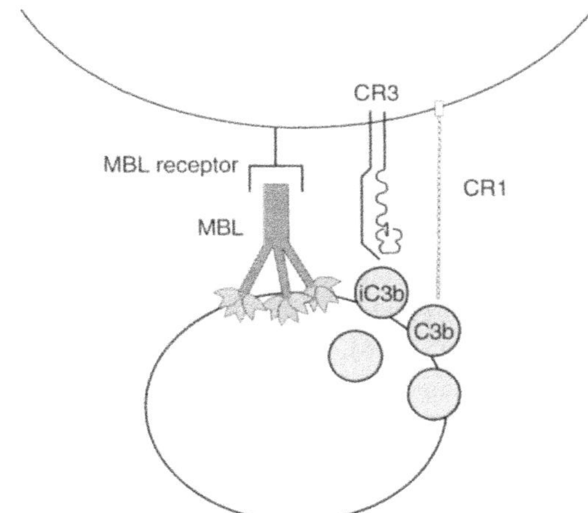

Figure 5. Opsonisation by MBL. MBL bound to microbial surfaces may enhance phagocytosis through direct interaction with a receptor and in addition can activate complement, resulting in surface bound iC3b interacting with CR3 or CR4, and C3b or C4b interacting with CR1. CR1, complement receptor 1; CR3, complement receptor 3.

However, there is no definitive report that C1q or SP-A or MBL directly interact with C1qRp, and recently the protein has been identified as the fetal stem cell marker AA4, which is a cell–cell adhesion molecule rather than an opsonic receptor (Dean *et al.*, 2000; Petrenko *et al.*, 1999). Therefore, antibodies against C1qRp inhibit many phagocytic phenomena by altering cell adhesion to the substratum, not by altering binding of opsonic ligands (McGreal *et al.*, 2002).

Another candidate receptor is complement receptor 1 (CR1), which mediates neutrophil and monocyte phagocytosis of particles coated with C3b and/or C4b. Ghiran *et al.* (2000) have demonstrated that MBL competes with C1q for binding to recombinant soluble CR1 and that this interaction, with an apparent dissociation constant of 5 nM, does not involve the CRD of MBL. In contrast, no interaction between MBL and B lymphocytes has been detected by Bajtay *et al.* (2000), although B cells are known to express CR1. There may be more than one receptor for C1q and MBL, and further characterization of such receptors is required to elucidate the importance of MBL receptor activities.

3.2. MBL-Dependent Cell-Mediated Cytotoxicity (MDCC)

While MBL does not bind normal mammalian cells, malignant transformations may modify oligosaccharide structures on cell surfaces and several glioma cell lines have been shown to bind MBL and activate the complement system (Fujita *et al.*, 1995). To test for such possible anti-tumor activity of MBL, nude mice were transplanted with an MBL-binding human colorectal carcinoma cell line and then injected with MBL-producing vaccinia virus, either by intratumoral injection or subcutaneous injection into adjacent non-tumor sites (Ma *et al.*, 1999). Upon intratumoral injection, the tumor size was observed to decrease, while subcutaneous injection resulted in inhibition, but not regression, of tumor growth. Similar results were obtained using vaccinia virus expressing the G54D-mutant form of MBL, and the *in vitro* treatment of carcinoma cells with MBL(G54D) and complement was shown to effect little cell lysis; hence, the antitumor activity observed was concluded to be independent of MBL-mediated complement activation, and this novel activity has been termed MBL-dependent cell-mediated cytotoxicity (MDCC)

(Ma *et al.*, 1999). A recent report based on an *in vitro* model supports such involvement of MBL in cell-mediated activity: the wild-type and G54D-mutant MBLs have been shown to bind to polymorphonuclear leukocytes (PMNs), which resulted in an increased leukocyte production of superoxide (Kawasaki *et al.*, 2000).

3.3. MBL Interaction with Microorganisms

MBL can bind to a wide range of clinically relevant microorganisms which display repetitive carbohydrate structures on their surfaces (Table 1) (for reviews, see Jack *et al.*, 2001; Turner, 1996).

The importance of surface structures of pathogens for MBL binding has long been evident. MBL was found to bind Ra, but not S or Re, chemotype strains of *Salmonella typhimurium* (Ihara *et al.*, 1982; Kawakami *et al.*, 1982). Indeed, recent studies support that lipopolysaccharide (LPS) or lipooligosaccharide (LOS) structure influences MBL binding: for example, sialylation of LOS of *Neisseria meningitidis* serogroup B was observed to inhibit its interaction with MBL (Jack *et al.*, 1998). While encapsulation of *N. meningitidis* had a minor effect on MBL binding in the same study, MBL interaction was significantly impaired by the presence of a capsule on *Cryptococcus neoformans*.

3.4. MBL and Disease Association

The immunological significance of MBL was first highlighted by Super *et al.* (1989), when children with an opsonic defect were identified as MBL-deficient. Since then, many studies have shown that low levels of MBL are associated with increased susceptibility to various infectious as well as autoimmune diseases (reviewed in Turner, 1998; and Turner and Hamvas, 2000) (Table 4).

Table 4. Disease association of MBL deficiency. The diseases listed below have been investigated for association with MBL deficiency.

Disease	References
Atopic dermatitis	Brandrup *et al.* (1999)
HIV	Nielsen *et al.* (1995); Garred *et al.* (1997a,b)
Recurrent miscarriage	Kilpartrick *et al.* (1995); Kruse *et al.* (2002)
Viral hepatitis	Thomas *et al.* (1996); Matsushita *et al.* (1998)
Atherosclerosis	Madsen *et al.* (1998b)
Malaria	Luty *et al.*, (1998)
Rheumatoid arthritis	Graudal *et al.* (1998); Saevarsdottir *et al.* (2001)
Cystic fibrosis	Garred *et al.* (1999)
Meningococcal disease	Hibberd *et al.* (1999)
Ischaemia-reperfusion injury	Collard *et al.* (2001)
Necrotizing pulmonary aspergillosis	Crosdale *et al.* (2001)
Systemic lupus erythematosus	Garred *et al.* (2001)
Acute respiratory infections	Koch *et al.* (2002)
Dermatomyositis	Werth *et al.* (2002).
Giant cell arteritis	Jacobsen *et al.* (2002)
Invasive pneumococcal disease	Roy *et al.* (2002)
Hepatitis B	Song *et al.* (2003)

Not all MBL-deficient individuals suffer from clinically significant recurrent infections, and Aittoniemi *et al.* (1998) propose that MBL deficiency alone is not an independent risk factor for infection but may require additional humoral immunodeficiency for phenotypic manifestation. Increased levels of infection were found in MBL-deficient individuals who are immunocompromised after chemotherapy (Peterslund *et al.*, 2001) or allogeneic transplantation (Mullighan *et al.* 2002); MBL replacement therapy, initially tested by Valdimarsson *et al.* (1998), may benefit such individuals. Analysis of sera from C2-deficient individuals indicated that the role of MBL was not only to opsonize the pathogens, but also to direct antibody isotype-switching from IgM to IgG (Selander *et al.*, 1999).

3.4.1. Infectious Diseases. The first large scale study of MBL deficiency and susceptibility to infectious disease was carried out by Garred *et al.* (1995), where a significantly higher frequency of homozygotes for the MBL structural mutations was found in the patient group (19 out of 228 patients), who suffered from clinical symptoms such as recurrent lung infections and septicemia, compared with a random control group (1 homozygote out of 123 controls). In support, another study demonstrated that the frequency of MBL mutation in children admitted to hospital with infection (e.g., meningitis, chest infections, and severe sepsis) (146 out of 345 children) was almost twice that of children admitted over the same period without infection (64 out of 272) (Summerfield *et al.*, 1997).

Subsequently, several studies have been carried out to examine a possible role of MBL in specific infections. MBL has been shown to bind glycoproteins from HIV-1 and -2 (Haurum *et al.*, 1993b), and *in vitro* infection of CD4+ lymphocytes by HIV has been inhibited by MBL (Ezekowitz *et al.*, 1989). Significantly more cases of MBL deficiency were detected in the HIV-positive patients than in controls (Nielsen *et al.*, 1995), and Garred *et al.* (1997a) reported a significantly higher frequency of homozygous individuals for MBL structural mutations in HIV-infected men than in the healthy controls and a greater level of MBL deficiency in HIV-positive patients than in HIV-negative controls in Tanzania.

3.4.2. Intracellular Pathogens. The high frequency of MBL structural mutations has been proposed to reflect selective advantages of MBL deficiency. Indeed, in the case of intracellular pathogens (such as *Mycobacteria, Leishmania*, and *Leigionella*), MBL deficiency might protect individuals against diseases. For example, the serum concentrations of MBL in 36 Ethiopian patients (with a median of 1688 ng/ml) infected with *M. leprae* were significantly higher than in 26 healthy Ethiopian blood donors (a median of 368 ng/ml) (Garred *et al.*, 1994); and the G57E mutation was found to be weakly associated with resistance to *M. tuberculosis* (Bellamy *et al.*, 1998). Therefore, it has been proposed that intracellular pathogens, which use C3 opsonization and C3 receptors to enter cells, would benefit from high levels of complement activation by MBL, and that MBL deficiency was maintained in populations because it decreased the infectivity of such parasites (Garred *et al.*, 1992a). MBL has also been implicated in direct opsonization of microorganisms and may provide direct entry to cells via an MBL receptor.

3.4.3. Autoimmune Disease. In addition to the infectious diseases mentioned above, MBL deficiency has been associated with autoimmune diseases, such as systemic lupus erythematosus (SLE) and rheumatoid arthritis (RA). Strong evidence has been accumulating for the correlation between MBL deficiency and susceptibility to SLE: higher frequencies of low plasma levels of MBL (Lau *et al.*, 1996), MBL structural mutations (Davies *et al.*, 1995, 1997; Lau *et al.*, 1996; Sullivan *et al.*, 1996) and the low-producing MBL promoter haplotype LX (Sullivan *et al.*, 1996; Ip *et al.*, 1998) have been found in SLE patients of different ethnic groups. Since the risk of SLE is further increased in the presence of C4B null alleles (Davies *et al.*, 1997), it is thought that the lectin pathway of complement activation may play a part in immune complex removal and impairment of this contributes to the development of SLE.

MBL has been shown to bind to agalactosyl IgG (IgG-G0), which is found in increased levels in RA, and to activate the complement system (Malhotra *et al.*, 1995). The authors postulated that since MBL and IgG-G0 were both found in the synovial fluids, the chronic inflammation of the synovial membrane may arise from interaction between these molecules and resulting complement activation. MBL may be a modulator of disease severity: for example, lower MBL levels appeared to be associated with earlier disease onset and more severe inflammatory manifestation (Graudal *et al.*, 1998). This was confirmed in a separate study by the same group, where the presence of MBL variant alleles was found associated with early disease onset. MBL may have a complex influence on RA, preventing disease onset at a young age yet enhancing inflammation in the late stages of the disease process.

3.5. Complement Activation via MBL

MBL is the only known collectin which can activate the complement system. It is found associated with serine proteases, the MASPs, which on activation cleave the downstream complement components C4 and C2 to initiate this pathway of complement activation, termed the MBLectin or lectin pathway. The nature of MBL ligands, the oligomeric structure of MBL, and the presence of structural mutations all affect the extent of complement activation via MBL.

3.5.1. The MASPs. MBL was first isolated from rabbit liver and serum (Kawasaki *et al.*, 1978; Kozutsumi *et al.*, 1980), and later identified as a bactericidal factor in non-immune mouse serum (and initially called Ra-reactive factor or RaRF), of which the activity against an Ra chemotype strain of *Salmonella* depended on the presence of guinea pig complement (Ihara *et al.*, 1982). Using purified C4 and C4-deficient guinea pig serum, Ikeda *et al.* (1987) later demonstrated that an MBL preparation from rat serum activated the complement system and lysed mannan-coated sheep erythrocytes in a dose-dependent manner. Since this complement-activating activity relied on the presence of complement component C4, the initiation of the complement activation via MBL was originally thought to proceed via a recruitment of proenzymic $C1r_2s_2$ by MBL on the activating surface (Ikeda *et al.*, 1987), just as C1q associates with proenzymic $C1r_2s_2$ to initiate the classical pathway of the complement system; in support of this model, purified $C1r_2s_2$ was demonstrated to interact with MBL on mannan-coated erythrocytes (Ohta *et al.*, 1990) and C1s was shown to be activated in C1q-deficient serum after zymosan treatment (Lu *et al.*, 1990).

However, Ji *et al.* (1988) observed that while the complement-activating capacity of MBL preparations was dependent on C4 and C2, it did not require C1q or C1 complex ($C1qr_2s_2$), thus suggesting the presence of a C4/C2-cleaving protease in the preparations. Subsequent analysis of MBL purified from human serum revealed that it was in fact associated with a novel C1r/s-like serine protease, termed MBL-associated serine protease (MASP) (Matsushita and Fujita, 1992); this was supported by studies on mouse RaRF, of which the carbohydrate-binding component was demonstrated to be identical to MBL (Kuge *et al.*, 1992) and the enzymatic component P100 of RaRF was shown to be a mouse equivalent of MASP (Takada *et al.*, 1993; Takayama *et al.*, 1994).

Biochemical analyses of human and mouse MASP demonstrated that they possessed proteolytic activity against C4 and C2, and proteolytic activation of C3 was also reported to occur at a very low rate (Matsushita and Fujita, 1995; Ogata *et al.*, 1995). It was then discovered that "MASP" was a mixture of two distinct but homologous proteases, now called MASP1 and MASP2, and that at least the C4-cleaving activity of "MASP" was due to MASP2, since C4 cleavage was observed in association with MASP2 but not with MASP1 (Thiel *et al.*, 1997). In addition, antiserum against the N-terminus of MASP2 identified a small protein of approximately 19 kDa in MBL-MASPs preparations (Thiel *et al.*, 1997); recently, this protein was determined to be an alternatively spliced product of the *MASP2* gene, called MBL-associated protein of 19 kDa (MAp19) (Stover *et al.*, 1999) or small MBL-associated protein (sMAP) (Takahashi *et al.*, 1999), that consisted of the first two domains of MASP2, followed by a unique sequence (EQSL), and hence lacked a serine protease domain. Furthermore, a third MASP was discovered and named MASP3, which is an alternatively spliced product of the *MASP1* gene and shares the same A-chain as MASP1 but has a different B-chain (Dahl *et al.*, 2001). Recently, an ascidian *Halocynthia roretzi* has been shown to express a protein with functional similarity to MBL (Sekine *et al.*, 2001), as well as proteins homologous to mammalian MASPs (Ji *et al.*, 1997). A C3- or C4-like protein was also discovered in the same species, and opsonic activity in the ascidian body fluid was inhibited by antibodies against this C3/4-like molecule (Nonaka *et al.*, 1999). Similarly, the coexistence of homologues of these key components was demonstrated in another protochordate *Clavelina picta* (Vasta *et al.*, 1999). Therefore, the lectin pathway is proposed to represent the primitive complement or opsonic system which played a pivotal role in innate immunity by enhancing phagocytosis before the emergence of the vertebrates. Hence, the activation of this lectin pathway and its control and the function of the pathway in human are of much interest.

Currently, the role of MASP-2 is uncontroversial. On binding of MBL to a target, MASP2 becomes activated and cleaves C4 and C2. MASP2 is therefore solely responsible for complement activation (Ambrus *et al.*, 2003; Vorup-Jensen *et al.*, 2000; Wong *et al.*, 1999). MASP1, despite earlier reports does not cleave active C3 *in vitro* (Ambrus *et al.*, 2003; Hajela *et al.*, 2002; Wong *et al.*, 1999). It does however cleave "dead" C3 *in vitro* (Hajela *et al.*, 2002), which is unlikely to be of biological significance. MASP-1 does however, cleave and activate fibrinogen and plasma transglutaminase *in vitro* at a high rate. This may be biologically relevant, but needs further study. The main function of MASP3 is also unlikely to complement activation (Dahl *et al.*, 2001).

4. REFERENCES

Aittoniemi, J., Baer, M., Soppi, E., Vesikari, T., and Miettinen, A., 1998, Mannan binding lectin deficiency and concomitant immunodefects, *Arch Dis Child.* 78:245–248.

Ambrus, G., Gal, P., Kojima, M., Szilagyi, K., Balczer, J., Antal, J., Graf, L., Laich, A., Moffatt, B.E., Schwaeble, W. *et al.*, 2003, Natural substrates and inhibitors of mannan-binding lectin-associated serine protease-1 and -2: a study on recombinant catalytic fragments, *J Immunol.* 170:1374–1382.

Arlaud, G.J. and Colomb, M.G., 1987, Modelling of C1, the first component of human complement: towards a consensus?, *Mol Immunol.* 24:317.

Arosa, F.A., de Jesus, O., Porto, G., Carmo, A.M., and de Sousa, M., 1999, Calreticulin is expressed on the cell surface of activated human peripheral blood T lymphocytes in association with major histocompatibility complex class I molecules, *J Biol Chem.* 274:16917–16922.

Bajtay, Z., Jozsi, M., Banki, Z., Thiel, S., Thielens, N., and Erdei, A., 2000, Mannan-binding lectin and C1q bind to distinct structures and exert differential effects on macrophages, *Eur J Immunol.* 30:1706–1713.

Bellamy, R., Ruwende, C., McAdam, K.P., Thursz, M., Sumiya, M., Summerfield, J., Gilbert, S.C., Corrah, T., Kwiatkowski, D., Whittle, H.C. *et al.*, 1998, Mannose binding protein deficiency is not associated with malaria, hepatitis B carriage nor tuberculosis in Africans, *Qjm.* 91:13–18.

Bhakdi, S. and Tranum-Jensen, J., 1991, Complement lysis: a hole is a hole. *Immunol Today.* 12:318–320; discussion 321.

Brandrup, F., Homburg, K.M., Wang, P., Garred, P., and Madsen, H.O., 1999, Mannan-binding lectin deficiency associated with recurrent cutaneous abscesses, prurigo and possibly atopic dermatitis. A family study, *Br J Dermatol.* 140:180–181.

Burton, D.R. and Woof, J.M., 1992, Human antibody effector function, *Adv Immunol.* 51:1–84.

Chen, C.B. and Wallis, R., 2001, Stoichiometry of complexes between mannose-binding protein and its associated serine proteases: defining functional units for complement activation, *J Biol Chem.* 276:25894–25902.

Collard, C.D., Montalto, M.C., Reenstra, W.R., Buras, J.A., and Stahl, G.L., 2001, Endothelial oxidative stress activates the lectin complement pathway: role of cytokeratin 1, *Am J Pathol.* 159:1045–1054.

Colley, K.J., Beranek, M.C., and Baenziger, J.U., 1988, Purification and characterization of the core-specific lectin from human serum and liver, *Biochem J.* 256:61–68.

Crosdale, D.J., Poulton, K.V., Ollier, W.E., Thomson, W., and Denning, D.W., 2001, Mannose-binding lectin gene polymorphisms as a susceptibility factor for chronic necrotizing pulmonary aspergillosis, *J Infect Dis.* 184:653–656.

Dahl, M.R., Thiel, S., Matsushita, M., Fujita, T., Willis, A.C., Christensen, T., Vorup-Jensen, T., and Jensenius, J.C., 2001, MASP-3 and its association with distinct complexes of the mannan-binding lectin complement activation pathway, *Immunity.* 15:127–135.

Davies, E.J., Snowden, N., Hillarby, M.C., Carthy, D., Grennan, D.M., Thomson, W., and Ollier, W.E., 1995, Mannose-binding protein gene polymorphism in systemic lupus erythematosus, *Arthritis Rheum.* 38:110–114.

Davies, E.J., Teh, L.S., Ordi-Ros, J., Snowden, N., Hillarby, M.C., Hajeer, A., Donn, R., Perez-Pemen, P., Vilardell-Tarres, M., and Ollier, W.E., 1997, A dysfunctional allele of the mannose binding protein gene associates with systemic lupus erythematosus in a Spanish population, *J Rheumatol.* 24:485–488.

Davies, J., Neth, O., Alton, E., Klein, N., and Turner, M., 2000, Differential binding of mannose-binding lectin to respiratory pathogens in cystic fibrosis, *Lancet.* 355:1885–1886.

Dean, Y.D., McGreal, E.P., Akatsu, H., and Gasque, P., 2000, Molecular and cellular properties of the rat AA4 antigen, a C-type lectin-like receptor with structural homology to thrombomodulin, *J Biol Chem.* 275:34382–34392.

Dempsey, P.W., Allison, M.E., Akkaraju, S., Goodnow, C.C., and Fearon, D.T., 1996, C3d of complement as a molecular adjuvant: bridging innate and acquired immunity, *Science.* 271:348–350.

Dodds, A.W., Sim, R.B., Porter, R.R., and Kerr, M.A., 1978, Activation of the first component of human complement (C1) by antibody-antigen aggregates, *Biochem J.* 175:383–390.

Esser, A.F., 1991, Big MAC attack: complement proteins cause leaky patches, *Immunol Today.* 12:316–318.

Ezekowitz, R.A., Kuhlman, M., Groopman, J.E., and Byrn, R.A., 1989, A human serum mannose-binding protein inhibits *in vitro* infection by the human immunodeficiency virus, *J Exp Med.* 169:185–196.

Fischer, P.B., Ellermann-Eriksen, S., Thiel, S., Jensenius, J.C., and Mogensen, S.C., 1994, Mannan-binding protein and bovine conglutinin mediate enhancement of herpes simplex virus type 2 infection in mice, *Scand J Immunol.* 39:439–445.

Fujita, T., Taira, S., Kodama, N., and Matsushita, M., 1995, Mannose-binding protein recognizes glioma cells: *in vitro* analysis of complement activation on glioma cells via the lectin pathway, *Jpn J Cancer Res.* 86:187–192.

Garred, P., Madsen, H.O., Kurtzhals, J.A., Lamm, L.U., Thiel, S., Hey, A.S., and Svejgaard, A., 1992a, Diallelic polymorphism may explain variations of the blood concentration of mannan-binding protein in Eskimos, but not in black Africans, *Eur J Immunogenet.* 19:403–412.

Garred, P., Thiel, S., Madsen, H.O., Ryder, L.P., Jensenius, J.C., and Svejgaard, A., 1992b, Gene frequency and partial protein characterization of an allelic variant of mannan binding protein associated with low serum concentrations, *Clin Exp Immunol.* 90:517–521.

Garred, P., Harboe, M., Oettinger, T., Koch, C., and Svejgaard, A., 1994, Dual role of mannan-binding protein in infections: another case of heterosis?, *Eur J Immunogenet.* 21:125–131.

Garred, P., Madsen, H.O., Hofmann, B., and Svejgaard, A., 1995, Increased frequency of homozygosity of abnormal mannan-binding-protein alleles in patients with suspected immunodeficiency, *Lancet.* 346:941–943.

Garred, P., Madsen, H.O., Balslev, U., Hofmann, B., Pedersen, C., Gerstoft, J., and Svejgaard, A., 1997a, Susceptibility to HIV infection and progression of AIDS in relation to variant alleles of mannose-binding lectin, *Lancet.* 349:236–240.

Garred, P., Richter, C., Andersen, A.B., Madsen, H.O., Mtoni, I., Svejgaard, A., and Shao, J., 1997b, Mannan-binding lectin in the sub-Saharan HIV and tuberculosis epidemics, *Scand J Immunol.* 46:204–208.

Garred, P., Pressler, T., Madsen, H.O., Frederiksen, B., Svejgaard, A., Hoiby, N., Schwartz, M., and Koch, C., 1999, Association of mannose-binding lectin gene heterogeneity with severity of lung disease and survival in cystic fibrosis, *J Clin Invest.* 104:431–437.

Garred, P., Voss, A., Madsen, H.O., and Junker, P., 2001, Association of mannose-binding lectin gene variation with disease severity and infections in a population-based cohort of systemic lupus erythematosus patients, *Genes Immun.* 2:442–450.

Ghiran, I., Barbashov, S.F., Klickstein, L.B., Tas, S.W., Jensenius, J.C., and Nicholson-Weller, A., 2000, Complement receptor 1/CD35 is a receptor for mannan-binding lectin. *J Exp Med.* 192:1797–1808.

Gigli, I., Porter, R.R., and Sim, R.B., 1976, The unactivated form of the first component of human complement, C1, *Biochem J.* 157:541–548.

Graudal, N.A., Homann, C., Madsen, H.O., Svejgaard, A., Jurik, A.G., Graudal, H.K., and Garred, P., 1998, Mannan binding lectin in rheumatoid arthritis. A longitudinal study, *J Rheumatol.* 25:629–635.

Green, P.J., Feizi, T., Stoll, M.S., Thiel, S., Prescott, A., and McConville, M.J., 1994, Recognition of the major cell surface glycoconjugates of Leishmania parasites by the human serum mannan-binding protein, *Mol Biochem Parasitol.* 66:319–328.

Hajela, K., Kojima, M., Ambrus, G., Wong, K.H., Moffatt, B.E., Ferluga, J., Hajela, S., Gal, P., and Sim, R.B., 2002, The biological functions of MBL-associated serine proteases (MASPs), *Immunobiology.* 205:467–475.

Hakansson, K. and Reid, K.B., 2000, Collectin structure: a review, *Protein Sci.* 9:1607–1617.

Hartshorn, K.L., Sastry, K., White, M.R., Anders, E.M., Super, M., Ezekowitz, R.A., and Tauber, A.I., 1993, Human mannose-binding protein functions as an opsonin for influenza A viruses, *J Clin Invest.* 91:1414–1420.

Haurum, J.S., Thiel, S., Haagsman, H.P., Laursen, S.B., Larsen, B., and Jensenius, J.C., 1993a, Studies on the carbohydrate-binding characteristics of human pulmonary surfactant-associated protein A and comparison with two other collectins: mannan-binding protein and conglutinin, *Biochem J.* 293:873–878.

Haurum, J.S., Thiel, S., Jones, I.M., Fischer, P.B., Laursen, S.B., and Jensenius, J.C., 1993b, Complement activation upon binding of mannan-binding protein to HIV envelope glycoproteins, *Aids.* 7:1307–1313.

Hibberd, M.L., Sumiya, M., Summerfield, J.A., Booy, R., and Levin, M., 1999, Association of variants of the gene for mannose-binding lectin with susceptibility to meningococcal disease, Meningococcal Research Group. *Lancet.* 353:1049–1053.

Holmskov, U., Malhotra, R., Sim, R.B., and Jensenius, J.C., 1994, Collectins: collagenous C-type lectins of the innate immune defense system, *Immunol Today.* 15:67–74.

Ihara, I., Harada, Y., Ihara, S., and Kawakami, M., 1982, A new complement-dependent bactericidal factor found in nonimmune mouse sera: specific binding to polysaccharide of Ra chemotype Salmonella, *J Immunol.* 128:1256–1260.

Ikeda, K., Sannoh, T., Kawasaki, N., Kawasaki, T., and Yamashina, I., 1987, Serum lectin with known structure activates complement through the classical pathway, *J Biol Chem.* 262:7451–7454.

Iobst, S.T., Wormald, M.R., Weis, W.I., Dwek, R.A., and Drickamer, K., 1994, Binding of sugar ligands to Ca^{2+}-dependent animal lectins. I. Analysis of mannose binding by site-directed mutagenesis and NMR, *J Biol Chem.* 269:15505–15511.

Ip, W.K., Chan, S.Y., Lau, C.S., and Lau, Y.L., 1998, Association of systemic lupus erythematosus with promoter polymorphisms of the mannose-binding lectin gene, *Arthritis Rheum.* 41:1663–1668.

Jack, D.L., Dodds, A.W., Anwar, N., Ison, C.A., Law, A., Frosch, M., Turner, M.W., and Klein, N.J., 1998, Activation of complement by mannose-binding lectin on isogenic mutants of *Neisseria meningitidis* serogroup B, *J Immunol.* 160:1346–13453.

Jack, D.L., Klein, N.J., and Turner, M.W., 2001, Mannose-binding lectin: targeting the microbial world for complement attack and opsonophagocytosis, *Immunol Rev.* 180:86–99.

Jacobsen, S., Baslund, B., Madsen, H.O., Tvede, N., Svejgaard, A., and Garred, P., 2002, Mannose-binding lectin variant alleles and HLA-DR4 alleles are associated with giant cell arteritis, *J Rheumatol.* 29, 2148–2153.

Janeway, C.A., Jr., 1989, Approaching the asymptote? Evolution and revolution in immunology, *Cold Spring Harb Symp Quant Biol.* 54:1–13.

Ji, X., Azumi, K., Sasaki, M., and Nonaka, M., 1997, Ancient origin of the complement lectin pathway revealed by molecular cloning of mannan binding protein-associated serine protease from a urochordate, the Japanese ascidian, *Halocynthia roretzi*, *Proc Natl Acad Sci USA.* 94:6340–6345.

Kahn, S.J., Wleklinski, M., Ezekowitz, R.A., Coder, D., Aruffo, A., and Farr, A., 1996, The major surface glycoprotein of *Trypanosoma cruzi* amastigotes are ligands of the human serum mannose-binding protein, *Infect Immun.* 64:2649–2656.

Kase, T., Suzuki, Y., Kawai, T., Sakamoto, T., Ohtani, K., Eda, S., Maeda, A., Okuno, Y., Kurimura, T., and Wakamiya, N., 1999, Human mannan-binding lectin inhibits the infection of influenza A virus without complement. *Immunology.* 97:385–392.

Kawakami, M., Ihara, I., Suzuki, A., and Harada, Y., 1982, Properties of a new complement-dependent bactericidal factor specific for Ra chemotype Salmonella in sera of conventional and germ-free mice, *J Immunol.* 129:2198–2201.

Kawasaki, N., Kawasaki, T., and Yamashina, I., 1983, Isolation and characterization of a mannan-binding protein from human serum, *J Biochem.* (Tokyo). 94:937–947.

Kawasaki, T., Etoh, R., and Yamashina, I., 1978, Isolation and characterization of a mannan-binding protein from rabbit liver, *Biochem Biophys Res Commun.* 81:1018–1024.

Kawasaki, T., Ma, T., Uemura, K., and Kawasaki, N., 2000, Mannan-binding protein (MBP)-dependent cell-mediated cytotoxicity (MDCC), *Mol Immunol.* 49:85.

Kelly, P., Jack, D.L., Naeem, A., Mandanda, B., Pollok, R.C., Klein, N.J., Turner, M.W., and Farthing, M.J., 2000, Mannose-binding lectin is a component of innate mucosal defense against *Cryptosporidium parvum* in AIDS, *Gastroenterology.* 119:1236–1242.

Kilpatrick, D.C., Bevan, B.H., and Liston, W.A., 1995, Association between mannan binding protein deficiency and recurrent miscarriage, *Hum Reprod.* 10:2501–2505.

Klabunde, J., Uhlemann, A.C., Tebo, A.E., Kimmel, J., Schwarz, R.T., Kremsner, P.G., and Kun, J.F., 2002, Recognition of *Plasmodium falciparum* proteins by mannan-binding lectin, a component of the human innate immune system, *Parasitol Res.* 88:113–117.

Koch, A., Melbye, M., Sorensen, P., Homoe, P., Madsen, H.O., Molbak, K., Hansen, C.H., Andersen, L.H., Hahn, G.W., and Garred, P., 2002 [In Process Citation], *Ugeskr Laeger.* 164:5635–5640.

Koppel, R. and Solomon, B., 2001, IgM detection via selective recognition by mannose-binding protein *J Biochem Biophys Methods.* 49:641–647.

Kozutsumi, Y., Kawasaki, T., and Yamashina, I., 1980, Isolation and characterization of a mannan-binding protein from rabbit serum, *Biochem Biophys Res Commun.* 95:658–664.

Kruse, C., Rosgaard, A., Steffensen, R., Varming, K., Jensenius, J.C., and Christiansen, O.B., 2002, Low serum level of mannan-binding lectin is a determinant for pregnancy outcome in women with recurrent spontaneous abortion, *Am J Obstet Gynecol.* 187:1313–1320.

Kuge, S., Ihara, S., Watanabe, E., Watanabe, M., Takishima, K., Suga, T., Mamiya, G., and Kawakami, M., 1992, cDNAs and deduced amino acid sequences of subunits in the binding component of mouse bactericidal factor, Ra-reactive factor: similarity to mannose-binding proteins, *Biochemistry.* 31:6943–6950.

Kuhlman, M., Joiner, K., and Ezekowitz, R.A., 1989, The human mannose-binding protein functions as an opsonin, *J Exp Med.* 169:1733–1745.

Lau, Y.L., Lau, C.S., Chan, S.Y., Karlberg, J., and Turner, M.W., 1996, Mannose-binding protein in Chinese patients with systemic lupus erythematosus, *Arthritis Rheum.* 39:706–708.

Lee, R.T., Ichikawa, Y., Kawasaki, T., Drickamer, K., and Lee, Y.C., 1992, Multivalent ligand binding by serum mannose-binding protein, *Arch Biochem Biophys.* 299:129–136.

Lipscombe, R.J., Lau, Y.L., Levinsky, R.J., Sumiya, M., Summerfield, J.A., and Turner, M.W., 1992a, Identical point mutation leading to low levels of mannose binding protein and poor C3b mediated opsonisation in Chinese and Caucasian populations, *Immunol Lett.* 32:253–257.

Lipscombe, R.J., Sumiya, M., Hill, A.V., Lau, Y.L., Levinsky, R.J., Summerfield, J.A., and Turner, M.W., 1992b, High frequencies in African and non-African populations of independent mutations in the mannose binding protein gene, *Hum Mol Genet.* 1:709–715.

Lipscombe, R.J., Sumiya, M., Summerfield, J.A., and Turner, M.W., 1995, Distinct physicochemical characteristics of human mannose binding protein expressed by individuals of differing genotype, *Immunology*. 85:660–667.

Lu, J.H., Thiel, S., Wiedemann, H., Timpl, R., and Reid, K.B., 1990, Binding of the pentamer/hexamer forms of mannan-binding protein to zymosan activates the proenzyme $C1r_2C1s_2$ complex, of the classical pathway of complement, without involvement of $C1_q$, *J Immunol*. 144:2287–2294.

Luty, A.J., Kun, J.F., and Kremsner, P.G., 1998, Mannose-binding lectin plasma levels and gene polymorphisms in *Plasmodium falciparum* malaria, *J Infect Dis*. 178:1221–1224.

Ma, Y., Shida, H., and Kawasaki, T., 1997, Functional expression of human mannan-binding proteins (MBPs) in human hepatoma cell lines infected by recombinant vaccinia virus: post-translational modification, molecular assembly, and differentiation of serum and liver MBP, *J Biochem*. (Tokyo). 122:810–818.

Ma, Y., Uemura, K., Oka, S., Kozutsumi, Y., Kawasaki, N., and Kawasaki, T., 1999, Antitumor activity of mannan-binding protein *in vivo* as revealed by a virus expression system: mannan-binding proteindependent cell-mediated cytotoxicity, *Proc Natl Acad Sci USA*. 96:371–375.

Madsen, H.O., Garred, P., Kurtzhals, J.A., Lamm, L.U., Ryder, L.P., Thiel, S., and Svejgaard, A., 1994, A new frequent allele is the missing link in the structural polymorphism of the human mannan-binding protein, *Immunogenetics*. 40:37–44.

Madsen, H.O., Garred, P., Thiel, S., Kurtzhals, J.A., Lamm, L.U., Ryder, L.P., and Svejgaard, A., 1995, Interplay between promoter and structural gene variants control basal serum level of mannan-binding protein, *J Immunol*. 155:3013–3020.

Madsen, H.O., Satz, M.L., Hogh, B., Svejgaard, A., and Garred, P., 1998a, Different molecular events result in low protein levels of mannan-binding lectin in populations from southeast Africa and South America, *J Immunol*. 161:3169–3175.

Madsen, H.O., Videm, V., Svejgaard, A., Svennevig, J.L., and Garred, P., 1998, Association of mannose-binding-lectin deficiency with severe atherosclerosis, *Lancet*. 352:959–960.

Malhotra, R., Thiel, S., Reid, K.B., and Sim, R.B., 1990, Human leukocyte C1q receptor binds other soluble proteins with collagen domains, *J Exp Med*. 172:955–959.

Malhotra, R., Haurum, J., Thiel, S., and Sim, R.B., 1992, Interaction of C1q receptor with lung surfactant protein A, *Eur J Immunol*. 22:1437–1445.

Malhotra, R., Haurum, J.S., Thiel, S., and Sim, R.B., 1994a, Binding of human collectins (SP-A and MBP) to influenza virus, *Biochem J*. 304:455–461.

Malhotra, R., Lu, J., Holmskov, U., and Sim, R.B., 1994b, Collectins, collectin receptors and the lectin pathway of complement activation, *Clin Exp Immunol*. 97, Suppl. 2:4–9.

Malhotra, R., Wormald, M.R., Rudd, P.M., Fischer, P.B., Dwek, R.A., and Sim, R.B., 1995, Glycosylation changes of IgG associated with rheumatoid arthritis can activate complement via the mannose-binding protein, *Nat Med*. 1:237–243.

Matsushita, M. and Fujita, T., 1992, Activation of the classical complement pathway by mannose-binding protein in association with a novel C1s-like serine protease, *J Exp Med*. 176:1497–1502.

Matsushita, M. and Fujita, T., 1995, Cleavage of the third component of complement (C3) by mannose-binding protein-associated serine protease (MASP) with subsequent complement activation, *Immunobiology*. 194:443–448.

Matsushita, M., Hijikata, M., Ohta, Y., and Mishiro, S., 1998, Association of mannose-binding lectin gene haplotype LXPA and LYPB with interferon-resistant hepatitis C virus infection in Japanese patients, *J Hepatol*. 29:695–700.

Matsushita, M., Endo, Y., and Fujita, T., 2000, Cutting edge: complement-activating complex of ficolin and mannose-binding lectin-associated serine protease, *J Immunol*. 164:2281–2284.

Matsushita, M., Endo, Y., Hamasaki, N., and Fujita, T., 2001, Activation of the lectin complement pathway by ficolins, *Int Immunopharmacol*. 1:359–363.

McGreal, E.P., Ikewaki, N., Akatsu, H., Morgan, B.P., and Gasque, P., 2002, Human C1qRp is identical with CD93 and the mNI-11 antigen but does not bind C1q, *J Immunol*. 168:5222–5232.

Morley, B.J. and Walport, M.J., 2000, *The Complement FactsBook*, Academic Press, London.

Mullighan, C.G., Heatley, S., Doherty, K., Szabo, F., Grigg, A., Hughes, T.P., Schwarer, A.P., Szer, J., Tait, B.D., Bik To, L. *et al*., 2002, Mannose-binding lectin gene polymorphisms are associated with major infection following allogeneic hemopoietic stem cell transplantation, *Blood*. 99:3524–3529.

Muto, S., Sakuma, K., Taniguchi, A., and Matsumoto, K., 1999, Human mannose-binding lectin preferentially binds to human colon adenocarcinoma cell lines expressing high amount of Lewis A and Lewis B antigens, *Biol Pharm Bull*. 22:347–352.

Neth, O., Jack, D.L., Dodds, A.W., Holzel, H., Klein, N.J., and Turner, M.W., 2000, Mannose-binding lectin binds to a range of clinically relevant microorganisms and promotes complement deposition, *Infect Immun.* 68:688–693.

Nevens, J.R., Mallia, A.K., Wendt, M.W., and Smith, P.K., 1992, Affinity chromatographic purification of immunoglobulin M antibodies utilizing immobilized mannan binding protein, *J Chromatogr.* 597:247–256.

Ng, K.K., Kolatkar, A.R., Park-Snyder, S., Feinberg, H., Clark, D.A., Drickamer, K., and Weis, W.I., 2002, Orientation of bound ligands in mannose-binding proteins. Implications for multivalent ligand recognition, *J Biol Chem.* 277:16088–16095.

Nicol, P.A. and Lachmann, P.J., 1973, The alternate pathway of complement activation. The role of C3 and its inactivator (KAF), *Immunology.* 24:259–275.

Nielsen, S.L., Andersen, P.L., Koch, C., Jensenius, J.C., and Thiel, S., 1995, The level of the serum opsonin, mannan-binding protein in HIV-1 antibody-positive patients, *Clin Exp Immunol.* 100:219–222.

Nonaka, M., Azumi, K., Ji, X., Namikawa-Yamada, C., Sasaki, M., Saiga, H., Dodds, A.W., Sekine, H., Homma, M.K., Matsushita, M. *et al.*, 1999, Opsonic complement component C3 in the solitary ascidian, *Halocynthia roretzi*, *J Immunol.* 162:387–391.

Ogata, R.T., Low, P.J., and Kawakami, M., 1995, Substrate specificities of the protease of mouse serum Ra-reactive factor, *J Immunol.* 154:2351–2357.

Ogden, C.A., deCathelineau, A., Hoffmann, P.R., Bratton, D., Ghebrehiwet, B., Fadok, V.A., and Henson, P.M., 2001, C1q and mannose binding lectin engagement of cell surface calreticulin and CD91 initiates macropinocytosis and uptake of apoptotic cells, *J Exp Med.* 194:781–795.

Ohta, M., Okada, M., Yamashina, I., and Kawasaki, T., 1990, The mechanism of carbohydrate-mediated complement activation by the serum mannan-binding protein, *J Biol Chem.* 265:1980–1984.

Ohtani, K., Suzuki, Y., Eda, S., Kawai, T., Kase, T., Yamazaki, H., Shimada, T., Keshi, H., Sakai, Y., Fukuoh, A. *et al.*, 1999, Molecular cloning of a novel human collectin from liver (CL-L1), *J Biol Chem.* 274:13681–13689.

Ollert, M.W., Kadlec, J.V., David, K., Petrella, E.C., Bredehorst, R., and Vogel, C.W., 1994, Antibody-mediated complement activation on nucleated cells. A quantitative analysis of the individual reaction steps, *J Immunol.* 153:2213–2221.

Pangburn, M.K. and Muller-Eberhard, H.J., 1980, Relation of putative thioester bond in C3 to activation of the alternative pathway and the binding of C3b to biological targets of complement, *J Exp Med.* 152:1102–1114.

Peterslund, N.A., Koch, C., Jensenius, J.C., and Thiel, S., 2001, Association between deficiency of mannose-binding lectin and severe infections after chemotherapy, *Lancet.* 358:637–638.

Petrenko, O., Beavis, A., Klaine, M., Kittappa, R., Godin, I., and Lemischka, I.R., 1999, The molecular characterization of the fetal stem cell marker AA4, *Immunity.* 10:691–700.

Polotsky, V.Y., Fischer, W., Ezekowitz, R.A., and Joiner, K.A., 1996, Interactions of human mannose-binding protein with lipoteichoic acids, *Infect Immun.* 64:380–383.

Roos, A., Bouwman, L.H., van Gijlswijk-Janssen, D.J., Faber-Krol, M.C., Stahl, G.L., and Daha, M.R., 2001, Human IgA activates the complement system via the mannan-binding lectin pathway, *J Immunol.* 167:2861–2868.

Roos, A., Bouwman, L.H., Munoz, J., Zuiverloon, T., Faber-Krol, M.C., Fallaux-van den Houten, F.C., Klar-Mohamad, N., Hack, C.E., Tilanus, M.G., and Daha, M.R., 2003, Functional characterization of the lectin pathway of complement in human serum, *Mol Immunol.* 39:655–668.

Roy, S., Knox, K., Segal, S., Griffiths, D., Moore, C.E., Welsh, K.I., Smarason, A., Day, N.P., McPheat, W.L., Crook, D.W., and Hill, A.V., 2002, MBL genotype and risk of invasive pneumococcal disease: a case-control study, *Lancet.* 359:1569–1573.

Rudd, P., Fortune, F., Lehner, T., Parekh, R., Patel, T., Wormald, M., Malhotra, R., Sim, R., and Dwek, R., 1995, Lectin–carbohydrate interactions in disease, T-cell recognition of IgA and IgD; mannose binding protein recognition of IgG0, *Adv Exp Med Biol.* 376:147–152.

Saevarsdottir, S., Vikingsdottir, T., Vikingsson, A., Manfredsdottir, V., Geirsson, A.J., and Valdimarsson, H., 2001, Low mannose binding lectin predicts poor prognosis in patients with early rheumatoid arthritis. A prospective study, *J Rheumatol.* 28:728–734.

Saifuddin, M., Hart, M.L., Gewurz, H., Zhang, Y., and Spear, G.T., 2000, Interaction of mannose-binding lectin with primary isolates of human immunodeficiency virus type 1, *J Gen Virol.* 81:949–955.

Schelenz, S., Malhotra, R., Sim, R.B., Holmskov, U., and Bancroft, G.J., 1995, Binding of host collectins to the pathogenic yeast *Cryptococcus neoformans*: human surfactant protein D acts as an agglutinin for acapsular yeast cells, *Infect Immun.* 63:3360–3336.

Sekine, H., Kenjo, A., Azumi, K., Ohi, G., Takahashi, M., Kasukawa, R., Ichikawa, N., Nakata, M., Mizuochi, T., Matsushita, M., *et al.* 2001, An ancient lectin-dependent complement system in an ascidian: novel lectin isolated from the plasma of the solitary ascidian, *Halocynthia roretzi*, *J Immunol.* 167:4504–4510.

Selander, B., Weintraub, A., Holmstrom, E., Sturfelt, G., Truedsson, L., Martensson, U., Jensenius, J.C., and Sjoholm, A.G., 1999, Low concentrations of immunoglobulin G antibodies to Salmonella serogroup C in C2 deficiency: suggestion of a mannan-binding lectin pathway-dependent mechanism, *Scand J Immunol.* 50:555–561.

Sheriff, S., Chang, C.Y., and Ezekowitz, R.A., 1994, Human mannose-binding protein carbohydrate recognition domain trimerizes through a triple alpha-helical coiled-coil, *Nat Struct Biol.* 1:789–794.

Sim, R.B., Twose, T.M., Paterson, D.S., and Sim, E., 1981, The covalent-binding reaction of complement component C3, *Biochem J.* 193:115–127.

Sim, R.B., 1993, *Activators and Inhibitors of Complement*, Kluwer Academic Publishers, Dordrecht.

Sim, R.B. and Malhotra, R., 1994, Interactions of carbohydrates and lectins with complement, *Biochem Soc Trans.* 22:106–111.

Sim, R.B. and Laich, A., 2000, Serine proteases of the complement system, *Biochem Soc Trans.* 28:545–550.

Sim, R.B., Moestrup, S.K., Stuart, G.R., Lynch, N.J., Lu, J., Schwaeble, W.J., and Malhotra, R., 1998, Interaction of C1q and the collectins with the potential receptors calreticulin (cC1qR/collectin receptor) and megalin, *Immunobiology.* 199:208–224.

Song le, H., Binh, V.Q., Duy, D.N., Juliger, S., Bock, T.C., Luty, A.J., Kremsner, P.G., and Kun, J.F., 2003, Mannose-binding lectin gene polymorphisms and hepatitis B virus infection in Vietnamese patients, *Mutat Res.* 522:119–125.

Stover, C.M., Thiel, S., Thelen, M., Lynch, N.J., Vorup-Jensen, T., Jensenius, J.C., and Schwaeble, W.J., 1999, Two constituents of the initiation complex of the mannan-binding lectin activation pathway of complement are encoded by a single structural gene, *J Immunol.* 162:3481–3490.

Sullivan, K.E., Wooten, C., Goldman, D., and Petri, M., 1996, Mannose-binding protein genetic polymorphisms in black patients with systemic lupus erythematosus, *Arthritis Rheum.* 39:2046–2051.

Sumiya, M., Super, M., Tabona, P., Levinsky, R.J., Arai, T., Turner, M.W., and Summerfield, J.A., 1991, Molecular basis of opsonic defect in immunodeficient children, *Lancet.* 337:1569–1570.

Summerfield, J.A., Sumiya, M., Levin, M., and Turner, M.W., 1997, Association of mutations in mannose binding protein gene with childhood infection in consecutive hospital series, *BMJ.* 314:1229–1232.

Super, M., Thiel, S., Lu, J., Levinsky, R.J., and Turner, M.W., 1989, Association of low levels of mannan-binding protein with a common defect of opsonisation, *Lancet.* 2:1236–1239.

Super, M., Gillies, S.D., Foley, S., Sastry, K., Schweinle, J.E., Silverman, V.J., and Ezekowitz, R.A., 1992, Distinct and overlapping functions of allelic forms of human mannose binding protein, *Nat Genet.* 2:50–55.

Swanson, A.F., Ezekowitz, R.A., Lee, A., and Kuo, C.C., 1998, Human mannose-binding protein inhibits infection of HeLa cells by *Chlamydia trachomatis*, *Infect Immun.* 66:1607–1612.

Tabona, P., Mellor, A., and Summerfield, J.A., 1995, Mannose binding protein is involved in first-line host defence: evidence from transgenic mice, *Immunology.* 85:153–159.

Takada, F., Takayama, Y., Hatsuse, H., and Kawakami, M., 1993, A new member of the C1s family of complement proteins found in a bactericidal factor, Ra-reactive factor, in human serum, *Biochem Biophys Res Commun.* 196:1003–1009.

Takahashi, M., Endo, Y., Fujita, T., and Matsushita, M., 1999, A truncated form of mannose-binding lectin-associated serine protease (MASP)-2 expressed by alternative polyadenylation is a component of the lectin complement pathway, *Int Immunol.* 11:859–863.

Takayama, Y., Takada, F., Takahashi, A., and Kawakami, M., 1994, A 100-kDa protein in the C4-activating component of Ra-reactive factor is a new serine protease having module organization similar to C1r and C1s, *J Immunol.* 152:2308–2316.

Tenner, A.J., Robinson, S.L., and Ezekowitz, R.A., 1995, Mannose binding protein (MBP) enhances mononuclear phagocyte function via a receptor that contains the 126,000 M(r) component of the C1q receptor, *Immunity.* 3:485–493.

Terai, I., Kobayashi, K., Fujita, T., and Hagiwara, K., 1993, Human serum mannose binding protein (MBP): development of an enzyme-linked immunosorbent assay (ELISA) and determination of levels in serum from 1085 normal Japanese and in some body fluids, *Biochem Med Metab Biol.* 50:111–119.

Terai, I., Kobayashi, K., Matsushita, M., and Fujita, T., 1997, Human serum mannose-binding lectin (MBL)-associated serine protease-1 (MASP-1): determination of levels in body fluids and identification of two forms in serum, *Clin Exp Immunol.* 110:317–323.

Thiel, S., Vorup-Jensen, T., Stover, C.M., Schwaeble, W., Laursen, S.B., Poulsen, K., Willis, A.C., Eggleton, P., Hansen, S., Holmskov, U. *et al.*, 1997, A second serine protease associated with mannan-binding lectin that activates complement, *Nature*. 386:506–510.

Thomas, H.C., Foster, G.R., Sumiya, M., McIntosh, D., Jack, D.L., Turner, M.W., and Summerfield, J.A., 1996, Mutation of gene of mannose-binding protein associated with chronic hepatitis B viral infection, *Lancet*. 348:1417–1419.

Townsend, R., Read, R.C., Turner, M.W., Klein, N.J., and Jack, D.L., 2001, Differential recognition of obligate anaerobic bacteria by human mannose-binding lectin, *Clin Exp Immunol.* 124:223–228.

Turner, M.W., 1996, Mannose-binding lectin: the pluripotent molecule of the innate immune system, *Immunol Today*. 17:532–540.

Turner, M.W., 1998, Mannose-binding lectin (MBL) in health and disease, *Immunobiology*. 199:327–339.

Turner, M.W., Dinan, L., Heatley, S., Jack, D.L., Boettcher, B., Lester, S., McCluskey, J., and Roberton, D., 2000, Restricted polymorphism of the mannose-binding lectin gene of indigenous Australians, *Hum Mol Genet.* 9:1481–1486.

Turner, M.W. and Hamvas, R.M., 2000, Mannose-binding lectin: structure, function, genetics and disease associations, *Rev Immunogenet.* 2:305–322.

Valdimarsson, H., Stefansson, M., Vikingsdottir, T., Arason, G.J., Koch, C., Thiel, S., and Jensenius, J.C., 1998, Reconstitution of opsonizing activity by infusion of mannan-binding lectin (MBL) to MBL-deficient humans, *Scand J Immunol.* 48:116–123.

van Emmerik, L.C., Kuijper, E.J., Fijen, C.A., Dankert, J., and Thiel, S., 1994, Binding of mannan-binding protein to various bacterial pathogens of meningitis, *Clin Exp Immunol.* 97:411–416.

Vasta, G.R., Quesenberry, M., Ahmed, H., and O'Leary, N., 1999, C-type lectins and galectins mediate innate and adaptive immune functions: their roles in the complement activation pathway, *Dev Comp Immunol.* 23:401–420.

Vorup-Jensen, T., Petersen, S.V., Hansen, A.G., Poulsen, K., Schwaeble, W., Sim, R.B., Reid, K.B., Davis, S.J., Thiel, S., and Jensenius, J.C., 2000, Distinct pathways of mannan-binding lectin (MBL)- and C1-complex autoactivation revealed by reconstitution of MBL with recombinant MBL-associated serine protease-2, *J Immunol.* 165:2093–2100.

Wallis, R. and Chen, J.Y., 1999, Molecular defects in variant forms of mannose-binding protein associated with immunodeficiency, *J Immunol.* 163:4953–4959.

Weis, W.I., Drickamer, K., and Hendrickson, W.A., 1992, Structure of a C-type mannose-binding protein complexed with an oligosaccharide, *Nature*. 360:127–134.

Weis, W.I. and Drickamer, K., 1994, Trimeric structure of a C-type mannose-binding protein, *Structure*. 2:1227–2240.

Werth, V.P., Berlin, J.A., Callen, J.P., Mick, R., and Sullivan, K.E., 2002, Mannose binding lectin (MBL) polymorphisms associated with low MBL production in patients with dermatomyositis, *J Invest Dermatol.* 119:1394–1399.

Wild, J., Robinson, D., and Winchester, B., 1983, Isolation of mannose-binding proteins from human and rat liver, *Biochem J.* 210:167–174.

Wong, N.K., Kojima, M., Dobo, J., Ambrus, G., and Sim, R.B., 1999, Activities of the MBL-associated serine proteases (MASPs) and their regulation by natural inhibitors, *Mol Immunol.* 36:853–861.

16

ANTI-INFLAMMATORY PROPERTIES OF SPECIFIC GLYCOFORMS OF HUMAN α_1-ACID GLYCOPROTEIN

Willem Van Dijk and Dennis C. W. Poland

Glycoimmunology Group
Department of Molecular Cell Biology & Immunology
VU Medical Center
Amsterdam, The Netherlands

1. INTRODUCTION

Upon damage to mammalian tissues, such as destruction due to infection, inflammation, neoplasia, burn wounds, and mechanical trauma, a complex series of events must follow in order to prevent further damage, and to restore homeostasis. The sum of these events is known as inflammation, of which the acute-phase response constitutes the set of immediate and early events. Local acute-phase reactions include aggregation of platelets, dilation and subsequent leakage of blood vessels, release of protease inhibitors, and release of cytokines and chemoattractant molecules. The latter attract leukocytes, which in turn, together with the fibroblasts and endothelial cells at the site of inflammation will produce more cytokines, like TNF-α, IL-1β and IL-6.

The hepatic acute-phase response is one of the systemic reactions. It represents large time-dependent fluctuations in the rate of hepatic synthesis and plasma concentrations of a number of plasma proteins, the so-called acute-phase proteins, which are induced by IL-1- and/or IL-6-type cytokines and glucocorticoids (Gabay and Kushner, 1999; Mackiewicz, 1997). The function of the acute-phase proteins is to keep the harmful effects of inflammation within narrow boundaries, for example by removing locally released proteases by protease inhibitors and by modulating the immune response. In chronic inflammatory diseases, like rheumatoid arthritis, where also changes in plasma levels of acute-phase proteins occur, the inflammatory response appears to be an ongoing process unable to be kept within the narrow boundaries of the acute-phase reaction.

The hepatic acute-phase response also results in characteristic changes in the glycosylation of the acute-phase proteins (De Graaf *et al.*, 1993; Havenaar *et al.*, 1997; Van Dijk and Mackiewicz, 1995; Van Dijk *et al.*, 1998). These regard the degree of

Glycobiology and Medicine, edited by John S. Axford
Kluwer Academic / Plenum Publishers, New York, 2003

branching of and the expression of sialyl Lewis^X (sLe^X) epitopes on their glycans. This occurs both in acute and chronic inflammatory processes for a variety of acute-phase proteins, although to different extents. Several studies have shown that also these changes are induced by inflammatory cytokines as well as hormones (Brinkman-Van der Linden *et al.*, 1996a, b; Pos *et al.*, 1989; Van Dijk and Mackiewicz, 1995). As is reviewed below for α_1-acid glycoprotein (AGP) the changes in glycosylation can have functional implications for anti-inflammatory properties of acute-phase proteins.

2. MOLECULAR PROPERTIES OF HUMAN PLASMA AGP

AGP, also known as orosomucoid, is an acute-phase protein varying in concentration from about 0.8 g/L in healthy controls to 3 g/L under inflamed conditions; the plasma level is decreased during pregnancy and oral oestrogen use (Brinkman-Van der Linden *et al.*, 1996a). Its apparent molecular weight as determined by SDS-PAGE under reducing conditions is 41500–43000 Dalton, of which 40% consists of carbohydrates. Human plasma AGP originates from the epithelial cells of the liver, but secretion of AGP by activated granulocytes (Poland *et al.*, 2002, submitted for publication) and B-cells (Gahmberg and Anderson, 1978) may contribute to its plasma level.

AGP contains five N-linked glycans that exhibit substantial heterogeneity in their structures when total AGP is analyzed. This heterogeneity results from the presence in plasma of various distinct AGP-glycoforms of which the plasma levels, and thus their relative occurrence, are dependent on the pathophysiological condition which is determined by cytokines and hormones (*cf.* Figure 1 and Table 1) (Brinkman-Van der Linden *et al.*,

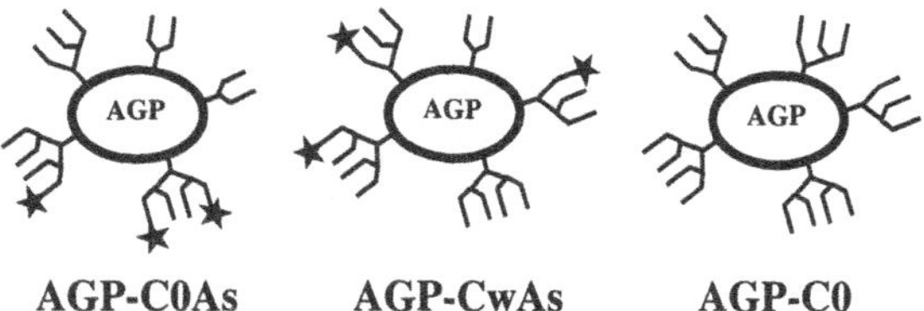

Figure 1. Examples of glycoforms that can occur in human blood plasma. See Table 1 for definition of glycoforms. The asterisks represent sialyl Lewis^X epitopes.

Table 1. AGP glycoforms detectable in plasma by lectin-affinity chromatography or electrophoresis.

Glycoform	Con A reactivity	AAL reactivity	Tri-/tetra-antennaes	Di-antennae(s)	α3-linked fucose	sLe^X groups
AGP-C0A0***	none (C0)	none (A0)	5	0	0	0
AGP-C0Aw	none (C0)	weak (Aw)	5	0	≤2	≤2
AGP-C0As**	none (C0)	strong (As)	5	0	≥3	≥3
AGP-CwA0	weak (Cw)	none (A0)	4	1	0	0
AGP-CwAw	weak (Cw)	weak (Aw)	4	1	≤2	≤2
AGP-CwAs**	weak (Cw)	strong (As)	4	1	≥3	≥3
AGP-CsA0*	strong (Cs)	none (A0)	≤3	≥2	0	0
AGP-CsAw*	strong (Cs)	weak (Aw)	≤3	≥2	≤2	≤2
AGP-CsAs*	strong (Cs)	strong (As)	≤3	≥2	≥3	≥3

*Increased in early acute-phase;
**Increased in late acute-phase and in chronic diseases;
***Increased in pregnancy.

1996a, b; De Graaf *et al.*, 1993; Havenaar *et al.*, 1997; Pos *et al.*, 1989; Schalkwijk *et al.*, 1999; Van den Heuvel *et al.*, 2000; Van Dijk and Mackiewicz, 1995; Van Dijk *et al.*, 1998). Methods employing concanavalin A (Con A) and *Aleuria aurantia* lectin (AAL) were used for the determination of the plasma concentrations as well as for the isolation of AGP-glycoforms that differ in diantennary glycan content and/or content of sLe^X (Table 1). It was established that the sLe^X content was determined by changes in the fucosylation, which in case of AGP occurs only on the branches of the glycans.

3. POSSIBLE FUNCTIONS OF HUMAN PLASMA AGP

The exact physiological function of AGP is not known, but the available data point to a carrier function for steroid hormones and lipophilic drugs. AGP has been suggested to belong to the class of retinol-binding proteins, the lipocalins (Pervaiz and Brew, 1987). Inflammation-induced increases in plasma AGP concentration will influence the free plasma concentrations of drugs for which AGP is the main carrier and consequently will affect their efficacies. For example, the increased plasma levels of AGP in depressed patients resulted in lowering the effective concentration of imipramine in these patients (Hervé *et al.*, 1996). AGP appeared to act in this way as an endogenous inhibitor for the binding of imipramine to the serotonin receptor and may influence the plasma-to-brain transport of this drug and comparable ones (Abraham *et al.*, 1987; Nemeroff *et al.*, 1990). A glycoform dependency for the binding activities of AGP described above has not been established up to now, and in fact has not been investigated for most of the compounds.

On the other hand a number of immunomodulatory activities have been described for AGP which in general are dependent on the composition of its glycans, that is, are dependent on the glycoform.

4. ANTI-INFLAMMATORY PROPERTIES OF AGP GLYCOFORMS

AGP-glycoforms expressing a high amount of sLe^X groups has been shown to be able to ameliorate neutrophil-mediated injuries in lung and intestine in a rat reperfusion model; AGP-glycoforms lacking sLe^X groups were clearly less active (Williams *et al.*, 1997). The sLe^X expressing AGP-glycoforms may have affected the selectin-mediated interaction of neutrophils in these tissues, because such AGP-glycoforms, in contrast to non-fucosylated AGP, can bind in a concentration- and Ca^{2+}-dependent manner to human E-selectin *in vitro* (Havenaar *et al.*, 1997; Williams *et al.*, 1997). In that sense, it must be realized that the primary interaction of neutrophils with activated endothelium occurs between sLe^X groups on their surface and endothelial selectins. The results of the reperfusion study are in support of our hypothesis (De Graaf *et al.*, 1993) that during an acute-phase response specific sLe^X-containing AGP glycoforms (see Table 1) are synthesized by the liver to provide in a feed-back inhibitor of selectin-dependent neutrophil adhesion to inflamed tissues (Figure 2).

The reperfusion studies, in addition, indicated that the sLe^X-containing AGP-glycoforms inhibited complement-mediated injuries in lung and intestine. *In vitro* studies

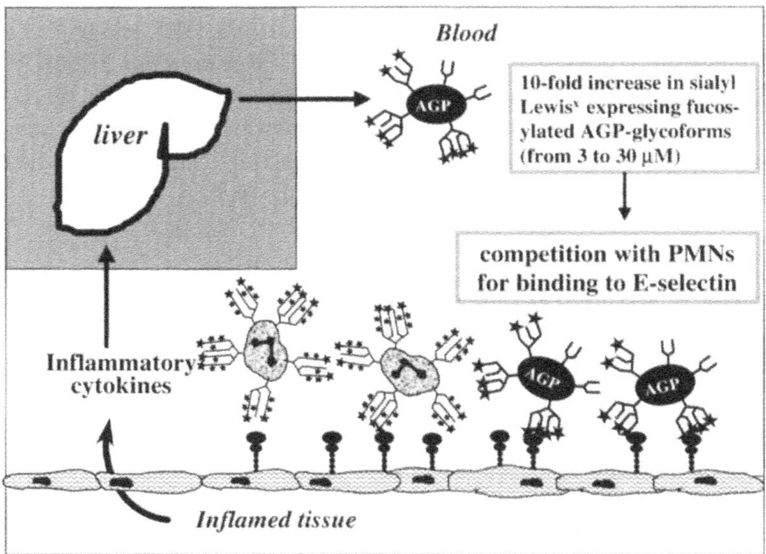

Figure 2. Model for the feed-back inhibition by acute-phase-induced AGP-glycoforms of the E-selectin mediated adhesion of neutrophils (PMN) to inflamed endothelium.

showed that strongly fucosylated AGP glycoforms can inhibit the alternative (Williams *et al.*, 1997) as well as the classical route of complement activation (Poland *et al.*, 2002, submitted for publication). The contribution of the branching of the glycans of AGP towards these effects are currently under investigation.

The extent of branching of the glycans have been shown to be of importance for other anti-inflammatory properties of AGP, like the ability to inhibit the proliferation of lymphocytes, and to induce an inhibitor of IL-1 co-mitogenic activity in macrophages (Bories *et al.*, 1990; Pos *et al.*, 1990). Also the degree of sialylation of AGP has been shown to be essential for its inhibitory effect on platelet aggregation (Costello *et al.*, 1979). Effects of fucosylation on these properties have not yet been investigated.

The great majority of the AGP-glycoforms present in normal or patient plasma are synthesized by the epithelial cells of the liver (reviewed in Van Dijk *et al.*, 1998). Recently, we detected that AGP is also synthesized by human neutrophils (Poland *et al.*, 2002 submitted). In addition we obtained immunohistochemical and biochemical evidence that during a human acute myocardial infarct (AMI) the infiltrated neutrophils release AGP into the infarct area (Poland *et al.*, 2002 submitted). The neutrophilic AGP clearly differed from normal plasma AGP in that all molecules are very strongly fucosylated and have extended branches resulting in a 10–20 kD higher molecular weight than for plasma AGP. Most probably this has made the neutrophilic AGP-glycoform a good ligand for E-selectin. It is tempting to speculate that the released AGP-glycoform will act as an endogenous feed-back inhibitor on E-selectin-mediated adhesion of neutrophils, in addition to the effect of the acute-phase induced plasma AGP (*cf.* Figure 1). In that way both effects would help to prevent excessive attraction of neutrophils to the inflamed area and thus prevent ongoing inflammation.

5. REFERENCES

Abraham, K.I., Ieni, J.R., and Meyerson, L.R., 1987, Purification and properties of a human plasma endogenous modulatoir for the platelet tricyclic binding/serotonin transport complex, *Biochim Biophys Acta.* 923:8–21.

Bories, P.N., Feger, J., Benbernou, N., Rouzeau, J.-D., Agneray, J., and Durand, G., 1990, Prevalence of tri- and tetraantennary glycans of human α_1-acid glycoprotein in release of macrophage inhibitor of interleukin-1 activity, *Inflammation.* 14:315–323.

Brinkman-Van der Linden, E.C.M., Havenaar, E.C., Van Ommen, E.C.R., Van Kamp, G.J., Gooren, L.J.G., and Van Dijk, W., 1996a, Oral estrogen treatment induced a decrease in expression of sialyl Lewis x on α_1-acid glycoprotein in females and male-to-female transsexuals, *Glycobiology.* 6:407–412.

Brinkman-Van der Linden, E.C.M., Mollicone, R., Oriol, R., Larson, G., Van den Eijnden, D.H., and Van Dijk, W., 1996b, A missense mutation in the FUT6 gene results in total absence of α3-fucosylation of human α_1-acid glycoprotein, *J Biol Chem.* 271:14492–14495.

Costello, M., Fiedel, B.A., and Gewurz, H., 1979, Inhibition of platelet aggregation by native and desialised α_1-acid glycoprotein, *Nature.* 281:677–678.

De Graaf, T.W., Van der Stelt, M.E., Anbergen, M.G., and Van Dijk, W., 1993, Inflammation-induced expression of sialyl Lewis X-containing glycan structures on α_1-acid glycoprotein (orosomucoid) in human sera, *J Exp Med.* 177:657–666.

Gabay, C. and Kushner, I., 1999, Acute-phase proteins and other systemic responses to inflammation, *New Engl J Med.* 340: 448–454.

Gahmberg, C.G. and Anderson, L.C., 1978, Leukocyte surface origin of human α_1-acid glycoprotein (orosomucoid). *J Exp Med.* 148:507–521.

Havenaar, E.C., Axford, J.S., Brinkman-Van der Linden, E.C.M., Alavi, A., Van Ommen, E.C.R., Van het Hof, B.J., Spector, T., Mackiewicz, A., and Van Dijk, W., 1998, Severe rheumatoid arthritis prohibits the pregnancy-induced decrease in α_3-fucosylation of α_1-acid glycoprotein, *Glycoconj J.* 15:723–729.

Havenaar, E.C., Dolhain, R.J.E.M., Turner, G.A., Goodarzi, M.T., Van Ommen, E.C.R., Breedveld, F.C., and Van Dijk, W., 1997a, Do synovial fluid acute phase proteins from patients with rheumatoid arthritis originate from serum, *Glycoconj J.* 14: 457–465.

Havenaar, E.C., Brinkman-Van der Linden, E.C.M., and Van Dijk, W., 1997b, Inflammation-induced α_1-acid glycoprotein expressing sialyl LewisX groups binds to E-selectin, *Glycoconj J.* 14:S85.

Hervé, F., Duche, J.C., Dathis, P. *et al.*, 1996, Binding of disopyramide, methadone, dipyridamol, chlorpromazine, lidocaine and progesterone to the two main genetic variants of human α_1-acid glycoprotein: evidence for drug-binding difference between the variants and for the presence of two separate drug-binding sites on α_1-acid glycoprotein, *Pharmacogenetics.* 6:403–415.

Mackiewicz, A., 1997, Acute-phase proteins and transformed cells, *Int Rev Cytol.* 170:225–300.

Nemeroff, C.B., Krishnan, K.R., Blazer, D.G. *et al.*, 1990, Elevated plasma concentrations of α_1-acid glycoprotein, a putative endogenous inhibitor of the tritiated imipramine binding site, in depressed patients, *Arch Gen Psychiatry.* 47:337–340.

Pervaiz, S. and Brew, K., 1987, Homology and structure-function correlations bewteen α_1-acid glycoprotein and serum retinol-binding protein and its relatives, *FASEB J.* 1:209–214.

Pos, O., Moshage, H.J., Yap, S.H., Snieders, J.P.M., Aarden, L.A., Van Gool, J., Boers, W., Brugman, A.M., and Van Dijk, W., 1989, Effects of monocytic products, recombinant interleukin-1 and recombinant interleukin-6 on the glycosylation of α_1-acid glycoprotein: studies with primary human hepatocyte cultures and rats, *Inflammation.* 13:415–424.

Pos, O., Oostendorp, R., Van der Stelt, M., Scheper, R., and Van Dijk, W., 1990, Con A-nonreactive human α_1-acid glycoprotein (AGP) is more effective in modulation of lymphocyte proliferation than Con A-reactive AGP serum variants, *Inflammation.* 14:133–141.

Sager, G. and Little, C., 1989, The effect of the plasticizers TBEP (Tris-(2-butoxyethyl)-phosphate) and DEHP (di-(2-ethylhexyl)phthalate) on beta-adrenergic ligand binding to α_1-acid glycoprotein and mononuclear leukocytes, *Biochem Pharmacol.* 38:2551–2557.

Schalkwijk, C.G., Poland, D.C.W., Van Dijk, W., Kok, A., Emeis, J.J., Drager, A.M., Doni, A., Van Hinsbergh, V.W.M., and Stehouwer, C.D.A., 1999, Plasma level of C-reactive protein is increased in IDDM patients without clinical macroangiopathy and correlates with markers of endothelial dysfunction: evidence for chronic inflammation, *Diabetologia.* 42:351–357.

Schmid, K., Kaufmann, H., Isemura, S., Bauer, F., Emura, J., Motoyama, T., Ishiguro, M., and Nanno, S., 1973, Structure of α_1-acid glycoprotein. The complete amino acid sequence, multiple amino acid substitutions, and homology with immunoglobulins. *Biochemistry.* 12:2711–2724.

Trovik, T.S., Jaeger, R., Jorde, R. *et al.*, 1992, Plasma protein binding of catecholamines, prazosin and propanolol in diabetes mellitus, *Eur J Clin Pharmacol.* 43:265–268.

Van den Heuvel, M.M., Poland, D.C.W., De Graaff, C.S., Hoefsmit, E.C.M., Postmus, P.E., Beelen, R.H.J., and Van Dijk, W., 2000, The degree of branching of α_1-acid glycoprotein in asthma, *Am J Respir Crit Care Med.* 161:1972–1978.

Van Dijk, W. and Mackiewicz, A., 1995, Interleukin-6-type cytokine-induced changes in acute phase protein glycosylation, *Ann NY Acad Sci US.* 762:319–330.

Van Dijk, W., Pos, O., VanderStelt, M.E., Moshage, H.J., Yap, S.H., Dente, L., Baumann, P., and Eap, C.B., 1991, Inflammation-induced changes in expression and glycosylation of genetic variants of α_1-acid glycoprotein (AGP): Studies with human sera, primary cultures of human hepatocytes and transgenic mice, *Biochem J.* 276:343–347.

Van Dijk, W., Brinkman-Van der Linden, E.C.M., and Havenaar, E.C., 1998, Glycosylation of alpha1-acid glycoprotein (orosomucoid) in health and disease: Occurrence, Regulation and possible functional implications, *Trens Glycosci Glycotechnol. (TIGG)* 10:235–245.

Williams, J.P., Weiser, M.R., Pechet, T.T.V., Kobzik, L., Moore, F.D., and Hechtmann, H.B., 1997, α_1-Acid glycoprotein reduces local and remote injuries after intestinal ischemia in the rat. *Am J Physiol* 273:G1031–G1053.

17

THE GLYCOSYLATION OF TYROSINASE IN MELANOMA CELLS AND THE EFFECT ON ANTIGEN PRESENTATION

Stefana M. Petrescu[1], Costin I. Popescu[1], Andrei J. Petrescu[1], and Raymond A. Dwek[2]

[1]Institute of Biochemistry
Romanian Academy
Splaiul Independentei 296
77700 Bucharest 17, Romania
[2]Glycobiology Institute
Department of Biochemistry
University of Oxford
South Parks Road, OX1 3QU, Oxford, UK

1. INTRODUCTION

Tyrosinase is constitutively expressed by normal melanocytes and melanoma cells. Both melanocytes and melanoma cells are proposed to differentiate from an early melanocyte precursor sharing most of their basic phenotypes. In these cells tyrosinase is transported through the secretory pathway to melanosomes, intracellular organelles specialized in melanin synthesis. Tyrosinase arrival in melanosomes initiates melanogenesis by catalyzing the first step of a cascade of oxidations, isomerical rearrangements and polymerizations starting with the oxidation of tyrosine to DOPA and ending with the formation of the insoluble polymer melanin.

Mammalian tyrosinases are glycoproteins displaying multiple occupied glycosylation sites shown to be highly homologous in all species. We and others have shown that these glycans are active players in the folding process of tyrosinase polypeptide. The importance of N-glycosylation for proper functioning of tyrosinase has been clearly established by the studies using N-glycosylation processing inhibitors (Petrescu *et al.*, 1997). Following the treatment of melanoma cells with α-glucosidase inhibitors, although correctly transported to melanosomes, tyrosinase was inactive. Synthesis of a functionally inactive tyrosinase was accompanied by the generation of amelanotic melanoma cells.

Glycobiology and Medicine, edited by John S. Axford
Kluwer Academic / Plenum Publishers, New York, 2003

To understand the role of glycans in tyrosinase function, a detailed characterization of the tyrosinase N-glycan composition and the identification of the site occupancy have been performed. By sequencing the mouse and hamster tyrosinases N-glycans it was found that both tyrosinases possess similar oligomannosidic series and sialylated complex bi- or triantennary structures. These data together with the high level of homology at the polypeptide level between mouse, hamster and human tyrosinase may reflect a possibly functionally relevant conservation of N-glycosylation between tyrosinases from different species.

The expression in CHO cells of mouse tyrosinase mutants lacking single N-glycosylation sites showed that sites N86, N230, N337, and N371 are fully occupied while sites N111 and N161 are unoccupied. Based on mutation experiments we found that individual N-glycans play distinctive roles depending on their location on the polypeptide chain. Moreover, we proposed that N-glycans influence tyrosinase activity by their interaction with the chaperones calnexin/calreticulin (CNX/CRT), which assist the correct folding of the nascent chain.

There is increasing evidence that the chaperones that bind tyrosinase during its folding act in conjunction with some of the ER associated degradation (ERAD) components in specific cellular conditions. Although little is known about the ERAD mechanisms, there is an obvious interrelationship between folding and degradation at the ER level which may influence the antigen presentation by MHC I.

2. TYROSINASE IN MELANOCYTES AND MELANOMA CELLS

Tyrosinase (monophenol, dihydroxyphenylalanine : oxygen oxidoreductase, EC 1.14.18.1) is a melanogenic enzyme that regulates pigment synthesis in mammals. Melanogenesis is a complex metabolic pathway in which L-tyrosine is processed to the final product, melanin, by a series of oxido-reduction and isomerization reactions catalyzed by a family of enzymes known as tyrosinase related proteins (TRPs) which are localized in the melanosomal membrane. Tyrosinase is responsible for catalyzing the first two steps of the melanin synthesis pathway: hydroxylation of tyrosine to dihydroxyphenylalanine (DOPA) and its subsequent oxidation to DOPA quinone (Prota *et al.*, 1976).

Tyrosinase is a type I membrane glycoprotein with 533 amino acids, 4 occupied N-glycosylation sites, 17 cysteine residues grouped in two cysteine rich domains and two copper binding domains, copper A and copper B (Hearing *et al.*, 1991) (Figure 1). Some of the structural motifs found in tyrosinase including the copper B domain appear to be highly conserved not only among tyrosinases from different species, but also among the melanogenic enzymes. For example, tyrosinase and tyrosinase related protein 1 (TRP1) share a significant level of homology in several regions including the catalytic domain and the potential N-glycosylation sites. Human and mouse tyrosinase share more than 95% homology at the primary sequence level. The enzymatic activity of tyrosinase is dependent upon the binding of two copper atoms in the copper A and copper B binding sites. At each of these sites three histidine residues coordinate the copper atom, and both of the copper atoms coordinate an O_2 molecule (Hearing, 1987). Site directed mutagenesis studies have revealed that three histidine residues, His 363, 367, and 389 are involved in the coordination of the copper binding in copper B binding domain.

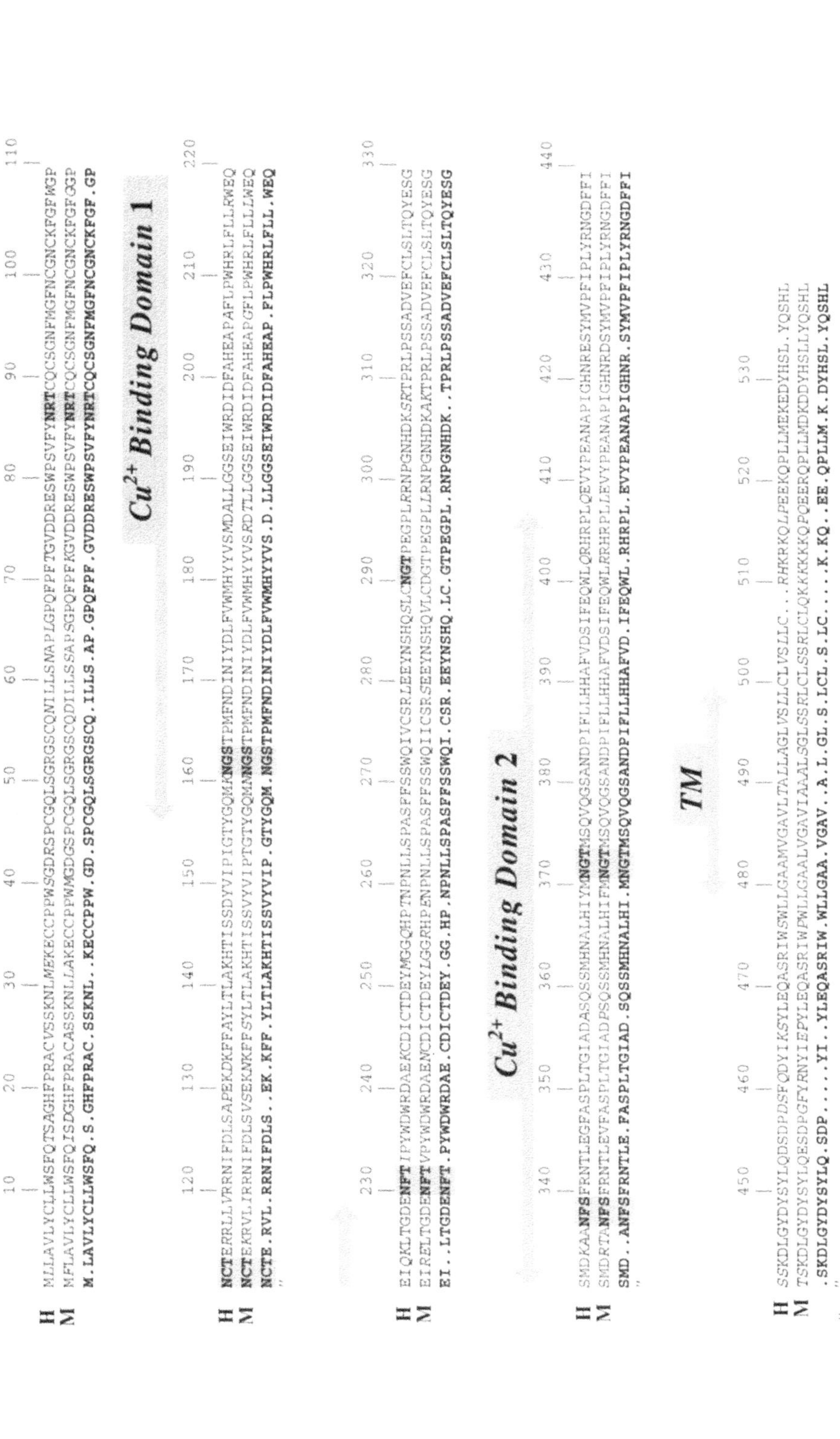

```
          10        20        30        40        50        60        70        80        90       100       110
           |         |         |         |         |         |         |         |         |         |         |
H  MLLAVLYCLLWSFQTSAGHFPRACVSSKNLMEKECCPPWSGDRSPCGQLSGRGSCQNILLSNAPLGPQFPFTGVDDRESWPSVFYNRTCQCSGNFMGFNCGNCKFGFWGP
M  MFLAVLYCLLWSFQISDGHFPRACASSKNLLAKECCPPWMGDGSPCGQLSGRGSCQDILLSSAPSGPQFPFKGVDDRESWPSVFYNRTCQCSGNFMGFNCGNCKFGFGGP
   M.LAVLYCLLWSFQ.S.GHFPRAC.SSKNL..KECCPPW.GD.SPCGQLSGRGSCQ.ILLS.AP.GPQFPF.GVDDRESWPSVFYNRTCQCSGNFMGFNCGNCKFGF.GP

                                                                                   Cu2+ Binding Domain 1

         120       130       140       150       160       170       180       190       200       210       220
           |         |         |         |         |         |         |         |         |         |         |
H  NCTERRLLVRRNIFDLSAPEKDKFFAYLTLAKHTISSDYVIPIGTYGQMKNGSTPMFNDINIYDLFVWMHYYVSMDALLGGSEIWRDIDFAHEAPAFLPWHRLFLLRWEQ
M  NCTEKRVLIRRNIFDLSVSEKNKFFSYLTLAKHTISSVYVIPTGTYGQMNNGSTPMFNDINIYDLFVWMHYYVSRDTLLGGSEIWRDIDFAHEAPGFLPWHRLFLLLWEQ
   NCTE.RVL.RRNIFDLS..EK.KFF.YLTLAKHTISSVYVIP.GTYGQM.NGSTPMFNDINIYDLFVWMHYYVS.D.LLGGSEIWRDIDFAHEAP.FLPWHRLFLL.WEQ

         230       240       250       260       270       280       290       300       310       320       330
           |         |         |         |         |         |         |         |         |         |         |
H  EIQKLTGDENFTIPYWDWRDAEKCDICTDEYMGGQHPTNPNLLSPASFFSSWQIVCSRLEEYNSHQSLCNGTPEGPLRRNPGNHDKSRTPRLPSSADVEFCLSLTQYESG
M  EIRELTGDENFTVPYWDWRDAENCDICTDEYLGGRHPENPNLLSPASFFSSWQIICSRSEEYNSHQVLCDGTPEGPLLRNPGNHDKAKTPRLPSSADVEFCLSLTQYESG
   EI..LTGDENFT.PYWDWRDAE.CDICTDEY.GG.HP.NPNLLSPASFFSSWQI.CSR.EEYNSHQ.LC.GTPEGPL.RNPGNHDK..TPRLPSSADVEFCLSLTQYESG

                    Cu2+ Binding Domain 2

         340       350       360       370       380       390       400       410       420       430       440
           |         |         |         |         |         |         |         |         |         |         |
H  SMDKAANFSFRNTLEGFASPLTGIADASQSSMHNALHIYMNGTMSQVQGSANDPIFLLHHAFVDSIFEQWLQRHRPLQEVYPEANAPIGHNRESYMVPFIPLYRNGDFFI
M  SMDRTANFSFRNTLEVFASPLTGIADPSQSSMHNALHIFMNGTMSQVQGSANDPIFLLHHAFVDSIFEQWLRRHRPLLEVYPEANAPIGHNRDSYMVPFIPLYRNGDFFI
   SMD..ANFSFRNTLE.FASPLTGIAD.SQSSMHNALHI.MNGTMSQVQGSANDPIFLLHHAFVD.IFEQWL.RHRPL.EVYPEANAPIGHNR.SYMVPFIPLYRNGDFFI

                                TM

         450       460       470       480       490       500       510       520       530
           |         |         |         |         |         |         |         |         |
H  SSKDLGYDYSYLQDSDPDSFQDYIKSYLEQASRIWSWLLGAAMVGAVLTALLAGLVSLLCLVSLLC...RHKRKQLPEEKQPLLMEKEDYHSL.YQSHL
M  TSKDLGYDYSYLQESDPGFYRNYIEPYLEQASRIWPWLLGAALVGAVIAAALSGLSSRLCLSSRLCLQKKKKKKQPQEERQPLLMDKDDYHSLLYQSHL
   .SKDLGYDYSYLQ.SDP.....YI..YLEQASRIW.WLLGAA.VGAV..A.L.GL.S.LCL.S.LC.....K.KQ..EE.QPLLM.K.DYHSL.YQSHL
```

Figure 1. Schematic representation of mouse and human tyrosinases. Human tyrosinase (H) and mouse tyrosinase (M) primary sequence together with the conserved residues (bold) and differences (dots) are depicted. The potential N-glycosylation sites are in gray cassettes. The copper domains, copper A (Cu^{2+} Binding Domain 1) and copper B (Cu^{2+} Binding Domain 2) and the transmembrane domain (TM) are also indicated.

Tyrosinase is the pigment enzyme synthesized in melanocytes. These cells are specialized in pigment production and export outside the cell to form the melanin deposits found in skin, hair bulbs, and eyes. Melanocytes synthesize two types of melanin, pheomelanins (yellow, red) and eumelanins (brown, black) (Figure 2) in ratios that are highly specific for each individual. Melanin has been shown to have a protective role against radiations and the absence of melanin yields oculocutaneous albinism (OCA), a recessive genetic disorder resulting in increased risk of skin cancer and visual disorders. Mutations in tyrosinase are responsible for many cases of albinism in which tyrosinase is retained in the ER and degraded. Tyrosinase is synthesized on ribosomes and translocated into the ER lumen where folding and maturation of the nascent chain occur. To reach its final destination, the melanosome, an intracellular organelle where melanin synthesis occurs, tyrosinase is transported from the ER through the secretory pathway.

As already mentioned, tyrosinase is a melanocyte protein and since melanoma cells are transformed melanocytes, this protein is constitutively expressed in melanoma cells. Interestingly, in advanced metastatic melanoma tyrosinase has been shown to behave like mutant proteins in normal melanocytes, being degraded in proteasomes following retention at the ER level. The rate of translation of tyrosinase has been reported to be faster in melanoma cells than in normal melanocytes, leading to incomplete N-glycan sequons occupancy. This was not specific for the synthesis of all melanoma proteins, since calnexin and TRP-1 were found to be synthesized at similar rates with those encountered in melanocytes.

In contrast to normal melanocytes, melanoma cells commonly express MHC class II molecules constitutively (Goodwin *et al.*, 2001). Interestingly, a melanoma cell line,

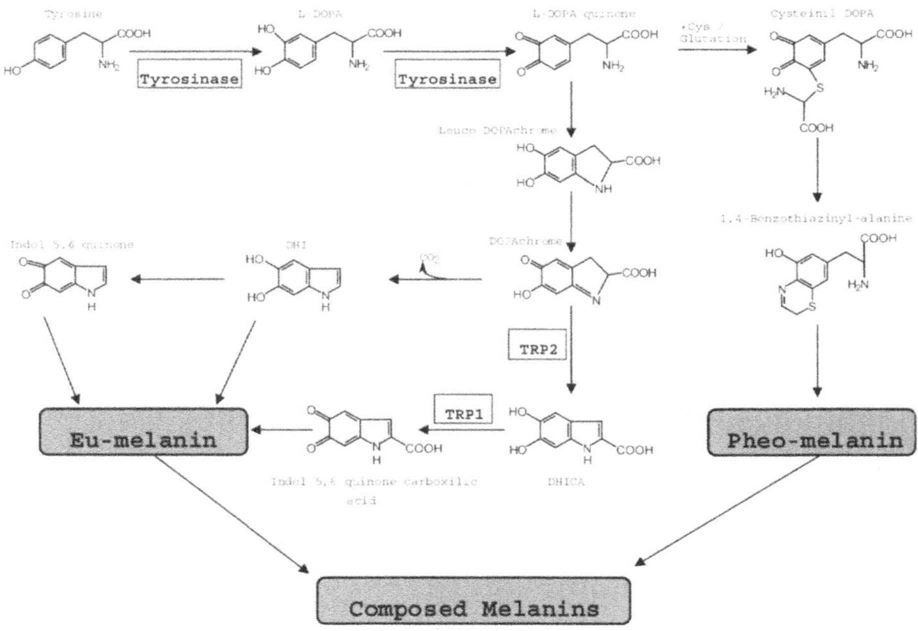

Figure 2. Biosynthetic pathways of melanin synthesis.

MelJuSo was the first non-professional antigen-presenting cell (APC) line studied in relation with its class II presentation. This phenotype is rather unusual for non-professional APCs that require IFN-g treatment in order to be induced to express MHC class II (Houghton *et al.*, 1982). Whilst class II expression is constitutive in B cells, most cells including tumor cells may express it upon IFN treatment.

Since clinical evidence suggested that class II expression in melanoma could be associated with tumor progression, the ability of metastatic melanoma to process and present antigens has been investigated. In all cell lines tested, melanoma cells were able to present peptide efficiently to CD4+ T cells, resulting in increased T-cell proliferation 5–26-fold over controls (Brady *et al.*, 1996). Furthermore, T-cell stimulation did not require CD28-mediated costimulation but rather CD54 (ICAM-1) which appears to be important in melanoma presentation, as opposed to B cells. Moreover, tissue culture experiments in which metastatic melanoma cells were co-cultured with CD4+ T cell revealed that melanoma cells were killed by the T cells secreted cytokines (Brady *et al.*, 2000). Immunotherapy, using primed CD4+ T cells and peptide, may be beneficial in patients whose tumors express HLA class II antigens.

3. TYROSINASE N-GLYCANS PROCESSING

By treating tissue culture melanoma cells with a drug that inhibits the normal glycan processing we realized that we are able to effectively bleach the normally black cells. The drug, an iminosugar bearing a hydrophobic side chain (NB-DNJ), a known inhibitor of glucosidases II, and I abolished the pigmentation of melanoma cells by inactivating tyrosinase (Petrescu *et al.*, 1997). This finding suggesting a role for glycans in tyrosinase function prompted us to investigate further the maturation and processing of tyrosinase in malignant melanoma.

By sequencing the mouse tyrosinase N-glycans and comparing them with the data on hamster (Toyofuku *et al.*, 1999) it was found that both tyrosinases possess similar oligomannosidic series and sialylated complex antennary structures. However, the ratio of high-mannose versus complex structures is 1:3 in hamster, as compared to 1:1 in mouse tyrosinase. These data together with the observation that mouse and human tyrosinase share 85% sequence identity and identical potential N-glycosylation sites indicate an interesting and possibly functionally relevant conservation of N-glycosylation between tyrosinases from different species (Figure 3).

We should note that the members of the TRP-family are all glycoproteins containing N-linked oligosaccharides. Their polypeptide chains have similar numbers of potential N-glycosylation sites and three of them are in well-conserved positions (Negroiu *et al.*, 1999). For instance, studies on TRP-1 and TRP-2 showed that carbohydrate moiety of TRP-2 seems to be similar to the one of hamster tyrosinase (Jimbow *et al.*, 1994) while TRP-1 is differently glycosylated. Both human and murine TRP-1 contain mixtures of high-mannose and complex structures. The carbohydrate analysis of TRP-1 from murine melanoma showed that only 16% of the glycans are of the high-mannose type, 16% are biantennary while 65% are processed to tri- and tetra-antennary structures. This is similar to that reported for TRP-1 from human melanocytes and melanoma cells (del Marmol *et al.*, 1996). In addition to differences between carbohydrate composition, there are substantial differences between the kinetics of maturation of tyrosinase (Vijayasaradhi *et al.*, 1995) and

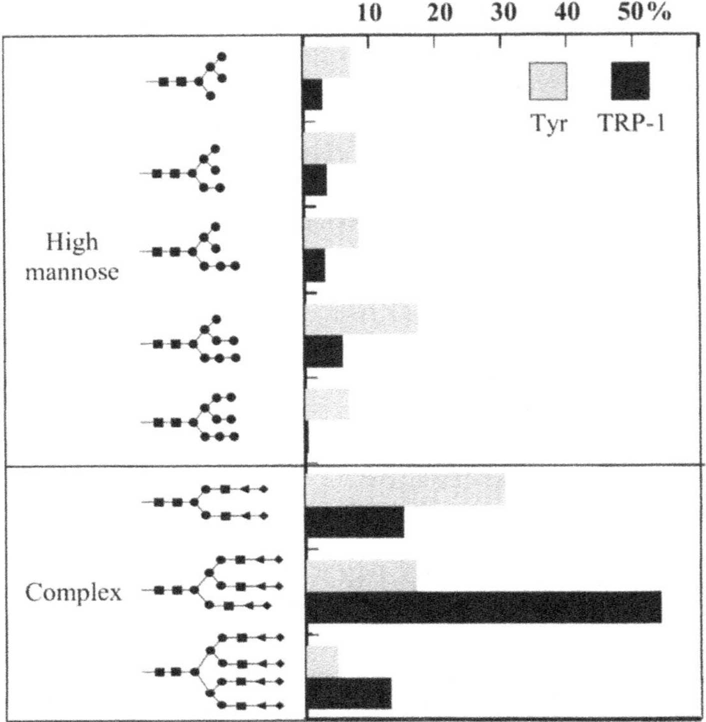

Figure 3. Structure of tyrosinase and TRP-1 N-glycans. The percentage molar ratio of total sugar chains released from each enzyme is indicated. σ—glucose; λ—mannose; ν—N-acetyl-glucosamine; τ—galactose; υ—sialic acid; H—fucose.

TRP-1 (Aroca *et al.*, 1992). TRP 1 is processed in less than one hour to its fully glycosylated form, while tyrosinase requires more than 3 hours for complete maturation. Within the ER and early Golgi TRP-1 is detected for ~30 minutes, whereas tyrosinase is present in the same compartments for at least 3 hours. Importantly, these results have been obtained by directly comparing the kinetics of N-glycan processing of the two glycoproteins in the same mouse melanoma cell line. The data indicate that differences in the overall processing time of the two glycoproteins are due primarily to their ER residency and hence to their folding process rather than to their transport through the secretory pathway.

Tyrosinase and TRP1 have been recently shown to behave differently in the presence of ER glucosidases inhibitors (Negroiu *et al.*, 1999). Tyrosinase from B16 cells treated with NB-DNJ contains oligosaccharides with Glc_3 $Man_{7\text{-}9}GlcNAc_2$ structure, which indicates that no further processing of N-glycans occurs in the presence of this inhibitor. By contrast, TRP-1 from the same cell line is able to overcome the inhibitory effect of the ER-glucosidases inhibitors and acquires oligosaccharides of complex type (Negroiu *et al.*, 2000). Similar results were observed with TRP-1 from SK Mel-19, a human melanoma cell line treated with castanospermine or 1-deoxynojirimycin, two other inhibitors of the ER-glucosidases I and II (del Marmol *et al.*, 1996).

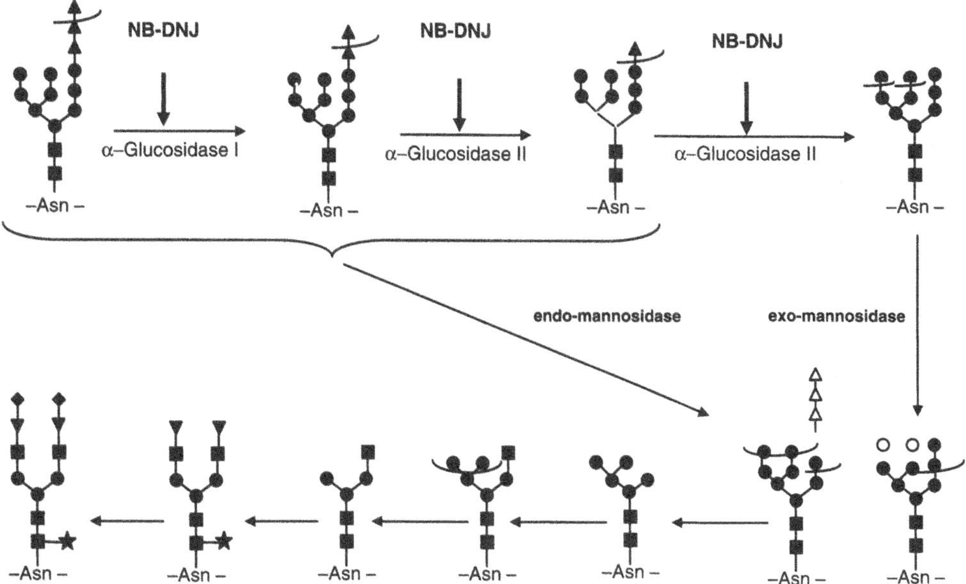

Figure 4. Early stages of N-glycan biosynthesis.

The glucosidase blockade can be bypassed by the use of the endomannosidase pathway, which acts in the Golgi and trims the three glucose and terminal mannose residues from the oligosaccharide moiety (Figure 4).

Therefore, these results suggest that tyrosinase is not a substrate for endomannosidase whereas TRP-1 is. This raises the interesting and intriguing question as to whether the Golgi endo-mannosidase only acts on those proteins, which are already correctly folded. It may well be that the misfolded proteins—as is tyrosinase in the presence of NB-DNJ—are such that the structure around the glycan moiety prohibits the access to the endo-mannosidase.

4. FOLDING AND MATURATION OF TYROSINASE

Nascent polypeptides are co-translationally N-glycosylated at the ER lumen in the presence of the multimeric enzyme oligosaccharyltransferase. This enzyme transfers an invariant triglucosylated oligomannosidic glycan to the Asn residue of an Asn-X-Ser(Thr) sequon on the polypeptide. Occurring almost simultaneously with the polypeptide translocation into the ER lumen, the glycosylation step results in nascent polypeptides sharing identical glycans attached to the chain. The covalently linked glycans are the unifying element allowing the unique ER quality control machinery to act on every single nascent chain. That this glycan generates recognition elements for the quality control is supported by the rapid trimming of the triglucosylated glycan to G2 and G1, process catalyzed by glucosidases I and II. The monoglucosylated glycan G1Man9 is a specific ligand for the lectins calnexin and calreticulin which recognize the glucose residue and the surface of the glucosylated α-1-6 arm (Petrescu *et al.*, 1997).

Calnexin and calreticulin are chaperones known to be components of the ER quality control (Zapun *et al.*, 1997). Both the membrane lectin calnexin and the highly homologous soluble lectin calreticulin act by driving the incompletely folded nascent chain into a glucosylation/deglucosylation cycle (Helenius *et al.*, 2001; Kowarik *et al.*, 2002). Whilst deglucosylation of the G1Man9 glycan occurs on a random basis by the action of glucosidase II, re-glucosylation of the glycan is one of the most tightly regulated process in the ER. Re-glucosylation of the Man9 glycan is catalyzed by glucosyltransferase, a soluble enzyme which recognize the high mannose glycan and transfers a glucose residue on its α 1-6 arm (Schrag *et al.*, 2003). Besides the glycan there is another important recognition element at the protein level, the enzyme interacting only with polypeptides exposing hydrophobic patches. Glucosyltransferase is the actual folding sensor in the ER acting on the incompletely folded polypeptides to facilitate their association with CNX/CRT. Why is the binding to CNX/CRT important for the nascent chain? Although they do not act directly on the chain, the two chaperones are thought to facilitate its interaction with ER folding factors. One example is the protein Erp57 involved in the disulfide bridges formation which has been found in ternary complexes with the chain and CNX (Schrag *et al.*, 2003). It is therefore likely that CNX association may help the stabilization of the folded structure by disulfide formation or rearrangements. On the other hand CNX binding redirect the polypeptide into the calnexin cycle where the polypeptide is again presented to glucosidase II for deglucosylation. Once folded, the chain stops being a substrate for GT and in the absence of the glucosyl tag it will be able to leave CNX cycle and exit the ER.

However, not the entire polypeptide population is able to reach the native state, this process being highly dependent on the efficiency of the folding. We have to bear in mind that the efficiency of the folding process varies from a protein to another. Moreover, some proteins are totally dependent on their association with the ER chaperones, whilst another category of proteins fold efficiently in the absence of chaperones. The first category of proteins are good candidates for the ER associated degradation process. It has been shown that polypeptide chains unable to reach the native conformation are directed by the ER chaperones into a retro-translocation pathway that drives the chain back into the cytosol through the translocon channel (Molinari and Helenius, 2000). Besides CNX/CRT, other chaperones such as PDI and BiP have been reported to associate with the chain before retro-translocation (Helenius *et al.*, 2001). The incompletely folded chain undergoes degradation into a specialized cytosolic complex, the proteasome, yielding peptides that can be presented by the MHC molecules.

Monitoring mouse tyrosinase maturation in melanoma cells we have reported a slow folding process of this membrane protein (Branza-Nichita *et al.*, 2000). Immunoprecipitation of the chain from pulse-labelled cells with antibodies directed against its C-terminal showed that tyrosinase folds in a two-stage process requiring more than 3 hours ER retention. By co-immunoprecipitation experiments we have demonstrated that calnexin association is required for the entire folding process (Petrescu *et al.*, 2000). Now we understand why the melanoma cells undergo a depigmentation process as a result of the inhibition of the early steps in glycan processing (Petrescu *et al.*, 1997). This inhibition abolishes calnexin interaction with tyrosinase which in turn yields a non-native protein unable to synthesize melanin. That calnexin is an absolute requirement for tyrosinase folding has been also demonstrated by the inactivation of human tyrosinase in the presence of glucosidase inhibitors in the culture medium (Popescu *et al.*, unpublished results).

Calreticulin, the soluble ER lectin interacts only transiently with mouse tyrosinase in the early stages of its maturation process (Petrescu *et al.*, 2000) and has been shown to associate with human tyrosinase only in cells treated with proteasome inhibitors (Halaban *et al.*, 1997). Although it is tempting to speculate that CRT is involved in the ER retention and retro-translocation of the misfolded tyrosinase into the cytosol, more evidence is required to make this statement.

Experiments with tyrosinase mutants lacking one or more glycans showed that individual glycans play different roles in tyrosinase folding and maturation. The presence of a specific pair of N-glycans located at sites 1 (N86), and 6 (N371) is crucial for correct folding of all the remaining mutants and sufficient to reproduce the wild type activity, regardless of the occupancy of the other N-glycosylation sites (Branza-Nichita *et al.*, 2000). Removal of one or both of the two glycan sites (N86 and N371) results in accelerated folding, shorter interaction time with calnexin and lower activity of the corresponding mutants, stressing the importance of site specific glycosylation of tyrosinase.

Based on these experiments, a local folding mechanism has been proposed for tyrosinase polypeptide chain which is regulated by the interaction with the calnexin/calreticulin cycle. In this model, individual N-glycans play distinctive roles depending on their location on the polypeptide chain. Regions of the protein local to some of the glycan sites may fold rapidly and correctly with limited chaperone assistance. The presence of glycans in regions of the protein structure that spontaneously fold less efficiently may be necessary for the correct function of the quality control system and hence may explain the conservation of glycosylation sites in tyrosinase. It is interesting to note that the removal of site six (N371) has the most significant effect on the activity of the expressed enzyme and this is the site at which mutations have been implicated in OCA.

Copper analysis of tyrosinase mutants with partial activity suggested the presence of populations of both active and inactive tyrosinase in differing amounts. Thus, different

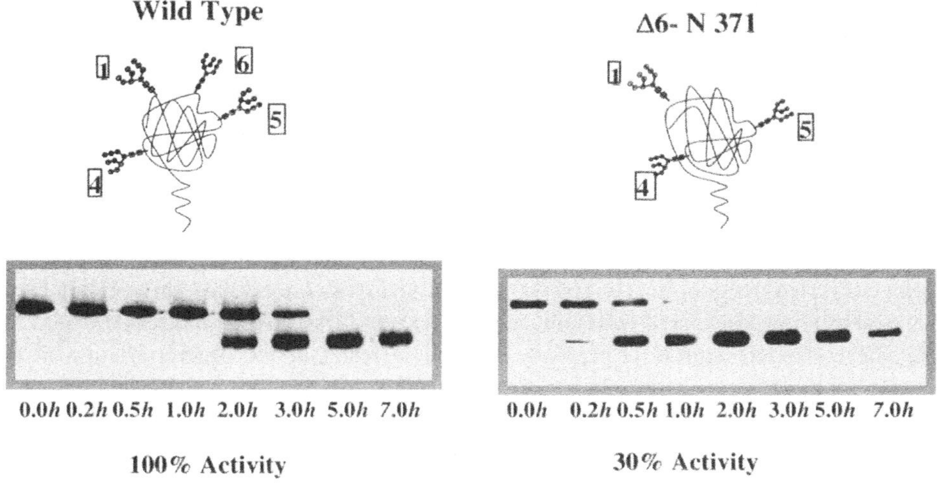

Figure 5. Folding and activity of tyrosinase mutant lacking N371 N-glycosylation sites. The lower panels illustrate typical immunoprecipitation experiments of pulse labelled melanoma cells chased for the indicated time points and analyzed by non-reducing SDS-PAGE.

N-glycosylation mutants show different yields of correctly folded enzyme. The population able to reach the native conformation increases with the time spent by the glycoprotein in the calnexin cycle.

This observation allows us to understand the OCA disease, which in 5% of cases is due to the absence of glycan site N371 in human tyrosinase. By monitoring the folding of mouse tyrosinase mutant we noticed that in the absence of calnexin, the mutant displays an accelerated folding and a dramatic decrease in tyrosinase activity as compared to the wild type (Figure 5). Considering the high homology between mouse and human tyrosinase and the ER retention reported for the human mutant (Halaban *et al.*, 2000) OCA type IA has been considered a tyrosinase folding disease (Halaban *et al.*, 1997; Petrescu *et al.*, 2000).

5. TYROSINASE DIFFERENTIATION ANTIGENS IN MELANOMA

Melanoma is the second lethal type of cancer after breast cancer. The estimated lifetime risk of an individual for developing melanoma is 1.4% (Blackwood *et al.*, 2002). It is now accepted that T-lymphocytes mediate tumor-specific protective immunity. In principal, T-lymphocytes are able to monitor and recognise any change or structural alteration occurring in somatic cells. T-cell-defined antigens on human melanomas are divided into three principal categories. These are individually distinct mutated antigens, cancer antigens, and melanocyte differentiation antigens. Cancer and differentiation antigens are not altered or mutated in tumor tissues. For a variety of reasons, tumor cells are not equipped to induce primary T-cell responses since among other factors they also release immuno suppressive molecules. However, tumor cells can be recognised by pre-activated cytotoxic T-lymphocytes which have been induced using differentiation antigens. This strategy essentially seeks to break tolerance and is the basis for developing active therapeutic immunization strategies. Since tyrosinase is a key enzyme in melanin synthesis, its normal differentiation antigens could potentially be used in melanoma therapy.

Melanoma reactive CTL (cytotoxic T lymphocytes) precursors are present in the immune repertoire. Despite the existing evidence that these T cells are activated *in vivo*, there is an yet unknown mechanism that reduces their efficacy in metastatic melanoma.

Since tyrosinase is a normal self protein produced by melanocytes in skin, hair, and eyes, the CTLs reactivity could be limited due to self-tolerance.

Interestingly, a tyrosinase peptide, YMNGTMSQV (N) corresponding to aminoacids 369–377 and including the N-linked glycosylation site 6, has been shown to be presented as the converted peptide, YMDGTMSQV (D) (Engelhard *et al.*, 2002) This peptide binds to the transporter associated with antigenic processing (TAP) which transports it into the ER. The converted peptide D probably arises as a result of the deglycosylation in the cytosol by the enzyme peptide: N-glycanase. This enzyme reaction also results in the conversion of Asn residues to Asp (Petrescu *et al.*, 2000). The D peptide has also been found to be presented by HLA-A0201 on cells expressing full-length tyrosinase. Following binding of the peptide to MHC, the complex passes through the Golgi where it is further glycosylated and then reach the cell surface. In general, deglycosylation of glycoproteins in the cytosol prior to degradation by the proteasome into peptides may provide a mechanism for limiting the number of peptides that can be presented. Therefore, the deglycosylation

Table 1. Tyrosinases antigens in melanoma.

Antigen	HLA restriction	Epitope	Sequence	Reference
Tyrosinase	A*01	243–251	KCDICTDEY	Kittlesen *et al.*, 1998
	A*01	146–156	SSDYVIPIGTY	Kawakami *et al.*, 1998
	A*0201	1–9	MLLAVLYCL	Wolfel *et al.*, 1994
	A*0201	369–377	YMNGTMSQV	Skipper *et al.*,1996
	A*0201	369–377	YMDGTMSQV	Boon *et al.*, 1994
	A*02402	206–214	AFLPWHRLF	Kang *et al.*, 1995
	B*44	192–205	SEIWRDIDDF	Brichard *et al.*, 1996
	DRβ1*0401	56–70	QNILLSNAPLGPQFP	Topalian *et al.*, 1996
	DRβ1*0401	448–462	DYSYKQDSDPDSFQD	Topalian *et al.*, 1996
	DRβ1*1501	386–406		Kobayashi *et al.*, 1998
Tyrosinase-related protein 1	A*31	1–9	MSLQRQFLR	Wang *et al.*, 1996
Tyrosinase-related protein 2	A*31	197–205	LLPGGRPYR	Wang *et al.*, 1996, 1998
	A*33	197–205	LLPGGRPYR	Castelli *et al.*, 1999
	A*0201	180–188	SVYDFFVWL	Parkhurst *et al.*, 1998
	C*0802	387–395	ANDPIFVVL	Castelli *et al.*, 1999
	A*68011	INT-2	EVISCKLIKR	Lupetti *et al.*, 1998

enzymes may play a crucial role in limiting the number of T-cell receptors required in the immune system.

6. CONCLUSIONS

Newly synthesized tyrosinase is translocated into the endoplasmic reticulum where folding and maturation of the nascent chain occurs. Folding of the translocated chain is regulated in the ER by multiple mechanisms destined to enable the polypeptide chain to acquire the conformation of the native glycoprotein. Properly folded tyrosinase achieves an export competent conformation by folding and refolding in the presence of the ER-resident molecular chaperones and exits the ER. Misfolded tyrosinase is retained in the ER and targeted to degradation, thus providing the cells with an efficient quality control mechanism. The incorrectly folded tyrosinase is retrotranslocated into the cytosol and degraded by the proteasomal proteases. Some of the resultant peptides are transported back into the endoplasmic reticulum where they associate with the MHC I complex to be presented to the immune system. Therefore, location and trafficking of tyrosinase rather than its expression level, may be an important factor in determining antigen presentation. Strategies for identifying antigen generating mutant tyrosinases will provide the basis for new generations of antigen-specific tumor vaccines.

7. REFERENCES

Aroca, P., Martinez-Liarte, J.H., Solano, F., Garcia-Borron, J.C., and Lozano, J.A., 1992, The action of glycosylases on dopachrome (2-carboxy-2,3-dihydroindole-5,6-quinone) tautomerase, *Biochem J.* 284:109–113.

Blackwood, M.A., Holmes, R., Synnestvedt, M., Young, M., George, C., Yang, H., Elder, D.E., Schuchter, L.M., Guerry, D., and Ganguly, A., 2002, Multiple primary melanoma revisited, *Cancer*. 94(8):2248–2255.

Brady, M.S., Eckels, D.D., Ree, S.Y., Schultheiss, K.E., and Lee, J.S., 1996, MHC class II-mediated antigen presentation by melanoma cells, *J Immunother Emphasis Tumor Immunol.* 19(6):387–397.

Brady, M.S., Lee, F., Petrie, H., Eckels, D.D., and Lee, J.S., 2000, CD4(+) T cells kill HLA-class-II-antigen-positive melanoma cells presenting peptide *in vitro*, *Cancer Immunol Immunother.* 48(11):621–626.

Branza-Nichita, N., Negroiu, G., Petrescu, A.J., Garman, E.F., Platt, F.M., Wormald, M.R., Dwek, R.A., and Petrescu, S.M., 2000, Mutations at critical N-glycosylation sites reduce tyrosinase activity by altering folding and quality control, *J Biol Chem.* 275(11):8169–8175.

Brichard, V.G., Herman, J., Van Pel, A., Wildmann, C., Gaugler, B., Wolfel, T., Boon, T., and Lethe, B., 1996, A tyrosinase nonapeptide presented by HLA-B44 is recognized on a human melanoma by autologous cytolytic T lymphocytes, *Eur J Immunol.* 26(1):224–223.

Castelli, C., Tarsini, P., Mazzocchi, A., Rini, F., Rivoltini, L., Ravagnani, F., Gallino, F., Belli, F., and Parmiani, G., 1999, Novel HLA-Cw8-restricted T cell epitopes derived from tyrosinase-related protein-2 and gp100 melanoma antigens, *J Immunol.* 162(3):1739–1748.

del Marmol, V., Beermann, F., 1996, Tyrosinase and related proteins in mammalian pigmentation. *FEBS Lett.* 381:165–168.

Engelhard, V.H., Brickner, A.G., and Zarling, A.L., 2002, Insights into antigen processing gained by direct analysis of the naturally processed class I MHC associated peptide repertoire, *Mol Immunol.* 39(3–4):127–137.

Goodwin, B.L., Xi, H., Tejiram, R., Eason, D.D., Ghosh, N., Wright, K.L., Nagarajan, U., Boss, J.M., and Blanck, G., 2001, Varying functions of specific major histocompatibility class II transactivator promoter III and IV elements in melanoma cell lines, *Cell Growth Differ.* 12(6):327–335.

Hearing, V.J. and Tsukamoto, K., 1991, Enzymatic control of pigmentation in mammals, *FASEB J.* 5:2902–2909.

Hearing, V.J., 1987, Mammalian monophenol monooxygenase (tyrosinase): purification, properties, and reactions catalyzed, *Methods Enzymol.* 142:154–165.

Halaban, R., Svedine, S., Cheng, E., Smicun, Y., Aron, R., and Hebert, D.N., 2000, Endoplasmic reticulum retention is a common defect associated with tyrosinase-negative albinism, *Proc Natl Acad Sci U S A.* 23;97(11):5889–5894.

Halaban, R., Cheng, E., Zhang, Y., Moellmann, G., Hanlon, D., Michalak, M., Setaluri, V., and Hebert, D.N., 1997, Aberrant retention of tyrosinase in the endoplasmic reticulum mediates accelerated degradation of the enzyme and contributes to the dedifferentiated phenotype of amelanotic melanoma cells, *Proc Natl Acad Sci USA*. 94(12):6210–6215.

Helenius, A. and Aebi, M., 2001, Intracellular functions of N-linked glycans, *Science*. 291(5512):2364–2369.

Kang, X., Kawakami, Y., el-Gamil, M., Wang, R., Sakaguchi, K., Yannelli, J.R., Appella, E., Rosenberg, S.A., and Robbins, P.F., 1995, Identification of a tyrosinase epitope recognized by HLA-A24-restricted, tumor-infiltrating lymphocytes, *J Immunol.* 155(3):1343–1348.

Kawakami, Y., Robbins, P.F., Wang, X., Tupesis, J.P., Parkhurst, M.R., Kang, X., Sakaguchi, K., Appella, E., and Rosenberg, S.A., 1998, Identification of new melanoma epitopes on melanosomal proteins recognized by tumor infiltrating T lymphocytes restricted by HLA-A1, -A2, and -A3 alleles, *J Immunol.* 161(12):6985–6992.

Kittlesen, D.J., Thompson, L.W., Gulden, P.H., Skipper, J.C., Colella, T.A., Shabanowitz, J., Hunt, D.F., Engelhard, V.H., Slingluff, C.L., Jr, and Shabanowitz, J.A., 1998, Human melanoma patients recognize an HLA-A1-restricted CTL epitope from tyrosinase containing two cysteine residues: implications for tumor vaccine development, *J Immunol.* 160(5):2099–2106.

Kobayashi, H., Kokubo, T., Sato, K., Kimura, S., Asano, K., Takahashi, H., Iizuka, H., Miyokawa, N., and Katagiri, M., 1998, CD4+ T cells from peripheral blood of a melanoma patient recognize peptides derived from nonmutated tyrosinase, *Cancer Res.* 58(2):296–301.

Kowarik, M., Kung, S., Martoglio, B., and Helenius, A., 2002, Protein folding during cotranslational translocation in the endoplasmic reticulum., *Mol Cell.* 10(4):769–778.

Lupetti, R., Pisarra, P., Verrecchia, A., Farina, C., Nicolini, G., Anichini, A., Bordignon, C., Sensi, M., Parmiani, G., and Traversari, C., 1998, Translation of a retained intron in tyrosinase-related protein (TRP) 2 mRNA generates a new cytotoxic T lymphocyte (CTL)-defined and shared human melanoma antigen not expressed in normal cells of the melanocytic lineage, *J Exp Med.* 188(6):1005–1016.

Molinari, M. and Helenius, A., 2000, Chaperone selection during glycoprotein translocation into the endoplasmic reticulum, *Science*. 288(5464):331–333.

Negroiu, G., Dwek, R.A., and Petrescu, S.M., 2000, Folding and maturation of tyrosinase-related protein-1 are regulated by the post-translational formation of disulfide bonds and by N-glycan processing, *J Biol Chem.* 275(41):32200–32207.

Negroiu, G., Branza-Nichita, N., Costin, G.E., Titu, H., Petrescu, A.J., Dwek, R.A., and Petrescu, S.M., 1999, Investigation of the intracellular transport of tyrosinase and tyrosinase related protein (TRP)-1. The effect of endoplasmic reticulum (ER)-glucosidases inhibition, *Cell Mol Biol* (Noisy-le-grand). 45(7):1001–1010.

Negroiu, G., Branza-Nichita, N., Petrescu, A.J., Dwek, R.A., and Petrescu, S.M., 1999, Protein specific N-glycosylation of tyrosinase and tyrosinase-related protein-1 in B16 mouse melanoma cells, *Biochem J.* 344(3):659–665.

Ohkura, T., Yamashita, K., Mishima, Y., and Kobata, A., 1984, Purification of hamster melanoma tyrosinases and structural studies of their asparagine-linked sugar chains, *Arch Biochem Biophys.* 235:63–77.

Parkhurst, M.R., Fitzgerald, E.B., Southwood, S., Sette, A., Rosenberg, S.A., and Kawakami, Y., 1998, Identification of a shared HLA-A*0201-restricted T-cell epitope from the melanoma antigen tyrosinase-related protein 2 (TRP2), *Cancer Res.* 58(21):4895–4901.

Petrescu, S.M., Branza-Nichita, N., Negroiu, G., Petrescu, A.J., and Dwek, R.A., 2000, Tyrosinase and glycoprotein folding: roles of chaperones that recognize glycans, *Biochemistry.* 39(18):5229–5237.

Petrescu, S.M., Petrescu, A.J., Titu, H.N., Dwek, R.A., and Platt, F.M., 1997, Inhibition of N-glycan processing in B16 melanoma cells results in inactivation of tyrosinase but does not prevent its transport to the melanosome, *J Biol Chem.* 272(25):15796–15803.

Prota, G. and Thomson, R.H., 1976, Melanin pigmentation in mammals, *Endeavor.* 35:32–38.

Roux, L. and Lloyd, K.O., 1986, Glycosylation characteristics of pigmentation-associated antigen (GP75): an intracellular glycoprotein of human melanocytes and malignant melanomas, *Arch Biochem Biophys.* 251:87–96.

Skipper, J.C., Hendrickson, R.C., Gulden, P.H., Brichard, V., Van Pel, A., Chen, Y., Shabanowitz, J., Wolfel, T., Slingluff, C.L., Jr, Boon, T., Hunt, D.F., and Engelhard, V.H., 1996, An HLA-A2-restricted tyrosinase antigen on melanoma cells results from posttranslational modification and suggests a novel pathway for processing of membrane proteins, *J Exp Med.* 183(2):527–534.

Schrag, J.D., Procopio, D.O., Cygler, M., Thomas, D.Y., and Bergeron, J.J., 2003, Lectin control of protein folding and sorting in the secretory pathway, *Trends Biochem Sci.* 28(1):49–57.

Topalian, S.L., Gonzales, M.I., Parkhurst, M., Li, Y.F., Southwood, S., Sette, A., Rosenberg, S.A., and Robbins, P.F., 1996, Melanoma-specific CD4+ T cells recognize nonmutated HLA-DR-restricted tyrosinase epitopes, *J Exp Med.* 183(5):1965–1971.

Toyofuku, K., Wada, I., Hirosaki, K., Park, J.S., Hori, Y., and Jimbow, K., 1999, Promotion of tyrosinase folding in COS 7 cells by calnexin. *J Biochem.* 125:82–89.

Ujvari, A., Aron, R., Eisenhaure, T., Cheng, E., Parag, H.A., Smicun, Y., Halaban, R., and Hebert, D.N., 2001, Translation rate of human tyrosinase determines its N-linked glycosylation level, *J Biol Chem.* 276(8):5924–5931.

Zapun, A., Petrescu, S.M., Rudd, P.M., Dwek, R.A., Thomas, D.Y., and Bergeron, J.J., 1997, Conformation-independent binding of monoglucosylated ribonuclease B to calnexin, *Cell.* 88(1):29–38.

van Vreeswijk, H., Ruiter, D.J., Brocker, E.B., Welvaart, K., and Ferrone, S., 1988, Differential expression of HLA-DR, DQ, and DP antigens in primary and metastatic melanoma, *Invest Dermatol.* 90(5):755–760.

Vijayasaradhi, S., Xu, Y., Bouchard, B., Houghton, A.N., Vijayasaradhi, S., Xu, Y., Bouchard, B., and Houghton, A.N., 1995, Intracellular sorting and targeting of melanosomal membrane proteins: identification of signals for sorting of the human brown locus protein, gp75, *J Cell Biol.* 130:807–820.

Wang, R.F., Parkhurst, M.R., Kawakami, Y., Robbins, P.F., and Rosenberg, S.A., 1996a, Utilization of an alternative open reading frame of a normal gene in generating a novel human cancer antigen, *J Exp Med.* 183(3):1131–1140.

Wang, R.F., Appella, E., Kawakami, Y., Kang, X., and Rosenberg, S.A., 1996b, Identification of TRP-2 as a human tumor antigen recognized by cytotoxic T lymphocytes, *J Exp Med.* 184(6):2207–2216.

Wang, R.F., Johnston, S.L., Southwood, S., Sette, A., and Rosenberg, S.A., 1998, Recognition of an antigenic peptide derived from tyrosinase-related protein-2 by CTL in the context of HLA-A31 and -A33, *J Immunol.* 160(2):890–897.

Wolfel, T., Van Pel, A., Brichard, V., Schneider, J., Seliger, B., Meyer zum Buschenfelde, K.H., and Boon, T., 1994, Two tyrosinase nonapeptides recognized on HLA-A2 melanomas by autologous cytolytic T lymphocytes, *Eur J Immunol.* 24(3):759–764.

18

GLYCOBIOLOGY OF THE RHEUMATIC DISEASES: AN UPDATE

Azita Alavi and John Axford

Academic Unit for Musculoskeletal Diseases
St George's Hospital Medical School
London SW17 ORE, UK

1. INTRODUCTION

Oligosaccharide components are an integral feature of glycoconjugates, and because of their wide spectrum of diversity, are the primary source of structural heterogeneity; which can profoundly affect the tertiary/quaternary structure and thus the overall biological function of the molecule (Varki, 1993; Kobata, 2000). Data generated over the past 17 years clearly indicate that glycosylation changes may significantly alter the biological activity and thus the overall role of a glycoconjugate in relation to the molecular mechanism of disease, for example, IgG and IgA1 agalactosylation in rheumatoid arthritis (RA) and IgA nephropathy/sLe^x expression in various diseases including RA (Amore *et al.*, 2001; Galon *et al.*, 1997; Goodarzi *et al.*, 1998; Jorgensen *et al.*, 1998).

The information encoded in the oligosaccharide structures may therefore provide new diagnostic and prognostic information, and a better understanding of the disease mechanisms involved, as for example, in RA (Malhotra *et al.*, 1995; Matsumoto *et al.*, 2000; van Zeben *et al.*, 1994; Wright *et al.*, 1998; Young *et al.*, 1991) and lead to advances in the design of novel oligosaccharide-based therapies (Chintalacharuvu *et al.*, 2001; Gorelik *et al.*, 2001; Gugliucci *et al.*, 2000).

2. OLIGOSACCHARIDE BIOSYNTHESIS

The mechanisms regulating the biosynthesis of the structures associated with most glycoconjugates is complex (Alavi, 1996; Dinter and Berger, 1995). The sequential action of glycosyltransferases produces oligosaccharide and glycan structures that reflect the constitution of the multi-glycosyltransferase system (van Zeben *et al.*, 1994) of a given cell type. The process occurs within the ER and the different Golgi cisternae

Glycobiology and Medicine, edited by John S. Axford
Kluwer Academic / Plenum Publishers, New York, 2003

(Alavi, 1996; Keusch *et al.*, 1998) and is governed by the sequential arrangement of the glycosidases and glycosyltransferases for example, β4-galactosyltransferase (GTase): a predominantly trans-Golgi glycoprotein that catalyzes the transfer of galactose to *N*-acetylglucosamine (Alavi, 1996).

$$\text{UDP-Gal} + \text{GlcNAc-R} \xleftrightarrow{\text{Mn}^{2+}} \text{Gal}\beta\text{1-4GlcNAc-R} + \text{UDP}$$

GTase has been extensively studied in relation to RA and is the principal enzyme involved in the galactosylation of human IgG (Keusch *et al.*, 1998).

3. IMMUNOGLOBULIN G

IgG is a glycoprotein consisting of two heavy and two light polypeptide chains forming the Fc and Fab moieties. The majority of sugars are located at a conserved site in the Fc region, at asparagine 297. This, together with the variable glycosylation in the Fab portion, gives an average of 2.8 oligosaccharide molecules per IgG molecule.

Detailed analysis of IgG has revealed extensive oligosaccharide micro-heterogeneity in the non-reducing termini, giving rise to more than 30 structures of the N-linked complex binary type (Routier *et al.*, 1998). This heterogeneity is the result of variations in outer arm sialic acid (SA) and galactose (Gal) content, as well as variation in bisecting *N*-acetylglucosamine (GlcNAc) and core linked fucose (Figure 1). This type of heterogeneity is associated with polyclonal human IgG as well as myeloma proteins (Youings *et al.*, 1996).

IgG oligosaccharide composition is not a static parameter and its synthesis may be governed by multiple factors (Alavi, 1996), including endocrines and various cytokines involved in the differentiation and proliferation of B cells (Chintalacharuvu, 2000). In this respect, it is not surprising that the levels of certain sugar residues vary to suit a given physiological situation (Kimura *et al.*, 2000), and that changes in the degree of IgG galactosylation have recently been shown to be a physiological feature of a normal antibody response (Lastra *et al.*, 1998).

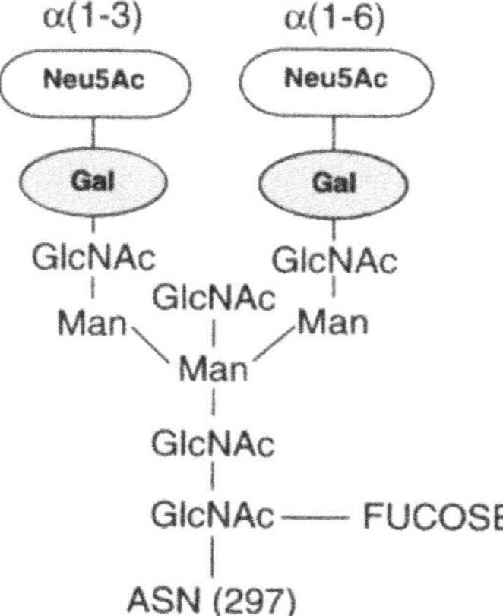

Figure 1. Composite structure of the complex N-linked biantennary oligosaccharide structure derived from human IgG. There is a common pentasaccharide core; two α-mannosyl (Man) residues attached to a β-mannosyl-di-*N*-acetylchitobiose (Man-GlcNAc-GlcNac—Asn) unit. Terminal sialic acid (Neu5Ac) is usually found in the Fab region of IgG, whereas terminal galactose (Gal) and *N*-acetylglucosamine (GlcNAc) are more common in the Fc region. Fucose and bisecting GlcNAc are relatively uncommon in the Fc sugars.

Glycosylation alters the structure of the IgG molecule with functional sequelae (Jefferis and Lund, 2002; Wright *et al.*, 2000; Wright and Morrison, 1993); the ability of IgG3 to trigger cell-lysis through its interaction with human killer cells is eliminated by de-glycosylation (Kumpel *et al.*, 1994). In addition to its effect on IgG effector functions, glycosylation changes have also been demonstrated to influence antibody-binding affinity (Wright and Morrison, 1993).

IgG glycosylation changes have been studied in a number of diseases, and disease associated changes have been observed in a restricted group including RA (Bond *et al.*, 1997; Gornik *et al.*, 1999; Watson *et al.*, 1999). In RA there is an increase in the number of oligosaccharide structures that lack the terminal galactose residues (Bond *et al.*, 1997) and there are indications that hypogalactosylation exists in other rheumatic diseases, for example, juvenile chronic arthritis (Watson *et al.*, 1999) and psoriatic arthritis (Martin *et al.*, 2001).

4. RHEUMATOID ARTHRITIS

Rheumatoid arthritis is a multisystem disorder in which immunological abnormalities characteristically result in a symmetrical joint inflammation, articular erosions, and extra-articular complications. It is the most common and disabling arthritis and affects approximately 1% of the adult population, with a 3 : 1 female to male ratio.

5. IgG GLYCOSYLATION IN RA

5.1. RA as a Glycosylation Disease

Following the initial observations of IgG glycosylation changes in patients with RA (Mullinax, 1975; published abstract) research in this field has followed an exponential path and has resulted in IgG-agalactosylation (IgG-G0) becoming almost synonymous with RA (Alavi *et al.*, 1998, 2000; Alavi and Axford, 1995a, b; Axford, 1999; Axford *et al.*, 1992, 2001; Axford and Alavi, 1995; Bond *et al.*, 1996; Delves *et al.*, 1990; Martin *et al.*, 2001; Routier *et al.*, 1998; Watson *et al.*, 1999).

Associations between IgG glycosylation changes and pathology have been suggested. In early synovitis, low levels of IgG galactose on patient presentation to the clinic were found to be a good predictor for determining which patients may go on to develop RA (Young *et al.*, 1991) and the degree of IgG hypogalactosylation in RA patients has been shown to correlate with disease activity (van Zeben *et al.*, 1994). Possible pathogenic mechanisms are suggested by *in vitro* data indicating, that the exposed terminal *N*-acetylglucosamine on agalactosylated IgG, may through its interaction with mannose-binding protein, be involved in the activation of complement (Malhotra *et al.*, 1995). Further evidence, in support of a pathogenic role for hypogalactosylated RA IgG, comes from a study which indicates that, high affinity rheumatoid factors; auto antibodies with specificity for the Fc region of IgG, may selectively bind IgG that is hypogalactosylated (Soltys *et al.*, 1995).

5.2. Analysis of IgG Sugars in RA

The oligosaccharide structures associated with IgG have been extensively characterized by a number of analytical approaches including various chromatographic and lectin-based techniques (Routier *et al.*, 1998). The most rapid and inexpensive of these techniques are:

(i) Lectin analysis; this technique uses intact IgG and utilizes *Ricinus communis* (RCA), *Bandeiraea simplicifolia II* (BSII) and *Sambucus nigra* (SNA) to detect terminal β1-4 galactose (Gal), *N*-acetylglucosamine (GlcNAc) and α2-6 sialic acid (SA) respectively (Bond *et al.*, 1997).

(ii) Fluorophore linked carbohydrate electrophoresis (FCE); this technique is combined with the enzymatic (PNGase-F) release of IgG oligosaccharides. The fluorescently (8 aminonaphthalene-1,3,6-trisulfonic acid) label biantennary saccharide moieties can be analysed using polyacrylamide gel electrophoresis (Martin *et al.*, 2001).

6. GLYCOSYLATION AND PREGNANCY INDUCED REMISSION OF DISEASE (ALAVI *et al.*, 2000)

Clinical studies have shown that up to 75% of women with RA experience a spontaneous clinical remission of their disease during pregnancy, which may last up to 40 weeks (Spector and Da Silva, 1992). Interestingly, the majority of these women subsequently suffer a clinical flare during the early post-partum period. Although many immunological and endocrinological explanations have been put forward, mechanisms responsible for these effects have yet to be firmly established (Spector and Da Silva, 1992).

Previous studies, examining pregnancy related IgG glycosylation, have on the whole been confined to changes that occur in relation to normal healthy pregnancies. A compositional analysis demonstrated an increase in IgG galactose in pregnancy (Pekelharing *et al.*, 1988), whilst another study using an antibody based (anti-GlcNAc Abs) analysis and fractionation, demonstrated a dramatic increase in the level of hypogalactosylated IgG, post partum (Rook *et al.*, 1991). In the latter study, a similar pattern of changes were noted in a single RA pregnancy with galactose levels mirroring disease activity and also in a further small group of RA patients, where only pre- and post-partum levels of IgG galactosylation changes were examined (Rook *et al.*, 1991).

6.1. Objective

To evaluate the extent to which pregnancy induced IgG glycosylation changes were associated with the clinical manifestations of RA during pregnancy.

6.2. Methods

Serum IgG glycosylation patterns were analyzed in 23 pregnant RA patients with active disease. Patients were randomly selected on the basis of whether they achieved spontaneous remission (n = 11) or did not remit (n = 12); of the latter group 6 patients experienced a relapse in disease activity. Serum IgG levels, of terminal galactose, *N*-acetylglucosamine and α-2,6 sialic acid were compared.

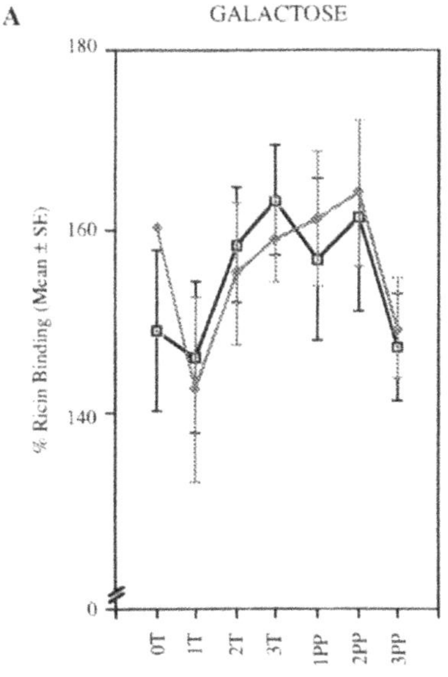

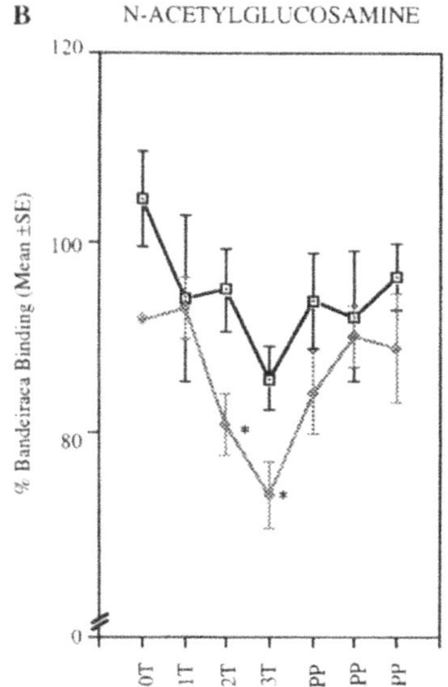

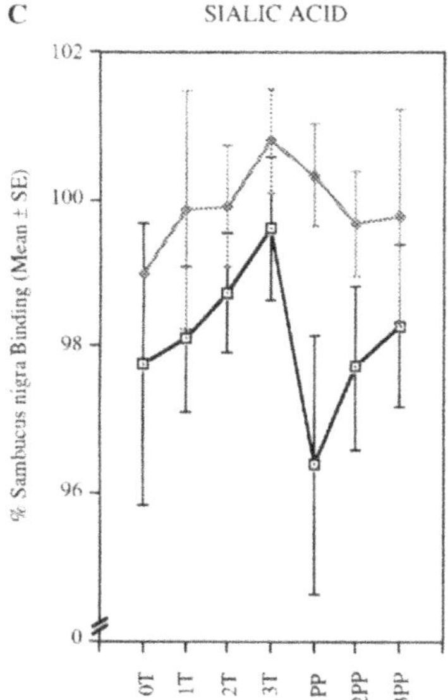

Figure 2. Oligosaccharide profiles of RA IgG glycoforms associated with the antenatal and post-partum period, in the remission -◆- and the no-remission -◻- group. Depicting the overall temporal changes (Mean ± SEM) in IgG glycoform microheterogeneity as governed by the levels of the terminal sugars; Galactose (A), *N*-acetylglucosamine (B), and α2-6 sialic acid.

*Represents significant ($p < 0.05$) differences between the two groups; remission and the no-remission, with respect to the mean levels of IgG GlcNAc.

6.3. Results

Exposed Gal levels increased ($p < 0.02$) and GlcNAc levels decreased ($p < 0.05$) in the antenatal period, and returned to preconception levels during post-partum. Interestingly, the GlcNAc rebound was instantaneous ($p < 0.005$), whereas Gal remained high for a further 10 weeks. SA did not undergo any major changes.

There were no significant Gal differences between the remission and the no-remission groups, in the antenatal period. However, remission was associated with an earlier and significantly greater antenatal reduction (2nd and 3rd trimester; $p < 0.02$) in GlcNAc (Figure 2). This unique and complex pattern of IgG oligosaccharide variation would imply that as well as Fc changes, there may be other variations involving Fab sugars and/or fluctuations in the levels of bisecting GlcNAc, that is site specific glycosylation changes.

Analysis of individual IgG samples during the 1st trimester revealed a significant negative correlation between Gal and GlcNAc in the remission group ($r = -0.80$; $p < 0.05$), which was opposite to that found in the relapse group ($r = +0.87$; $p < 0.03$).

There were no significant differences between the groups with regard to the timing and/or incidence of the post-partum flare of the disease.

6.4. Conclusion

This is the first time that such sugar changes have been reported in relation to the various clinical manifestations of RA during pregnancy. In addition to demonstrating glycosylation changes that are integrally associated with remission, our results also indicate that the observed temporal changes are due to differential site-specific glycosylation, that is Fc and Fab (Bond *et al.*, 1997; Youings *et al.*, 1996).

Changes in the oligosaccharide composition of IgG are integrally associated with the clinical manifestation of RA in pregnancy. The complex temporal changes in these parameters and the time frame at which they occur, are the critical discerning factors associated with the clinical outcome of the disease during pregnancy.

7. CLINICAL UTILITY (MARTIN *et al.*, 2001)

There is no specific diagnostic test for RA. Pilot investigations indicate potential to use sugar printing as a means of differentiating RA from other more benign arthritic disorders (Martin *et al.*, 2001). IgG glycosylation has been studied in a number of different rheumatic diseases (Bond *et al.*, 1997; Ichikawa *et al.*, 1998; Watson *et al.*, 1999). The data indicates that disease associated changes are not only confined to the degree of galactosylation, but that each disease may be associated with a specific mechanism that gives rise to a particular permutation(s) in the oligosaccharide structures associated with IgG. These permutations may be of relevance in not only the differentiation of rheumatic diseases, but also in distinguishing disease associated pathogenic processes.

7.1. Objective

To investigate FCE as a method of analysis of IgG glycosylation changes and to carry out a comparative study of IgG sugar changes in RA, Psoriatic arthritis (PsA) and ankylosing spondylitis (AS).

7.2. Methods

Serum IgG glycosylation patterns were analyzed in patients with RA ($n = 21$), AS ($n = 20$), PsA ($n = 20$), and healthy adults ($n = 36$). Fluorophore linked carbohydrate electrophoresis was performed using the FCE N-linked oligosaccharide profiling kit (Glyco Inc, UK). Each oligosaccharide band was characterized further by sequential enzymatic degradation using the FCE oligosaccharide sequencing kit (Glyco Inc, USA).

7.3. Results

A total of six bands were obtained (Table 1).

All three diseases (Figure 3) were associated with a significant increase in G0f ($p < 0.001$) and a reciprocal decrease in G2f structures ($p = 0.001$).

Significant quantitative differences ($p = 0.001$) were observed in the relative intensity of 5/6 of the bands; G2f, G1f, G0f, a2f (Band 5), and a2, between the disease groups studied. Each disease was associated with a particular pattern of glycosylation changes (Table 2).

7.4. Conclusion

FCE analysis is a useful tool for analysing IgG sugar profiles. Analysis of the different bands for each rheumatic disease group, demonstrated significantly different profiles for each group and in comparison to healthy adults. This technique may therefore prove to be a useful clinical tool.

Table 1. Composite structure of the biantennary oligosaccharides, corresponding to bands 1–6, obtained by fluorophore linked carbohydrate electrophoresis of IgG derived sugars. Key: Asn—, common heptasaccharide core attached to Asparagine; Fucosylated, core fucose-(α1–6); Diagalacto, fully galactosylated; Disialo, fully sialylated.

Band	Oligosaccharide-structure	Abbreviation
1	Asn——core Fucosylated – Digalacto – Asialo	G2f
2	Asn——core Fucosylated – monogalacto – Asialo	G1f
3	Asn——core Fucosylated – digalacto – Disialo (α*2,3 linked SA*)	a2f
4	Asn——core Fucosylated – Agalacto – Asialo	G0f
5	Asn——core Fucosylated – Digalacto – Disialo (α*2,6 linked SA*)	a2f
6	Asn——core Fucosylated – Digalacto – Disialo	a2

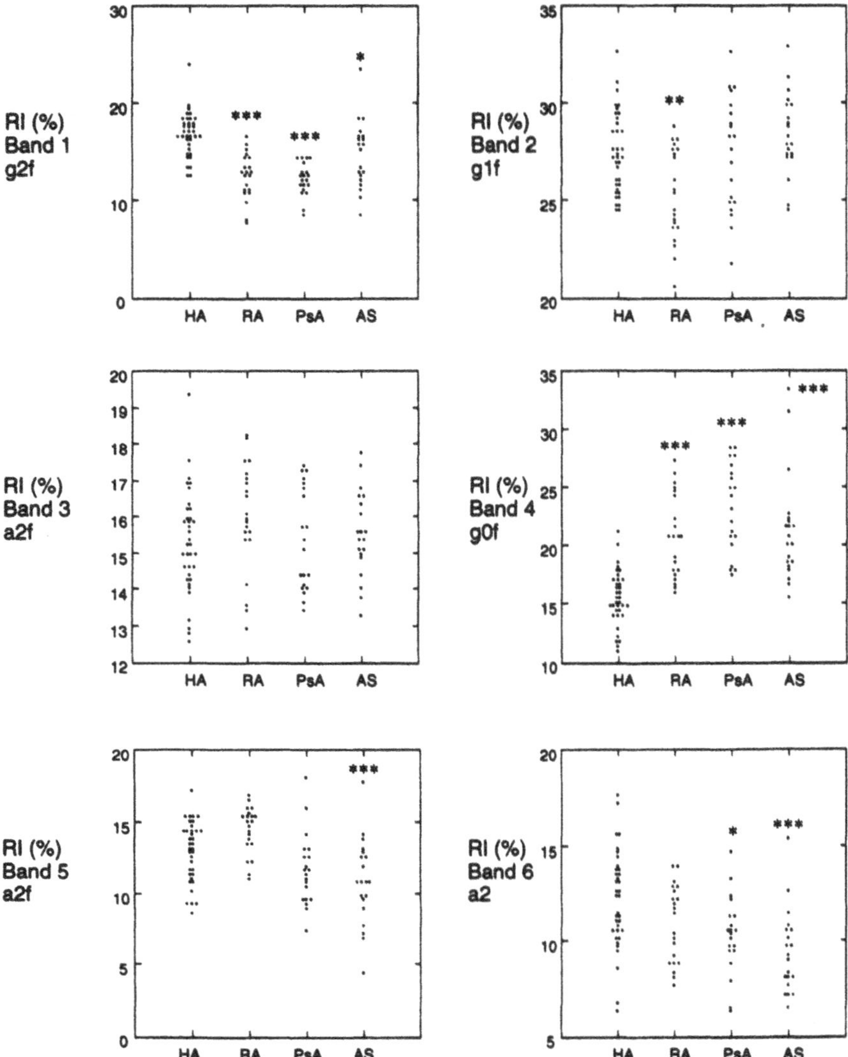

Figure 3. Graphical representation of the data; relative intensity (RI) obtained from the FCE N-linked oligosaccharide profiling of serum IgG samples obtained from healthy adults (HA, $n = 36$), rheumatoid arthritis (RA, $n = 21$), psoriatic arthritis (PsA, $n = 20$), and ankylosing spondylitis (AS, $n = 20$).
$^{*}p < 0.002$, $^{**}p < 0.01$, $^{***}p < 0.001$.

Table 2. Percentage change in band intensity when compared to the IgG sugar profile obtained for healthy individuals.

Disease	G2f	G1f	a2f	G0f	a2f	a2
RA	−26% ***	−8% **	+4%	+34% ***	+12%	−10%
PsA	−27% ***	0	0	+51% ***	−10%	−15% *
AS	−12% *	+3%	+1%	+38% ***	−16% ***	−23% ***

$^{***}p < 0.001$, $^{**}p < 0.01$ and $^{*}p < 0.02$.

8. CONCLUSION

Glycosylation has been shown to play a central role in the pathogenesis of RA and has the potential to be of further benefit with regard to diagnosis and differentiation of rheumatic diseases.

REFERENCES

Alavi, A. and Axford, J.S., 1996, The glycosyltransferases. In: Isenberg, D.A., Rademacker, T.W., editors. Abnormalities of IgG glycosylation and Immunological disorders. London: John Wiley & Sons; 149–169.

Alavi, A., Arden, N., Spector, T.D., and Axford, J.S., 2000, Immunoglobulin G glycosylation and clinical outcome in rheumatoid arthritis during pregnancy, *J Rheumatol.* 27(6):1379–1385.

Alavi, A. and Axford, J., 1995a, Evaluation of beta 1,4-galactosyltransferase in rheumatoid arthritis and its role in the glycosylation network associated with this disease, *Glycoconj J.* 12(3):206–210.

Alavi, A. and Axford, J., 1995b, Beta 1,4-galactosyltransferase variations in rheumatoid arthritis. *Adv Exp Med Biol.* 376:185–192.

Alavi, A., Axford, J.S., Hay, F.C., and Jones, M.G., 1998, Tissue-specific galactosyltransferase abnormalities in an experimental model of rheumatoid arthritis, *Ann Med Interne (Paris).* 149(5):251–260.

Amore, A., Cirina, P., Conti, G., Brusa, P., Peruzzi, L., and Coppo, R., 2001, Glycosylation of circulating IgA in patients with IgA nephropathy modulates proliferation and apoptosis of mesangial cells, *J Am Soc Nephrol.* 12(9):1862–1871.

Axford, J.S., 1999, Glycosylation and rheumatic disease, *Biochim Biophys Acta.* 1455(2–3):219–229.

Axford, J., 2001, The impact of glycobiology on medicine, *Trends Immunol.* 22(5):237–239.

Axford, J.S. and Alavi, A., 1995, An introduction to glycosylation and rheumatic disease. What is the current state of play?, *Adv. Exp. Med. Biol.* 376:171–177.

Axford, J., Kieda, C., and van Dijk, W., 2001, Meeting report-Jenner 5: glycobiology and medicine, *Glycobiology.* 11(2):5G–7G.

Axford, J.S., Sumar, N., Alavi, A., Isenberg, D.A., Young, A., and Bodman, K.B., 1992, Changes in normal glycosylation mechanisms in autoimmune rheumatic disease, *J Clin Invest.* 89(3):1021–1031.

Bond, A., Alavi, A., Axford, J.S., Bourke, B.E., Bruckner, F.E., and Kerr, M.A., 1997, A detailed lectin analysis of IgG glycosylation, demonstrating disease specific changes in terminal galactose and *N*-acetylglucosamine, *J Autoimmun.* 10(1):77–85.

Bond, A., Alavi, A., Axford, J.S., Youinou, P., and Hay, F.C., 1996, The relationship between exposed galactose and *N*-acetylglucosamine residues on IgG in rheumatoid arthritis (RA), juvenile chronic arthritis (JCA) and Sjogren's syndrome (SS), *Clin Exp Immunol.* 105(1):99–103.

Chintalacharuvu, S.R. and Emancipator, S.N., 2000, Differential glycosylation of two glycoproteins synthesized by murine b cells in response to IL-4 plus IL-5, *Cytokine.* 12(8):1182–1188.

Chintalacharuvu, S.R., Urankar-Nagy, N., Petersilge, C.A., Abdul-Karim, F.W., and Emancipator, S.N., 2001, Treatment of collagen induced arthritis by proteolytic enzymes: immunomodulatory and disease modifying effects, *J Rheumatol.* 28(9):2049–2059.

Delves, P.J., Lund, T., Axford, J.S., Alavi-Sadrieh, A., Lydyard, P.M., and MacKenzie, L., 1990, Polymorphism and expression of the galactosyltransferase-associated protein kinase gene in normal individuals and galactosylation-defective rheumatoid arthritis patients, *Arthritis Rheum.* 33(11):1655–1664.

Dinter, A. and Berger, E.G., 1995, The regulation of cell- and tissue-specific expression of glycans by glycosyltransferases, *Adv Exp Med Biol.* 376:53–82.

Galon, J., Robertson, M.W., Galinha, A., Mazieres, N., Spagnoli, R., and Fridman, W.H., 1997, Affinity of the interaction between Fc gamma receptor type III (Fc gammaRIII) and monomeric human IgG subclasses. Role of Fc gammaRIII glycosylation, *Eur J Immunol.* 27(8):1928–1932.

Goodarzi, M.T., Axford, J.S., Varanasi, S.S., Alavi, A., Cunnane, G., and Fitzgerald, O., 1998, Sialyl Lewis(x) expression on IgG in rheumatoid arthritis and other arthritic conditions: a preliminary study, *Glycoconj J.* 15(12):1149–1154.

Gorelik, E., Galili, U., and Raz, A., 2001, On the role of cell surface carbohydrates and their binding proteins (lectins) in tumor metastasis, *Cancer Metastasis Rev.* 20(3–4):245–277.

Gornik, I., Maravic, G., Dumic, J., Flogel, M., and Lauc, G., 1999, Fucosylation of IgG heavy chains is increased in rheumatoid arthritis, *Clin Biochem.* 32(8):605–608.

Gugliucci, A., 2000, Glycation as the glucose link to diabetic complications, *J Am Osteopath Assoc.* 100(10):621–634.

Ichikawa, Y., Yamada, C., Horiki, T., Hoshina, Y., Uchiyama, M., and Yamada, Y., 1998, Anti-agalactosyl IgG antibodies and isotype profiles of rheumatoid factors in Sjogren's syndrome and rheumatoid arthritis, *Clin Exp Rheumatol.* 16(6):709–715.

Jefferis, R. and Lund, J., 2002, Interaction sites on human IgG-Fc for FcgammaR: current models, *Immunol Lett.* 82(1–2):57–65.

Jorgensen, H.G., Elliott, M.A., Priest, R., and Smith, K.D., 1998, Modulation of sialyl Lewis X dependent binding to E-selectin by glycoforms of alpha-1-acid glycoprotein expressed in rheumatoid arthritis, *Biomed Chromatogr.* 12(6):343–349.

Keusch, J., Lydyard, P.M., and Delves P.J., 1998, The effect on IgG glycosylation of altering beta1,4-galactosyltransferase-1 activity in B cells, *Glycobiology.* 8(12):1215–1220.

Kimura, S., Numaguchi, M., Kaizu, T., Kim, D., Takagi, Y., and Gomi, K., 2000, High galactosylation of oligosaccharides in umbilical cord blood IgG, and its relationship to placental function, *Clin Chim Acta.* 299(1–2):169–177.

Kobata, A., 2000, A journey to the world of glycobiology. *Glycoconj J.* 17(7–9):443–464.

Kumpel, B.M., Rademacher, T.W., Rook, G.A., Williams, P.J., and Wilson, I.B., 1994, Galactosylation of human IgG monoclonal anti-D produced by EBV-transformed B-lymphoblastoid cell lines is dependent on culture method and affects Fc receptor-mediated functional activity, *Hum Antibodies Hybridomas.* 5(3–4):143–151.

Lastra, G.C., Thompson, S.J., Lemonidis, A.S., and Elson, C.J., 1998, Changes in the galactose content of IgG during humoral immune responses, *Autoimmunity.* 28(1):25–30.

Malhotra, R., Wormald, M.R., Rudd, P.M., Fischer, P.B., Dwek, R.A., and Sim, R.B., 1995, Glycosylation changes of IgG associated with rheumatoid arthritis can activate complement via the mannose-binding protein, *Nat Med.* 1(3):237–243.

Martin, K., Talukder, R., Hay, F.C., and Axford, J.S., 2001, Characterization of changes in IgG associated oligosaccharide profiles in rheumatoid arthritis, psoriatic arthritis, and ankylosing spondylitis using fluorophore linked carbohydrate electrophoresis. *J Rheumatol.* 28(7):1531–1536.

Matsumoto, A., Shikata, K., Takeuchi, F., Kojima, N., and Mizuochi, T., 2000, Autoantibody activity of IgG rheumatoid factor increases with decreasing levels of galactosylation and sialylation, *J Biochem (Tokyo).* 128(4):621–628.

Pekelharing, J.M., Hepp, E., Kamerling, J.P., Gerwig, G.J., and Leijnse, B., 1988, Alterations in carbohydrate composition of serum IgG from patients with rheumatoid arthritis and from pregnant women, *Ann Rheum Dis.* 47(2):91–95.

Rook, G.A., Steele, J., Brealey, R., Whyte, A., Isenberg, D., and Sumar, N., 1991, Changes in IgG glycoform levels are associated with remission of arthritis during pregnancy, *J Autoimmun.* 4(5):779–794.

Routier, F.H., Hounsell, E.F., Rudd, P.M., Takahashi, N., Bond, A., and Hay, F.C., 1998, Quantitation of the oligosaccharides of human serum IgG from patients with rheumatoid arthritis: a critical evaluation of different methods, *J Immunol Methods.* 213(2):113–130.

Soltys, A.J., Bond, A., Westwood, O.M., and Hay, F.C., 1995, The effects of altered glycosylation of IgG on rheumatoid factor-binding and immune complex formation, *Adv Exp Med Biol.* 376:155–160.

Spector, T.D. and Da Silva, J.A., 1992, Pregnancy and rheumatoid arthritis: an overview, *Am J Reprod Immunol.* 28(3–4):222–225.

van Zeben, D., Rook, G.A., Hazes, J.M., Zwinderman, A.H., Zhang, Y., and Ghelani, S., 1994, Early agalactosylation of IgG is associated with a more progressive disease course in patients with rheumatoid arthritis: results of a follow-up study, *Br J Rheumatol.* 33(1):36–43.

Varki, A., 1993, Biological roles of oligosaccharides: all of the theories are correct, *Glycobiology.* 3(2):97–130.

Watson, M., Rudd, P.M., Bland, M., Dwek, R.A., and Axford, J.S., 1999, Sugar printing rheumatic diseases: a potential method for disease differentiation using immunoglobulin G oligosaccharides, *Arthritis Rheum.* 42(8):1682–1690.

Wright, A. and Morrison, S.L., 1993, Antibody variable region glycosylation: biochemical and clinical effects, *Springer Semin Immunopathol.* 15(2–3):259–273.

Wright, A. and Morrison, S.L., 1998, Effect of C2-associated carbohydrate structure on Ig effector function: studies with chimeric mouse-human IgG1 antibodies in glycosylation mutants of Chinese hamster ovary cells, *J Immunol.* 160(7):3393–3402.

Wright, A., Sato, Y., Okada, T., Chang, K., Endo, T., and Morrison, S., 2000, *In vivo* trafficking and catabolism of IgG1 antibodies with Fc associated carbohydrates of differing structure, *Glycobiology*, 10(12):1347–1355.

Youings, A., Chang, S.C., Dwek, R.A., and Scragg, I.G., 1996, Site-specific glycosylation of human immunoglobulin G is altered in four rheumatoid arthritis patients, *Biochem J.* 314(Pt 2):621–630.

Young, A., Sumar, N., Bodman, K., Goyal, S., Sinclair, H., and Roitt, I., 1991, Agalactosyl IgG: an aid to differential diagnosis in early synovitis, *Arthritis Rheum.* 34(11):1425–1429.

INDEX

The manufacturer's authorised representative in the EU is Springer Nature Customer Service Centre GmbH, Europaplatz 3, 69115 Heidelberg, Germany. If you have any concerns regarding our products, please contact ProductSafety@springernature.com

Printed and bound by CPI Group (UK) Ltd, Croydon, CR0 4YY

23/04/2026

02095627-0001